Svetlin G. Georgiev (Ed.)

Dynamic Calculus and Equations on Time Scales

Also of Interest

Numerical Analysis on Time Scales
Svetlin G. Georgiev, Inci M. Erhan, 2022
ISBN 978-3-11-078725-2, e-ISBN (PDF) 978-3-11-078732-0

Integral Inequalities on Time Scales
Svetlin G. Georgiev, 2020
ISBN 978-3-11-070550-8, e-ISBN (PDF) 978-3-11-070555-3

Functional Analysis with Applications
Svetlin G. Georgiev, Khaled Zennir, 2019
ISBN 978-3-11-065769-2, e-ISBN (PDF) 978-3-11-065772-2

Sequences and Series in Calculus
Joseph D. Fehribach, 2023
ISBN 978-3-11-076835-0, e-ISBN (PDF) 978-3-11-076839-8

Nonlinear Dynamics
Axelle Amon, Marc Lefranc, 2023
ISBN 978-3-11-067786-7, e-ISBN (PDF) 978-3-11-067787-4

Dynamic Calculus and Equations on Time Scales

Edited by
Svetlin G. Georgiev

DE GRUYTER

Mathematics Subject Classification 2020
Primary: 39A10, 34A60, 34B37; Secondary: 34C10, 34K20

Editor
Prof. Dr. Svetlin G. Georgiev
Department of Differential Equations
Faculty of Mathematics and Informatics
Kliment Ohridski University of Sofia
1126 Sofia
Bulgaria
svetlingeorgiev1@gmail.com

ISBN 978-3-11-118289-6
e-ISBN (PDF) 978-3-11-118297-1
e-ISBN (EPUB) 978-3-11-118519-4

Library of Congress Control Number: 2023938954

Bibliographic information published by the Deutsche Nationalbibliothek
The Deutsche Nationalbibliothek lists this publication in the Deutsche Nationalbibliografie; detailed bibliographic data are available on the Internet at http://dnb.dnb.de.

Cover image: piranka / E+ / Getty Images
Typesetting: VTeX UAB, Lithuania
Printing and binding: CPI books GmbH, Leck

www.degruyter.com

Preface

Time scale theory was first initiated by Stefan Hilger in 1988 in his PhD thesis to unify both approaches of dynamic modeling, namely difference and differential equations. Similar ideas have been used before and go back to the introduction of the Riemann–Stieltjes integral which unifies sums and integrals. Many results to differential equations carry over easily to corresponding results for difference equations, while other results seem to be totally different in nature. Because of these reasons, the theory of dynamic equations is an active area of research. The time scale calculus can be applied to any fields in which dynamic processes are described by discrete- or continuous-time models. So, the calculus of time scales has various applications involving discontinuous domains such as certain bug populations, phytoremediation of metals, wound healing, maximization problems in economics and traffic problems.

This book presents some recent developments in the area of dynamic calculus and dynamic equations on time scales. The book contains ten chapters. Chapter 1 presents a projector analysis of dynamic systems on time scales. The linear time-varying dynamic systems are introduced and they are classified into those of the first, second, third, and fourth kind. The considered systems are investigated in the case when they are regular with tractability index 1. Jets of a function of one independent time scale variable and jets of a function of n independent real variables and one independent time scale variable are defined. Jet spaces are introduced and some of their properties are given. In the chapter, differentiable functions and total derivatives are defined. Nonlinear dynamic systems on arbitrary time scales are considered. Properly involved derivatives, constraints and consistent initial values for the considered equations are investigated. A linearization for nonlinear dynamic systems is introduced and the total derivative for regular linearized equations with tractability index 1 is investigated. Chapter 2 deals with the fundamental properties of the Muckenhoupt and Gehring weights on time scales. Some results related to the self-improving properties of the Muckenhoupt and Gehring classes and some higher integrability results for nonincreasing functions on time scales are presented. The main approach is based on proving some properties of integral operators with powers, the Hölder inequality, chain rules, as well as some connecting relations between Muckenhoupt and Gehring classes on time scales. In Chapter 3, a new definition of periodicity on isolated time scales introduced by Bohner, Mesquita, and Streipert is applied to the study of Ulam stability. If the graininess (step size) of an isolated time scale is bounded by a finite constant, then the linear 1- and 2-periodic dynamic equations are Ulam stable if and only if the exponential function has modulus different from unity. If the graininess increases at least linearly to infinity, the 1- and 2-periodic dynamic equations are not Ulam stable. Applying these results, several examples of the first-order linear 1- or 2-periodic dynamic equations on specific isolated time scales are given in the chapter, such as h-difference equations, q-difference equations, triangular equations, Fibonacci equations, and harmonic equations. In some cases the minimum Ulam stability

https://doi.org/10.1515/9783111182971-201

constant is found. A multivalued logarithm on time scales recently introduced for delta-differentiable functions that never vanish is covered in Chapter 4. This is accomplished using an extended definition of the cylinder transformation from which the definition of exponential functions on time scales arose. The definition of a logarithm function on arbitrary time scales with familiar and useful properties then follows. Chapter 5 deals with the existence and stability results for hybrid fuzzy differential equations with tempered Ξ-Hilfer fractional derivatives on time scales. Further, sufficient conditions for the existence and uniqueness of solutions by using hybrid fixed point theorem are obtained. In addition, it demonstrates Ulam-type stability. Finally, a suitable example to illustrate the main results is given. In Chapter 6, a new class of Ambartsumian equations of the fractional type with tempered Ξ-Hilfer fractional derivative with boundary conditions is examined. The provided problem is transformed into an equivalent fixed point problem, which is then resolved by using the Banach and Krasnosel'skii fixed point theorems. Ulam stability is investigated. An example is included to verify the theoretical results. In Chapter 7, a series solution method on general time scales is introduced. The derivation of the method and its application to dynamic and integral equations is discussed in detail. Several examples illustrating the method are presented. In Chapter 8, the dual results, delta and nabla inequalities, and their special cases, continuous and discrete inequalities, are unified into diamond alpha case, and new forms of such results as well as new diamond alpha Bennett–Leindler-type dynamic inequalities are established by developing a novel method, which does not require the integration by parts formula and the fundamental theorem of calculus. These theorems are standard arguments in the proofs of similar theorems in the delta and nabla approaches but do not follow naturally in the diamond-alpha calculus. In Chapter 9, a de la Vallée Poussin-type inequality for impulsive dynamic equations on time scales is derived. This inequality is often used in conjunction with disconjugacy and/or (non)oscillation. Hence, it appears to be a very useful tool for the qualitative study of dynamic equations. In this chapter, generalizing the classical de la Vallée Poussin inequality for impulsive dynamic equations on arbitrary time scales, disconjugacy criteria and some results on nonoscillation are obtained. In Chapter 10, the divided differences and σ-divided differences on time scales are introduced. The Newton and σ-Newton interpolation polynomial are constructed. In addition, the Hermite interpolation polynomial on time scales is constructed by using the divided differences table. Examples are presented to illustrate the theoretical results.

This book is addressed to a wide audience of specialists such as mathematicians, physicists, engineers, and biologists. It can be used as a textbook at the graduate level and as a reference book for several disciplines.

Paris, May 2023 Svetlin G. Georgiev

Contents

List of Contributing Authors

Douglas R. Anderson
Department of Mathematics
Concordia College
Moorhead
MN 56562
USA
E-mail: andersod@cord.edu

Martin Bohner
Department of Mathematics and Statistics
Missouri University of Science and Technology
Rolla
MO 65409-0020
USA
E-mail: bohner@mst.edu

Sibel Doğru Akgöl
Department of Mathematics
Atilim University
06830 Incek
Ankara
Turkey
E-mail: sibel.dogruakgol@atilim.edu.tr

İnci M. Erhan
Department of Computer Engineering
Aydın Adnan Menderes University
09010 Aydın
Turkey
E-mail: incierhan@gmail.com

Svetlin G. Georgiev
Department of Mathematics
Sorbonne University
Paris
France
E-mail: svetlingeorgiev1@gmail.com

Najlaa Jaddoa
Department of Mathematics
Gazi University
06500 Beşevler
Ankara
Turkey
E-mail: nabd.jaddoa@gazi.edu.tr

K. Kanagarajan
Department of Mathematics
Sri Ramakrishna Mission Vidyalaya College of Arts and Science
Coimbatore 641020
India
E-mail: kanagarajank@gmail.com

Zeynep Kayar
Department of Mathematics
Van Yüzüncü Yıl University
Van
Turkey
E-mail: zeynepkayar@yyu.edu.tr

Billur Kaymakçalan
Department of Mathematics
Çankaya University
06810 Ankara
Turkey
E-mail: billurkaymakcalan@gmail.com

S. Manikandan
Department of Mathematics
Sri Ramakrishna Mission Vidyalaya College of Arts and Science
Coimbatore 641020
TamilnaduIndia
E-mail: p.s.manikandanshanmugam@gmail.com

Masakazu Onitsuka
Department of Applied Mathematics
Okayama University of Science
Okayama
700-0005
Japan
E-mail: onitsuka@ous.ac.jp

Abdullah Özbekler
Department of Mathematics
Atilim University
06830 Incek
Ankara
Turkey
E-mail: aozbekler@gmail.com

Neslihan Nesliye Pelen
Department of Mathematics
Ondokuz Mayıs University
Samsun
Turkey
E-mail: nesliyeaykir@gmail.com

Samir H. Saker
Department of Mathematics
Mansoura University
Mansoura 35516
Egypt
E-mail: shsaker@mans.edu.eg

Rezan Sevinik-Adıgüzel
Department of Mathematics
Atilim University
06830 Incek
Ankara
Turkey
E-mail: rezan.adiguzel@atilim.edu.tr

S. Sivasundaram
Department of Mathematics
College of Natural Science, Engineering and Mathematics
Daytona Beach
FL 32114
USA
E-mail: seenithi@gmail.com

D. Vivek
Department of Mathematics
PSG College of Arts & Science
Coimbatore 641014
India
E-mail: peppyvivek@gmail.com

R. Vivek
Department of Mathematics
Sri Ramakrishna Mission Vidyalaya College of Arts and Science
Coimbatore 641020
India
E-mail: vivekmaths1997@gmail.com

Guo-Cheng Wu
Data Recovery Key Laboratory of Sichuan Province
College of Mathematics and Information Science, Neijiang Normal University
Neijiang 641100
PR China
E-mail: wuguocheng@gmail.com

Svetlin G. Georgiev

1 Projector analysis of dynamic systems on time scales

Abstract: This chapter presents a projector analysis of dynamic systems on time scales. We investigate the linear time-varying dynamic systems and classify them into those of the first, second, third, and fourth kind. The considered systems are investigated in the case when they are regular with tractability index 1. Then, we define jets of a function of one independent time scale variable and jets of a function of n independent real variables and one independent time scale variable. We introduce jet spaces and give some of their properties. In the chapter, we also define differentiable functions and total derivatives. We consider nonlinear dynamic systems on arbitrary time scales. We define properly involved derivatives, constraints, and consistent initial values for the considered equations. We introduce a linearization for nonlinear dynamic systems and investigate the total derivative for regular linearized equations with tractability index 1.

1.1 Linear time-varying dynamic-algebraic equations

This chapter is devoted to linear time-varying dynamic-algebraic equations. We classify them into those of the first, second, third and fourth kind. We investigate them in the case when they are regular with tractability index 1.

Suppose $\mathbb{T}$ is a time scale with forward jump operator and delta differentiation operator σ and Δ, respectively. Let $I \subseteq \mathbb{T}$.

1.1.1 Linear time-varying dynamic-algebraic equations of the first kind

In this section, we will investigate the following linear time-varying dynamic-algebraic equation:

$$A^\sigma(t)(Bx)^\Delta(t) = C^\sigma(t)x^\sigma(t) + f(t), \quad t \in I, \tag{1.1}$$

where $A : I \to \mathscr{M}_{n\times m}$, $B : I \to \mathscr{M}_{m\times n}$, $C : I \to \mathscr{M}_{n\times n}$, and $f : I \to \mathbb{R}^n$ are given. Here, with $\mathscr{M}_{p\times q}$ we denote the set of all $p \times q$ real matrices.

Definition 1.1. Equation (1.1) is said to be a linear time-varying dynamic-algebraic equation of the first kind.

Svetlin G. Georgiev, Department of Mathematics, Sorbonne University, Paris, France, e-mail: svetlingeorgiev1@gmail.com

https://doi.org/10.1515/9783111182971-001

We will consider the solutions of (1.1) within the space $\mathscr{C}^1_B(I)$. Below, we remove the explicit dependence on t for the sake of notational simplicity.

1.1.1.1 A particular case

Suppose that $A, C : I \to \mathscr{M}_{n\times n}$. Consider the equation

$$A^\sigma x^\Delta = C^\sigma x^\sigma + f. \tag{1.2}$$

We will show that equation (1.2) can be reduced to equation (1.1). Suppose that P is a $\mathscr{C}^1$-projector along $\ker A^\sigma$. Then

$$A^\sigma P = A^\sigma$$

and

$$A^\sigma x^\Delta = A^\sigma P x^\Delta = A^\sigma (Px)^\Delta - A^\sigma P^\Delta x^\sigma.$$

Hence, equation (1.2) takes the form

$$A^\sigma (Px)^\Delta - A^\sigma P^\Delta x^\sigma = C^\sigma x^\sigma + f,$$

or

$$A^\sigma (Px)^\Delta = (A^\sigma P^\Delta + C^\sigma) x^\sigma + f.$$

Set

$$C_1^\sigma = A^\sigma P^\Delta + C^\sigma.$$

Thus, (1.2) takes the form

$$A^\sigma (Px)^\Delta = C_1^\sigma x^\sigma + f, \tag{1.3}$$

i. e., equation (1.2) is a particular case of equation (1.1).

Example. Let

$$\mathbb{T} = 2^{\mathbb{N}_0}, \quad A(t) = \begin{pmatrix} 1 & 0 & 0 \\ 0 & -t & 1 \\ 0 & 0 & 0 \end{pmatrix}, \quad C(t) = \begin{pmatrix} -t & 1 & t \\ 0 & 1 & 2t \\ t & 0 & 1 \end{pmatrix}, \quad t \in \mathbb{T}. \tag{1.4}$$

We have

$$\sigma(t) = 2t, \quad t \in \mathbb{T},$$

and

$$A^\sigma(t) = \begin{pmatrix} 1 & 0 & 0 \\ 0 & -2t & 1 \\ 0 & 0 & 0 \end{pmatrix}, \quad C^\sigma(t) = \begin{pmatrix} -2t & 1 & 2t \\ 0 & 1 & 4t \\ 2t & 0 & 1 \end{pmatrix}, \quad t \in \mathbb{T}.$$

We will find a vector

$$y(t) = \begin{pmatrix} y_1(t) \\ y_2(t) \\ y_3(t) \end{pmatrix}, \quad t \in \mathbb{T},$$

so that

$$A^\sigma(t)y(t) = \begin{pmatrix} 0 \\ 0 \\ 0 \end{pmatrix}, \quad t \in \mathbb{T}.$$

We have

$$\begin{pmatrix} 0 \\ 0 \\ 0 \end{pmatrix} = \begin{pmatrix} 1 & 0 & 0 \\ 0 & -2t & 1 \\ 0 & 0 & 0 \end{pmatrix} \begin{pmatrix} y_1(t) \\ y_2(t) \\ y_3(t) \end{pmatrix} = \begin{pmatrix} y_1(t) \\ -2ty_2(t) + y_3(t) \\ 0 \end{pmatrix}, \quad t \in \mathbb{T},$$

whereupon

$$\begin{pmatrix} y_1(t) \\ y_2(t) \\ y_3(t) \end{pmatrix} = \begin{pmatrix} 0 \\ 1 \\ 2t \end{pmatrix}, \quad t \in \mathbb{T},$$

and the null projector to $A^\sigma(t)$, $t \in \mathbb{T}$, is

$$Q(t) = \begin{pmatrix} 0 & 0 & 0 \\ 0 & 0 & 1 \\ 0 & 0 & 2t \end{pmatrix}, \quad t \in \mathbb{T}.$$

Hence,

$$P(t) = I - Q(t) = \begin{pmatrix} 1 & 0 & 0 \\ 0 & 1 & 0 \\ 0 & 0 & 1 \end{pmatrix} - \begin{pmatrix} 0 & 0 & 0 \\ 0 & 0 & 1 \\ 0 & 0 & 2t \end{pmatrix} = \begin{pmatrix} 1 & 0 & 0 \\ 0 & 1 & -1 \\ 0 & 0 & 1-2t \end{pmatrix}, \quad t \in \mathbb{T},$$

is a projector along $\ker A^\sigma$. Note that

$$
\begin{aligned}
P^{\Delta}(t) &= \begin{pmatrix} 0 & 0 & 0 \\ 0 & 0 & 0 \\ 0 & 0 & -2 \end{pmatrix}, \\
C_1^{\sigma}(t) &= A^{\sigma}(t)P^{\Delta}(t) + C^{\sigma}(t) \\
&= \begin{pmatrix} 1 & 0 & 0 \\ 0 & -2t & 1 \\ 0 & 0 & 0 \end{pmatrix} \begin{pmatrix} 0 & 0 & 0 \\ 0 & 0 & 0 \\ 0 & 0 & -2 \end{pmatrix} + \begin{pmatrix} -2t & 1 & 2t \\ 0 & 1 & 4t \\ 2t & 0 & 1 \end{pmatrix} \\
&= \begin{pmatrix} 0 & 0 & 0 \\ 0 & 0 & -2 \\ 0 & 0 & 0 \end{pmatrix} + \begin{pmatrix} -2t & 1 & 2t \\ 0 & 1 & 4t \\ 2t & 0 & 1 \end{pmatrix} \\
&= \begin{pmatrix} -2t & 1 & 2t \\ 0 & 1 & 4t-2 \\ 2t & 0 & 1 \end{pmatrix}.
\end{aligned}
$$

Equation (1.2) can be written as follows:

$$
\begin{pmatrix} 1 & 0 & 0 \\ 0 & -2t & 1 \\ 0 & 0 & 0 \end{pmatrix} \begin{pmatrix} x_1^{\Delta}(t) \\ x_2^{\Delta}(t) \\ x_3^{\Delta}(t) \end{pmatrix} = \begin{pmatrix} -2t & 1 & 2t \\ 0 & 1 & 4t \\ 2t & 0 & 1 \end{pmatrix} \begin{pmatrix} x_1^{\sigma}(t) \\ x_2^{\sigma}(t) \\ x_3^{\sigma}(t) \end{pmatrix} + \begin{pmatrix} f_1(t) \\ f_2(t) \\ f_3(t) \end{pmatrix}, \quad t \in \mathbb{T},
$$

or

$$
\begin{aligned}
x_1^{\Delta}(t) &= -2tx_1^{\sigma}(t) + x_2^{\sigma}(t) + 2tx_3^{\sigma}(t) + f_1(t), \\
-2tx_2^{\Delta}(t) + x_3^{\Delta}(t) &= x_2^{\sigma}(t) + 4tx_3^{\sigma}(t) + f_2(t), \\
0 &= 2tx_1^{\sigma}(t) + x_3^{\sigma}(t) + f_3(t), \quad t \in \mathbb{T}.
\end{aligned}
$$

This system, using (1.3), can be rewritten in the form

$$
\begin{aligned}
&\begin{pmatrix} 1 & 0 & 0 \\ 0 & -2t & 1 \\ 0 & 0 & 0 \end{pmatrix} \left(\begin{pmatrix} 1 & 0 & 0 \\ 0 & 1 & -1 \\ 0 & 0 & 1-2t \end{pmatrix} \begin{pmatrix} x_1(t) \\ x_2(t) \\ x_3(t) \end{pmatrix} \right)^{\Delta} \\
&= \begin{pmatrix} -2t & 1 & 2t \\ 0 & 1 & 4t-2 \\ 2t & 0 & 1 \end{pmatrix} \begin{pmatrix} x_1^{\sigma}(t) \\ x_2^{\sigma}(t) \\ x_3^{\sigma}(t) \end{pmatrix} + \begin{pmatrix} f_1(t) \\ f_2(t) \\ f_3(t) \end{pmatrix}, \quad t \in \mathbb{T},
\end{aligned}
$$

or

$$
\begin{pmatrix} 1 & 0 & 0 \\ 0 & -2t & 1 \\ 0 & 0 & 0 \end{pmatrix} \begin{pmatrix} x_1(t) \\ x_2(t) - x_3(t) \\ (1-2t)x_3(t) \end{pmatrix}^{\Delta} = \begin{pmatrix} -2tx_1^{\sigma}(t) + x_2^{\sigma}(t) + 2tx_3^{\sigma}(t) + f_1(t) \\ x_2^{\sigma}(t) + (4t-2)x_3^{\sigma}(t) + f_2(t) \\ 2tx_1^{\sigma}(t) + x_3^{\sigma}(t) + f_3(t) \end{pmatrix},
$$

$t \in \mathbb{T}$,

or

$$\begin{pmatrix} 1 & 0 & 0 \\ 0 & -2t & 1 \\ 0 & 0 & 0 \end{pmatrix} \begin{pmatrix} x_1^\Delta(t) \\ x_2^\Delta(t) - x_3^\Delta(t) \\ (1-4t)x_3^\Delta(t) - 2x_3(t) \end{pmatrix} = \begin{pmatrix} -2tx_1^\sigma(t) + x_2^\sigma(t) + 2tx_3^\sigma(t) + f_1(t) \\ x_2^\sigma(t) + (4t-2)x_3^\sigma(t) + f_2(t) \\ 2tx_1^\sigma(t) + x_3^\sigma(t) + f_3(t) \end{pmatrix},$$

$t \in \mathbb{T}$,

or

$$\begin{aligned} x_1^\Delta(t) &= -2tx_1^\sigma(t) + x_2^\sigma(t) + 2tx_3^\sigma(t) + f_1(t), \\ -2t(x_2^\Delta(t) - x_3^\Delta(t)) + (1-4t)x_3^\Delta(t) - 2x_3(t) &= x_2^\sigma(t) + (4t-2)x_3^\sigma(t) + f_2(t), \\ 0 &= -2tx_1^\sigma(t) + x_3^\sigma(t) + f_3(t), \quad t \in \mathbb{T}, \end{aligned}$$

or

$$\begin{aligned} x_1^\Delta(t) &= -2tx_1^\sigma(t) + x_2^\sigma(t) +_2 tx_3^\sigma(t) + f_1(t), \\ -2tx_2^\Delta(t) + (1-2t)x_3^\Delta(t) &= x_2^\sigma(t) + (4t-2)x_3^\sigma(t) + 2x_3(t) + f_2(t), \\ 0 &= 2tx_1^\sigma(t) + x_3^\sigma(t) + f_3(t), \quad t \in \mathbb{T}. \end{aligned}$$

1.1.1.2 Standard form index 1 problems

In this section, we will investigate the equation

$$A^\sigma (Px)^\Delta = C^\sigma x^\sigma + f, \tag{1.5}$$

where $\ker A$ is a $\mathscr{C}^1$-space, $C \in \mathscr{C}(I)$, P is a $\mathscr{C}^1$-projector along $\ker A$. Then

$$AP = A.$$

Assume in addition that

$$Q = I - P$$

and

(B1) the matrix

$$A_1 = A + CQ$$

is invertible.

Definition 1.2. Equation (1.5) is said to be regular with tractability index 1.

We will start our investigations with the following useful lemma.

Lemma 1.1. *Suppose that* (B1) *holds. Then*

$$A_1^{-1}A = P$$

and

$$A_1^{-1}CQ = Q.$$

Proof. We have

$$A_1P = (A + CQ)P = AP + CQP = A.$$

Since $Q = I - P$ and $\ker P = \ker A$, we have $\operatorname{im} Q = \ker A$ and

$$AQ = 0.$$

Then

$$A_1Q = (A + CQ)Q = AQ + CQQ = CQ.$$

This completes the proof. □

Now, we multiply equation (1.5) by $(A_1^{-1})^\sigma$ and get

$$(A_1^{-1})^\sigma A^\sigma (Px)^\Delta = (A_1^{-1})^\sigma C^\sigma x^\sigma + (A_1^{-1})^\sigma f.$$

Now, we employ the first equation of Lemma 1.1 and get

$$P^\sigma (Px)^\Delta = (A_1^{-1})^\sigma C^\sigma x^\sigma + (A_1^{-1})^\sigma f. \tag{1.6}$$

We decompose x in the following way:

$$x = Px + Qx.$$

Then equation (1.6) takes the following form:

$$\begin{aligned} P^\sigma (Px)^\Delta &= (A_1^{-1})^\sigma C^\sigma (P^\sigma x^\sigma + Q^\sigma x^\sigma) + (A_1^{-1})^\sigma f \\ &= (A_1^{-1})^\sigma C^\sigma P^\sigma x^\sigma + (A_1^{-1})^\sigma C^\sigma Q^\sigma x^\sigma + (A_1^{-1})^\sigma f. \end{aligned}$$

Using the second equation of Lemma 1.1, the latter equation can be rewritten as follows:

$$P^\sigma (Px)^\Delta = (A_1^{-1})^\sigma C^\sigma P^\sigma x^\sigma + Q^\sigma x^\sigma + (A_1^{-1})^\sigma f. \tag{1.7}$$

We multiply equation (1.7) with the projector P^σ and, using

$$PP = P, \quad PQ = 0,$$

we find

$$P^\sigma P^\sigma (Px)^\Delta = P^\sigma (A_1^{-1})^\sigma C^\sigma P^\sigma x^\sigma + P^\sigma Q^\sigma x^\sigma + P^\sigma (A_1^{-1})^\sigma f,$$

or

$$P^\sigma (Px)^\Delta = P^\sigma (A_1^{-1})^\sigma C^\sigma P^\sigma x^\sigma + P^\sigma (A_1^{-1})^\sigma f. \tag{1.8}$$

Note that

$$P^\sigma (Px)^\Delta = (PPx)^\Delta - P^\Delta Px = (Px)^\Delta - P^\Delta Px.$$

Hence using (1.8), we find

$$(Px)^\Delta - P^\Delta Px = P^\sigma (A_1^{-1})^\sigma C^\sigma P^\sigma x^\sigma + P^\sigma (A_1^{-1})^\sigma f,$$

or

$$(Px)^\Delta = P^\Delta Px + P^\sigma (A_1^{-1})^\sigma C^\sigma P^\sigma x^\sigma + P^\sigma (A_1^{-1})^\sigma f. \tag{1.9}$$

Now, we multiply equation (1.7) by Q^σ and find

$$Q^\sigma P^\sigma (Px)^\Delta = Q^\sigma (A_1^{-1}) C^\sigma B^{-\sigma} B^\sigma P^\sigma x^\sigma + Q^\sigma Q^\sigma x^\sigma + Q^\sigma (A_1^{-1})^\sigma f,$$

or

$$0 = Q^\sigma (A_1^{-1})^\sigma C^\sigma B^{-\sigma} B^\sigma P^\sigma x^\sigma + Q^\sigma x^\sigma + Q^\sigma (A_1^{-1})^\sigma f. \tag{1.10}$$

Set

$$u = Px, \quad v = Qx.$$

Then, by (1.9) and (1.10), we get the system

$$\begin{aligned} u^\Delta &= P^\Delta u + P^\sigma (A_1^{-1})^\sigma C^\sigma B^{-\sigma} u^\sigma + P^\sigma (A_1^{-1})^\sigma f, \\ v^\sigma &= -Q^\sigma (A_1^{-1})^\sigma C^\sigma B^{-\sigma} u^\sigma - Q^\sigma (A_1^{-1})^\sigma f. \end{aligned} \tag{1.11}$$

We find $u \in \mathscr{C}^1(I)$ from the first equation of the system (1.11), and then we find $v^\sigma \in \mathscr{C}(I)$ from the second equation of the system (1.11). Hence, for the solution x of equation (1.5), we have the following representation:

$$x^\sigma = u^\sigma + v^\sigma = P^\sigma x^\sigma + Q^\sigma x^\sigma.$$

Example. Let

$$\mathbb{T} = \mathbb{N}, \qquad A(t) = \begin{pmatrix} -1 & t+1 & -1 \\ 0 & 0 & 0 \\ 0 & 2t+2 & -2 \end{pmatrix},$$
$$C(t) = \begin{pmatrix} 0 & 0 & 1 \\ 0 & -t & 1 \\ 0 & 2 & 1 \end{pmatrix}, \quad P(t) = \begin{pmatrix} 1 & 0 & 0 \\ 0 & 0 & 0 \\ 0 & -(t+1) & 1 \end{pmatrix}, \quad t \in \mathbb{T}. \tag{1.12}$$

Here

$$\sigma(t) = t+1, \quad t \in \mathbb{T}.$$

We have

$$A(t)P(t) = \begin{pmatrix} -1 & t+1 & -1 \\ 0 & 0 & 0 \\ 0 & 2t+2 & -2 \end{pmatrix} \begin{pmatrix} 1 & 0 & 0 \\ 0 & 0 & 0 \\ 0 & -(t+1) & 1 \end{pmatrix} = \begin{pmatrix} -1 & t+1 & -1 \\ 0 & 0 & 0 \\ 0 & 2t+2 & -2 \end{pmatrix},$$
$$t \in \mathbb{T}.$$

Therefore P is a projector along $\ker A$. Next,

$$\begin{aligned} Q(t) &= I - P(t) \\ &= \begin{pmatrix} 1 & 0 & 0 \\ 0 & 0 & 0 \\ 0 & -(t+1) & 1 \end{pmatrix} - \begin{pmatrix} 1 & 0 & 0 \\ 0 & 0 & 0 \\ 0 & -(t+1) & 1 \end{pmatrix} \\ &= \begin{pmatrix} 0 & 0 & 0 \\ 0 & 1 & 0 \\ 0 & t+1 & 0 \end{pmatrix}, \quad t \in \mathbb{T}, \end{aligned}$$

and

$$\begin{aligned} A_1(t) &= A(t) + C(t)Q(t) \\ &= \begin{pmatrix} -1 & t+1 & -1 \\ 0 & 0 & 0 \\ 0 & 2t+2 & -2 \end{pmatrix} + \begin{pmatrix} 0 & 0 & 1 \\ 0 & -t & 1 \\ 0 & 2 & 1 \end{pmatrix} \begin{pmatrix} 0 & 0 & 0 \\ 0 & 1 & 0 \\ 0 & t+1 & 0 \end{pmatrix} \\ &= \begin{pmatrix} -1 & t+1 & -1 \\ 0 & 0 & 0 \\ 0 & 2t+2 & -2 \end{pmatrix} + \begin{pmatrix} 0 & t+1 & 0 \\ 0 & 1 & 0 \\ 0 & t+3 & 0 \end{pmatrix} \\ &= \begin{pmatrix} -1 & 2(t+1) & -1 \\ 0 & 1 & 0 \\ 0 & 3t+5 & -1 \end{pmatrix}, \quad t \in \mathbb{T}. \end{aligned}$$

Note that

$$\det A_1(t) = 2 \neq 0, \quad t \in \mathbb{T}.$$

Thus, A_1 is invertible. We will find its cofactors. We compute

$$a_{11}(t) = \begin{vmatrix} 1 & 0 \\ 3t+5 & -2 \end{vmatrix} = -2, \quad a_{12}(t) = -\begin{vmatrix} 0 & 0 \\ 0 & -1 \end{vmatrix} = 0, \quad a_{13}(t) = \begin{vmatrix} 0 & 1 \\ 0 & 3t+5 \end{vmatrix} = 0,$$

$$a_{21}(t) = -\begin{vmatrix} 2(t+1) & -1 \\ 3t+5 & -2 \end{vmatrix} = -(-4(t+1) + 3t + 5) = -(-4t - 4 + 3t + 5) = t - 1,$$

$$a_{22}(t) = \begin{vmatrix} -1 & -1 \\ 0 & -2 \end{vmatrix} = 2, \quad a_{23}(t) = -\begin{vmatrix} -1 & 2(t+1) \\ 0 & 3t+5 \end{vmatrix} = 3t + 5,$$

$$a_{31}(t) = \begin{vmatrix} 2(t+1) & -1 \\ 1 & 0 \end{vmatrix} = 1, \quad a_{32}(t) = -\begin{vmatrix} -1 & -1 \\ 0 & 0 \end{vmatrix} = 0,$$

$$a_{33}(t) = \begin{vmatrix} -1 & 2(t+1) \\ 0 & 1 \end{vmatrix} = -1, \quad t \in \mathbb{T}.$$

Consequently,

$$A_1^{-1}(t) = \frac{1}{2}\begin{pmatrix} -2 & t-1 & 1 \\ 0 & 2 & 0 \\ 0 & 3t+5 & -1 \end{pmatrix} = \begin{pmatrix} -1 & \frac{t-1}{2} & \frac{1}{2} \\ 0 & 1 & 0 \\ 0 & \frac{3t+5}{2} & -\frac{1}{2} \end{pmatrix}, \quad t \in \mathbb{T}.$$

Hence,

$$A^\sigma(t) = \begin{pmatrix} -1 & t+2 & -1 \\ 0 & 0 & 0 \\ 0 & 2t+4 & -2 \end{pmatrix}, \quad (A_1^{-1})^\sigma(t) = \begin{pmatrix} -1 & \frac{t}{2} & \frac{1}{2} \\ 0 & 1 & 0 \\ 0 & \frac{3t+8}{2} & -\frac{1}{2} \end{pmatrix},$$

$$C^\sigma(t) = \begin{pmatrix} 0 & 0 & 1 \\ 0 & -t-1 & 1 \\ 0 & 2 & 1 \end{pmatrix}, \qquad P^\sigma(t) = \begin{pmatrix} 1 & 0 & 0 \\ 0 & 0 & 0 \\ 0 & -(t+2) & 1 \end{pmatrix},$$

$$Q^\sigma(t) = \begin{pmatrix} 0 & 0 & 0 \\ 0 & 1 & 0 \\ 0 & t+2 & 0 \end{pmatrix}, \qquad P^\Delta(t) = \begin{pmatrix} 0 & 0 & 0 \\ 0 & 0 & 0 \\ 0 & -1 & 0 \end{pmatrix}, \quad t \in \mathbb{T}.$$

Also,

$$P^\sigma(t)(A^{-1})^\sigma(t) = \begin{pmatrix} 1 & 0 & 0 \\ 0 & 1 & -1 \\ 0 & 0 & -t-1 \end{pmatrix}\begin{pmatrix} -1 & \frac{t}{2} & \frac{1}{2} \\ 0 & 1 & 0 \\ 0 & \frac{3t+8}{2} & -\frac{1}{2} \end{pmatrix} = \begin{pmatrix} -1 & \frac{t}{2} & \frac{1}{2} \\ 0 & 0 & 0 \\ 0 & \frac{t+4}{2} & -\frac{1}{2} \end{pmatrix},$$

$$P^\sigma(t)(A_1^{-1})^\sigma(t)C^\sigma(t) = \begin{pmatrix} -1 & \frac{t}{2} & \frac{1}{2} \\ 0 & 0 & 0 \\ 0 & \frac{t+4}{2} & -\frac{1}{2} \end{pmatrix}\begin{pmatrix} 0 & 0 & 1 \\ 0 & -t-1 & 1 \\ 0 & 2 & 1 \end{pmatrix}$$
$$= \begin{pmatrix} 0 & -\frac{(t+2)(t-1)}{2} & \frac{t-1}{2} \\ 0 & 0 & 0 \\ 0 & -\frac{(t+2)(t+3)}{2} & \frac{t+3}{2} \end{pmatrix},$$
$$Q^\sigma(t)(A_1^{-1})^\sigma(t) = \begin{pmatrix} 0 & 0 & 0 \\ 0 & 0 & 1 \\ 0 & 0 & t+2 \end{pmatrix}\begin{pmatrix} -1 & \frac{t}{2} & \frac{1}{2} \\ 0 & 1 & 0 \\ 0 & \frac{3t+8}{2} & -\frac{1}{2} \end{pmatrix}$$
$$= \begin{pmatrix} 0 & 0 & 0 \\ 0 & \frac{3t+8}{2} & -\frac{1}{2} \\ 0 & \frac{(3t+8)(t+2)}{2} & -\frac{t+2}{2} \end{pmatrix},$$
$$Q^\sigma(t)(A_1^{-1})^\sigma(t)C^\sigma(t) = \begin{pmatrix} 0 & 0 & 0 \\ 0 & \frac{3t+8}{2} & -\frac{1}{2} \\ 0 & \frac{(3t+8)(t+2)}{2} & -\frac{t+2}{2} \end{pmatrix}\begin{pmatrix} 0 & 0 & 1 \\ 0 & -t-1 & 1 \\ 0 & 2 & 1 \end{pmatrix}$$
$$= \begin{pmatrix} 0 & 0 & 0 \\ 0 & -\frac{(3t+5)(t+2)}{2} & \frac{3(t+2)}{2} \\ 0 & -\frac{(3t+5)(t+2)^2}{2} & \frac{(3t+7)(t+2)}{2} \end{pmatrix}, \quad t \in \mathbb{T}.$$

Then, equation (1.5) can be rewritten as follows:

$$\begin{pmatrix} -1 & t+2 & -1 \\ 0 & 0 & 0 \\ 0 & 2t+4 & -2 \end{pmatrix}\left(\begin{pmatrix} 1 & 0 & 0 \\ 0 & 0 & 0 \\ 0 & -(t+1) & 1 \end{pmatrix}\begin{pmatrix} x_1(t) \\ x_2(t) \\ x_3(t) \end{pmatrix}\right)^\Delta$$
$$= \begin{pmatrix} 0 & 0 & 1 \\ 0 & -1-t & 1 \\ 0 & 2 & 1 \end{pmatrix}\begin{pmatrix} x_1^\sigma(t) \\ x_2^\sigma(t) \\ x_3^\sigma(t) \end{pmatrix} + \begin{pmatrix} f_1(t) \\ f_2(t) \\ f_3(t) \end{pmatrix}, \quad t \in \mathbb{T},$$

or

$$\begin{pmatrix} -1 & t+2 & -1 \\ 0 & 0 & 0 \\ 0 & 2t+4 & -2 \end{pmatrix}\begin{pmatrix} x_1(t) \\ 0 \\ -(t+1)x_2(t) + x_3(t) \end{pmatrix}^\Delta = \begin{pmatrix} x_1^\sigma(t) \\ -(t+1)x_2^\sigma(t) + x_3^\sigma(t) \\ 2x_2^\sigma(t) + x_3^\sigma(t) \end{pmatrix}$$
$$+ \begin{pmatrix} f_1(t) \\ f_2(t) \\ f_3(t) \end{pmatrix}, \quad t \in \mathbb{T},$$

or

$$\begin{pmatrix} -1 & t+2 & -1 \\ 0 & 0 & 0 \\ 0 & 2t+4 & -2 \end{pmatrix} \begin{pmatrix} x_1^\Delta(t) \\ 0 \\ -x_2^\sigma(t) - (t+1)x_2^\Delta(t) + x_3^\Delta(t) \end{pmatrix} = \begin{pmatrix} x_1^\sigma(t) \\ -(t+1)x_2^\sigma(t) + x_3^\sigma(t) \\ 2x_2^\sigma(t) + x_3^\sigma(t) \end{pmatrix} + \begin{pmatrix} f_1(t) \\ f_2(t) \\ f_3(t) \end{pmatrix}, \quad t \in \mathbb{T},$$

or

$$\begin{pmatrix} -x_1^\Delta(t) + x_2^\sigma(t) + (t+1)x_2^\Delta(t) - x_3^\Delta(t) \\ 0 \\ 2x_2^\sigma(t) + 2(t+1)x_2^\Delta(t) - 2x_3^\Delta(t) \end{pmatrix} = \begin{pmatrix} x_1^\sigma(gt) + f_1(t) \\ -(t+1)x_2^\sigma(t) + x_3^\sigma(t) + f_2(t) \\ 2x_2^\sigma(t) + x_3^\sigma(t) + f_3(t) \end{pmatrix},$$
$$t \in \mathbb{T},$$

or

$$\begin{aligned} -x_1^\Delta(t) + x_2^\sigma(t) + (t+1)x_2^\Delta(t) - x_3^\Delta(t) &= x_1^\sigma(t) + f_1(t), \\ 0 &= -(t+1)x_2^\sigma(t) + x_3^\sigma(t) + f_2(t), \\ 2x_2^\sigma(t) + 2(t+1)x_2^\Delta(t) - 2x_3^\Delta(t) &= 2x_2^\sigma(t) + x_3^\sigma(t) + f_3(t), \quad t \in \mathbb{T}, \end{aligned}$$

or

$$\begin{aligned} -x_1^\Delta(t) + (t+1)x_2^\Delta(t) - x_3^\Delta(t) &= x_1^\sigma(t) - x_2^\sigma(t) + f_1(t), \\ 0 &= -(t+1)x_2^\sigma(t) + x_3^\sigma(t) + f_2(t), \\ 2(t+1)x_2^\Delta(t) - 2x_3^\Delta(t) &= x_3^\sigma(t) + f_3(t), \quad t \in \mathbb{T}. \end{aligned}$$

The system (1.11) can be rewritten as follows:

$$\begin{pmatrix} u_1^\Delta(t) \\ u_2^\Delta(t) \\ u_3^\Delta(t) \end{pmatrix} = \begin{pmatrix} 0 & 0 & 0 \\ 0 & 0 & 0 \\ 0 & -1 & 0 \end{pmatrix} \begin{pmatrix} u_1(t) \\ u_2(t \\ u_3(t) \end{pmatrix} + \begin{pmatrix} 0 & -\frac{(t-1)(t+2)}{2} & \frac{t-1}{2} \\ 0 & 0 & 0 \\ 0 & -\frac{(t+2)(t+3)}{2} & \frac{t+3}{2} \end{pmatrix} \begin{pmatrix} u_1^\sigma(t) \\ u_2^\sigma(t) \\ u_3^\sigma(t) \end{pmatrix} + \begin{pmatrix} -1 & \frac{t}{2} & \frac{1}{2} \\ 0 & 0 & 0 \\ 0 & \frac{t+4}{2} & -\frac{1}{2} \end{pmatrix} \begin{pmatrix} f_1(t) \\ f_2(t) \\ f_3(t) \end{pmatrix},$$

$$\begin{pmatrix} 0 \\ 0 \\ v^\sigma(t) \end{pmatrix} = -\begin{pmatrix} 0 & 0 & 0 \\ 0 & -\frac{(3t+5)(t+1)}{2} & \frac{3(t+2)}{2} \\ 0 & -\frac{(3t+5)(t+2)^2}{2} & \frac{(3t+7)(t+2}{2} \end{pmatrix} \begin{pmatrix} u_1^\sigma(t) \\ u_2^\sigma(t) \\ u_3^\sigma(t) \end{pmatrix} - \begin{pmatrix} 0 & 0 & 0 \\ 0 & \frac{3t+8}{2} & -\frac{1}{2} \\ 0 & \frac{(3t+8)(t+2)}{2} & -\frac{t+2}{2} \end{pmatrix} \begin{pmatrix} f_1(t) \\ f_2(t) \\ f_3(t) \end{pmatrix}, \quad t \in \mathbb{T},$$

or

$$\begin{pmatrix} u_1^\Delta(t) \\ u_2^\Delta(t) \\ u_3^\Delta(t) \end{pmatrix} = \begin{pmatrix} 0 \\ 0 \\ -u_2^\sigma(t) \end{pmatrix} + \begin{pmatrix} -\frac{(t-1)(t+2)}{2} u_2^\sigma(t) + \frac{t-1}{2} u_3^\sigma(t) \\ 0 \\ -\frac{(t+2)(t+3)}{2} u_2^\sigma(t) + \frac{t+3}{2} u_3^\sigma(t) \end{pmatrix} + \begin{pmatrix} -f_1(t) + \frac{t}{2} f_2(t) + \frac{1}{2} f_3(t) \\ 0 \\ \frac{t+4}{2} f_2(t) - \frac{1}{2} f_3(t) \end{pmatrix},$$

$$\begin{pmatrix} 0 \\ 0 \\ v^\sigma(t) \end{pmatrix} = \begin{pmatrix} 0 \\ -\frac{(3t+5)(t+2)}{2} u_2^\sigma(t) + \frac{3(t+2)}{2} u_3^\sigma(t) \\ -\frac{(3t+5)(t+2)^2}{2} u_2^\sigma(t) + \frac{(3t+7)(t+2)}{2} u_3^\sigma(t) \end{pmatrix} - \begin{pmatrix} 0 \\ \frac{3t+8}{2} f_2(t) - \frac{1}{2} f_3(t) \\ \frac{(3t+8)(t+2)}{2} f_2(t) - \frac{t+2}{2} f_3(t) \end{pmatrix}, \quad t \in \mathbb{T},$$

or

$$\begin{aligned} u_1^\Delta(t) &= -\frac{(t-1)(t+2)}{2} u_2^\sigma(t) + \frac{t-1}{2} u_3^\sigma(t) - f_1(t) + \frac{t}{2} f_2(t) + \frac{1}{2} f_3(t), \\ u_2^\Delta(t) &= 0, \\ u_3^\Delta(t) &= -u_2(t) - \frac{(t+2)(t+3)}{2} u_2^\sigma(t) + \frac{t+3}{2} u_3^\sigma(t) + \frac{t+4}{2} f_2(t) - \frac{1}{2} f_3(t), \\ 0 &= -\frac{(3t+5)(t+2)}{2} u_2^\sigma(t) + \frac{3(t+2)}{2} u_3^\sigma(t) + \frac{3t+8}{2} f_2(t) - \frac{1}{2} f_3(t), \\ v^\sigma(t) &= -\frac{(3t+5)(t+2)^2}{2} u_2^\sigma(t) + \frac{(3t+7)(t+2)}{2} u_3^\sigma(t) \\ &\quad + \frac{(3t+8)(t+2)}{2} f_2(t) - \frac{t+2}{2} f_3(t), \quad t \in \mathbb{T}. \end{aligned}$$

1.1.2 Linear time-varying dynamic-algebraic equations of the second kind

In this chapter we will investigate the following linear time-varying dynamic-algebraic equation:

$$A^\sigma(t)(Bx)^\Delta(t) = C(t)x(t) + f(t), \quad t \in I, \tag{1.13}$$

where $A, B, C : I \to \mathscr{M}_{m\times m}$ and $f : I \to \mathbb{R}^m$ are given. Equation (1.13) will be said to be a linear time-varying dynamic-algebraic equation of the second kind. We will consider the solutions of (1.13) within the space $\mathscr{C}_B^1(I)$. Below, we remove the explicit dependence on t for the sake of notational simplicity.

1.1.2.1 A particular case

Consider the equation

$$A^{\sigma}x^{\Delta} = Cx + f. \tag{1.14}$$

We will show that equation (1.14) can be reduced to equation (1.13). Suppose that P is a $\mathscr{C}^1$-projector along $\ker A$. Then

$$AP = A$$

and

$$A^{\sigma}x^{\Delta} = A^{\sigma}P^{\sigma}x^{\Delta} = A^{\sigma}(Px)^{\Delta} - A^{\sigma}P^{\Delta}x.$$

Hence, equation (1.14) takes the form

$$A^{\sigma}(Px)^{\Delta} - A^{\sigma}P^{\Delta}x = Cx + f,$$

or

$$A^{\sigma}(Px)^{\Delta} = (A^{\sigma}P^{\Delta} + C)x + f.$$

Set

$$C_1 = A^{\sigma}P^{\Delta} + C.$$

Thus, (1.14) takes the form

$$A^{\sigma}(Px)^{\Delta} = C_1x + f, \tag{1.15}$$

i. e., equation (1.14) is a particular case of equation (1.13).

Example. Let $\mathbb{T} = 2^{\mathbb{N}_0}$ and the matrices A and C be as in (1.4). Then

$$\sigma(t) = 2t, \quad t \in \mathbb{T},$$

and

$$A^{\sigma}(t) = \begin{pmatrix} 1 & 0 & 0 \\ 0 & -2t & 1 \\ 0 & 0 & 0 \end{pmatrix}, \quad t \in \mathbb{T}.$$

We will find a vector

$$y(t) = \begin{pmatrix} y_1(t) \\ y_2(t) \\ y_3(t) \end{pmatrix}, \quad t \in \mathbb{T},$$

so that

$$A(t)y(t) = \begin{pmatrix} 0 \\ 0 \\ 0 \end{pmatrix}, \quad t \in \mathbb{T}.$$

We have

$$\begin{pmatrix} 0 \\ 0 \\ 0 \end{pmatrix} = \begin{pmatrix} 1 & 0 & 0 \\ 0 & -t & 1 \\ 0 & 0 & 0 \end{pmatrix} \begin{pmatrix} y_1(t) \\ y_2(t) \\ y_3(t) \end{pmatrix} = \begin{pmatrix} y_1(t) \\ -ty_2(t) + y_3(t) \\ 0 \end{pmatrix}, \quad t \in \mathbb{T},$$

whereupon

$$\begin{pmatrix} y_1(t) \\ y_2(t) \\ y_3(t) \end{pmatrix} = \begin{pmatrix} 0 \\ 1 \\ t \end{pmatrix}, \quad t \in \mathbb{T},$$

and the null projector to $A(t)$, $t \in \mathbb{T}$, is

$$Q(t) = \begin{pmatrix} 0 & 0 & 0 \\ 0 & 0 & 1 \\ 0 & 0 & t \end{pmatrix}, \quad t \in \mathbb{T}.$$

Hence,

$$P(t) = I - Q(t) = \begin{pmatrix} 1 & 0 & 0 \\ 0 & 1 & 0 \\ 0 & 0 & 1 \end{pmatrix} - \begin{pmatrix} 0 & 0 & 0 \\ 0 & 0 & 1 \\ 0 & 0 & t \end{pmatrix} = \begin{pmatrix} 1 & 0 & 0 \\ 0 & 1 & -1 \\ 0 & 0 & 1-t \end{pmatrix}, \quad t \in \mathbb{T},$$

is a projector along $\ker A$. Observe that

$$P^{\Delta}(t) = \begin{pmatrix} 0 & 0 & 0 \\ 0 & 0 & 0 \\ 0 & 0 & -1 \end{pmatrix},$$

$$\begin{aligned} C_1(t) = A^{\sigma}(t)P^{\Delta}(t) + C(t) &= \begin{pmatrix} 1 & 0 & 0 \\ 0 & -2t & 1 \\ 0 & 0 & 0 \end{pmatrix} \begin{pmatrix} 0 & 0 & 0 \\ 0 & 0 & 0 \\ 0 & 0 & -1 \end{pmatrix} + \begin{pmatrix} -t & 1 & t \\ 0 & 1 & 2t \\ t & 0 & 1 \end{pmatrix} \\ &= \begin{pmatrix} 0 & 0 & 0 \\ 0 & 0 & -1 \\ 0 & 0 & 0 \end{pmatrix} + \begin{pmatrix} -t & 1 & t \\ 0 & 1 & 2t \\ t & 0 & 1 \end{pmatrix} = \begin{pmatrix} -t & 1 & t \\ 0 & 1 & 2t-1 \\ t & 0 & 1 \end{pmatrix}. \end{aligned}$$

Equation (1.2) can be written as follows:

$$\begin{pmatrix} 1 & 0 & 0 \\ 0 & -2t & 1 \\ 0 & 0 & 0 \end{pmatrix}\begin{pmatrix} x_1^\Delta(t) \\ x_2^\Delta(t) \\ x_3^\Delta(t) \end{pmatrix} = \begin{pmatrix} -t & 1 & t \\ 0 & 1 & 2t \\ t & 0 & 1 \end{pmatrix}\begin{pmatrix} x_1(t) \\ x_2(t) \\ x_3(t) \end{pmatrix} + \begin{pmatrix} f_1(t) \\ f_2(t) \\ f_3(t) \end{pmatrix}, \quad t \in \mathbb{T},$$

or

$$\begin{aligned} x_1^\Delta(t) &= -tx_1(t) + x_2(t) + tx_3(t) + f_1(t), \\ -2tx_2^\Delta(t) + x_3^\Delta(t) &= x_2(t) + 2tx_3(t) + f_2(t), \\ 0 &= tx_1(t) + x_3(t) + f_3(t), \quad t \in \mathbb{T}. \end{aligned}$$

This system, using (1.15), can be rewritten in the form

$$\begin{aligned} &\begin{pmatrix} 1 & 0 & 0 \\ 0 & -2t & 1 \\ 0 & 0 & 0 \end{pmatrix}\left(\begin{pmatrix} 1 & 0 & 0 \\ 0 & 1 & -1 \\ 0 & 0 & 1-t \end{pmatrix}\begin{pmatrix} x_1(t) \\ x_2(t) \\ x_3(t) \end{pmatrix}\right)^\Delta \\ &= \begin{pmatrix} -t & 1 & t \\ 0 & 1 & 2t-1 \\ t & 0 & 1 \end{pmatrix}\begin{pmatrix} x_1(t) \\ x_2(t) \\ x_3(t) \end{pmatrix} + \begin{pmatrix} f_1(t) \\ f_2(t) \\ f_3(t) \end{pmatrix}, \quad t \in \mathbb{T}, \end{aligned}$$

or

$$\begin{pmatrix} 1 & 0 & 0 \\ 0 & -2t & 1 \\ 0 & 0 & 0 \end{pmatrix}\begin{pmatrix} x_1(t) \\ x_2(t) \\ x_3(t) \end{pmatrix}^\Delta = \begin{pmatrix} -tx_1(t) + x_2(t) + tx_3(t) \\ x_2(t) + (2t-1)x_3(t) \\ tx_1(t) + x_3(t) \end{pmatrix} + \begin{pmatrix} f_1(t) \\ f_2(t) \\ f_3(t) \end{pmatrix}, \quad t \in \mathbb{T},$$

or

$$\begin{pmatrix} 1 & 0 & 0 \\ 0 & -2t & 1 \\ 0 & 0 & 0 \end{pmatrix}\begin{pmatrix} x_1^\Delta(t) \\ x_2^\Delta(t) \\ x_3^\Delta(t) \end{pmatrix} = \begin{pmatrix} -tx_1(t) + x_2(t) + tx_3(t) + f_1(t) \\ x_2(t) + (2t-1)x_3(t) + f_2(t) \\ tx_1(t) + x_3(t) + f_3(t) \end{pmatrix}, \quad t \in \mathbb{T},$$

or

$$\begin{aligned} x_1^\Delta(t) &= -tx_1(t) + x_2(t) + tx_3(t) + f_1(t), \\ -2t(x_2^\Delta(t) - x_3^\Delta(t)) + (1-2t)x_3^\Delta(t) - x_3(t) &= x_2(t) + (2t-1)x_3(t) + f_2(t), \\ 0 &= tx_1(t) + x_3(t) + f_3(t), \quad t \in \mathbb{T}, \end{aligned}$$

or

$$\begin{aligned} x_1^\Delta(t) &= -tx_1(t) + x_2(t) + tx_3(t) + f_1(t), \\ -2tx_2^\Delta(t) + x_3^\Delta(t) &= x_2(t) + 2tx_3(t) + f_2(t), \\ 0 &= tx_1(t) + x_3(t) + f_3(t), \quad t \in \mathbb{T}. \end{aligned}$$

1.1.2.2 Standard form index 1 problems

In this section, we will investigate the equation

$$A^{\sigma}(Px)^{\Delta} = Cx + f, \tag{1.16}$$

where $\ker A$ is a $\mathscr{C}^1$-space, $C \in \mathscr{C}(I)$, P is a $\mathscr{C}^1$-projector along $\ker A$. Then

$$AP = A.$$

Assume in addition that

$$Q = I - P$$

and

(C1) the matrix

$$A_1^{\sigma} = A + CQ$$

is invertible.

Definition 1.3. Equation (1.16) is said to be regular with tractability index 1.

We will start our investigations with the following useful lemma.

Lemma 1.2. *Suppose that* (C1) *holds. Then*

$$A_1^{\sigma-1}A = P$$

and

$$A_1^{\sigma-1}CQ = Q.$$

Proof. We have

$$A_1^{\sigma}P = (A + CQ)P = AP + CQP = A.$$

Since $Q = I - P$ and $\ker P = \ker A$, we have $\operatorname{im} Q = \ker A$ and

$$AQ = 0.$$

Then

$$A_1^{\sigma}Q = (A + CQ)Q = AQ + CQQ = CQ.$$

This completes the proof. □

Now, we multiply equation (1.16) with $A_1^{\sigma-1}$ and get

$$A_1^{\sigma-1}A^\sigma(Px)^\Delta = A_1^{\sigma-1}Cx + A_1^{\sigma-1}f.$$

Now, we employ the second equation of Lemma 1.2 and get

$$P(Px)^\Delta = A_1^{\sigma-1}Cx + A_1^{\sigma-1}f. \tag{1.17}$$

We decompose x in the following way:

$$x = Px + Qx.$$

Then equation (1.17) takes the following form:

$$\begin{aligned} P(Px)^\Delta &= A_1^{\sigma-1}C(Px + Qx) + A_1^{\sigma-1}f \\ &= A_1^{\sigma-1}CPx + A_1^{\sigma-1}CQx + A_1^{\sigma-1}f. \end{aligned}$$

Using the second equation of Lemma 1.2, the latter equation can be rewritten as follows:

$$P(Px)^\Delta = A_1^{\sigma-1}CPx + Qx + A_1^{\sigma-1}f. \tag{1.18}$$

We multiply equation (1.18) with the projector P and, using

$$PP = P, \quad PQ = 0,$$

we find

$$PP(Px)^\Delta = PA_1^{\sigma-1}CPx + PQx + PA_1^{\sigma-1}f,$$

or

$$P(Px)^\Delta = PA_1^{\sigma-1}CPx + PA_1^{\sigma-1}f. \tag{1.19}$$

Note that

$$P(Px)^\Delta = (PPx)^\Delta - P^\Delta P^\sigma x^\sigma = (Px)^\Delta - P^\Delta P^\sigma x^\sigma.$$

Hence by (1.19), we find

$$(Px)^\Delta - P^\Delta P^\sigma x^\sigma = PA_1^{\sigma-1}CPx + PA_1^{\sigma-1}f,$$

or

$$(Px)^\Delta = P^\Delta P^\sigma x^\sigma + PA_1^{\sigma-1}CPx + PA_1^{\sigma-1}f. \tag{1.20}$$

Now, we multiply equation (1.18) by Q and find

$$QP(Px)^{\Delta} = QA_1^{\sigma-1}CPx + QQx + QA_1^{\sigma-1}f,$$

or

$$0 = QA_1^{\sigma-1}CPx + Qx + QA_1^{\sigma-1}f. \tag{1.21}$$

Set

$$u = Px, \quad v = Qx.$$

Then, by (1.20) and (1.21), we get the system

$$\begin{aligned} u^{\Delta} &= P^{\Delta}u^{\sigma} + PA_1^{\sigma-1}Cu + PA_1^{\sigma-1}f, \\ v &= -QA_1^{\sigma-1}Cu - QA_1^{\sigma-1}f. \end{aligned} \tag{1.22}$$

We find $u \in \mathscr{C}^1(I)$ from the second equation of the system (1.22), and then we find $v \in \mathscr{C}(I)$ from the second equation of the system (1.22). Hence the solution of the equation (1.16) is given by

$$x = u + v = Px + Qx.$$

Example. Let $\mathbb{T} = \mathbb{N}$ and A, P and C be as in (1.12). Then, equation (1.16) can be rewritten as follows:

$$\begin{aligned} &\begin{pmatrix} -1 & t+2 & -1 \\ 0 & 0 & 0 \\ 0 & 2t+4 & -2 \end{pmatrix} \left(\begin{pmatrix} 1 & 0 & 0 \\ 0 & 1 & -1 \\ 0 & 0 & -1 \end{pmatrix} \begin{pmatrix} x_1(t) \\ x_2(t) \\ x_3(t) \end{pmatrix} \right)^{\Delta} \\ &= \begin{pmatrix} 0 & 0 & 1 \\ 0 & -t & 1 \\ 0 & 2 & 1 \end{pmatrix} \begin{pmatrix} x_1(t) \\ x_2(t) \\ x_3(t) \end{pmatrix} + \begin{pmatrix} f_1(t) \\ f_2(t) \\ f_3(t) \end{pmatrix}, \quad t \in \mathbb{T}, \end{aligned}$$

or

$$\begin{pmatrix} -1 & t+2 & -1 \\ 0 & 0 & 0 \\ 0 & 2t+4 & -2 \end{pmatrix} \begin{pmatrix} x_1(t) \\ x_2(t) \\ x_3(t) \end{pmatrix}^{\Delta} = \begin{pmatrix} x_1(t) \\ -tx_2(t) + x_3(t) \\ 2x_2(t) + x_3(t) \end{pmatrix} + \begin{pmatrix} f_1(t) \\ f_2(t) \\ f_3(t) \end{pmatrix}, \quad t \in \mathbb{T},$$

or

$$\begin{pmatrix} -1 & t+2 & -1 \\ 0 & 0 & 0 \\ 0 & 2t+4 & -2 \end{pmatrix} \begin{pmatrix} x_1^{\Delta}(t) \\ x_2^{\Delta}(t) \\ x_3^{\Delta}(t) \end{pmatrix} = \begin{pmatrix} x_1(t) \\ -tx_2(t) + x_3(t) \\ 2x_2(t) + x_3(t) \end{pmatrix} + \begin{pmatrix} f_1(t) \\ f_2(t) \\ f_3(t) \end{pmatrix}, \quad t \in \mathbb{T},$$

or

$$\begin{aligned} -x_1^\Delta(t) + (t+2)(x_2^\Delta(t) - x_3^\Delta(t)) + x_3^\sigma(t) + x_3^\Delta(t) &= x_1(t) + f_1(t), \\ 0 &= -tx_2(t) + x_3(t) + f_2(t), \\ (2t+4)(x_2^\Delta(t) - x_3^\Delta(t)) + 2x_3^\sigma(t) + 2x_3^\Delta(t) &= 2x_2(t) + x_3(t) + f_3(t), \quad t \in \mathbb{T}, \end{aligned}$$

or

$$\begin{aligned} -x_1^\Delta(t) + (t+2)x_2^\Delta(t) - (t+1)x_3^\Delta(t) &= x_1(t) - x_3(t) + f_1(t), \\ 0 &= -tx_2(t) + x_3(t) + f_2(t), \\ (2t+4)x_2^\Delta(t) - 2(t+1)x_3^\Delta(t) &= 2x_2(t) - x_3(t) + f_3(t), \quad t \in \mathbb{T}. \end{aligned}$$

Next, we will rewrite the system (1.22). We have

$$\begin{aligned} P(t)(A_1^{-1}(t))^\sigma &= \begin{pmatrix} 1 & 0 & 0 \\ 0 & 1 & -1 \\ 0 & 0 & -t \end{pmatrix} \begin{pmatrix} -1 & \frac{2t+1}{2} & \frac{1}{2} \\ 0 & -\frac{1}{2(t+2)} & \frac{1}{2(t+1)} \\ 0 & -1 & 0 \end{pmatrix} \\ &= \begin{pmatrix} -1 & \frac{2t+1}{2} & \frac{1}{2} \\ 0 & \frac{2t+3}{2(t+2)} & \frac{1}{2(t+2)} \\ 0 & t & 0 \end{pmatrix}, \\ P(t)(A^{-1}(t))^\sigma C(t) &= \begin{pmatrix} -1 & \frac{2t+1}{2} & \frac{1}{2} \\ 0 & \frac{2t+3}{2(t+2)} & \frac{1}{2(t+2)} \\ 0 & t & 0 \end{pmatrix} \begin{pmatrix} 0 & 0 & 1 \\ 0 & -t-1 & 1 \\ 0 & 2 & 1 \end{pmatrix} \\ &= \begin{pmatrix} 0 & -t(t+2) & \frac{2t+1}{2} \\ 0 & -\frac{(2t+3)(t+1)^2-2}{2(t+2)} & \frac{2t^2+7t+7}{2(t+2)} \\ 0 & -t(t+1) & t \end{pmatrix}, \\ Q(t)(A_1^{-1}(t))^\sigma &= \begin{pmatrix} 0 & 0 & 0 \\ 0 & 0 & 1 \\ 0 & 0 & t+1 \end{pmatrix} \begin{pmatrix} -1 & \frac{2t+1}{2} & \frac{1}{2} \\ 0 & -\frac{1}{2(t+2)} & \frac{1}{2(t+2)} \\ 0 & -1 & 0 \end{pmatrix} \\ &= \begin{pmatrix} 0 & 0 & 0 \\ 0 & -1 & 0 \\ 0 & -(t+1) & 0 \end{pmatrix}, \\ Q(t)(A^{-1}(t))^\sigma C(t) &= \begin{pmatrix} 0 & 0 & 0 \\ 0 & -1 & 0 \\ 0 & -(t+1) & 0 \end{pmatrix} \begin{pmatrix} 0 & 0 & 1 \\ 0 & -t-1 & 1 \\ 0 & 2 & 1 \end{pmatrix} \\ &= \begin{pmatrix} 0 & 0 & 0 \\ 0 & t+1 & -1 \\ 0 & (t+1)^2 & -(t+1) \end{pmatrix}, \quad t \in \mathbb{T}. \end{aligned}$$

Hence, the system (1.22) takes the form

$$\begin{pmatrix} u_1^\Delta(t) \\ u_2^\Delta(t) \\ u_3^\Delta(t) \end{pmatrix} = \begin{pmatrix} 0 & 0 & 0 \\ 0 & 0 & 0 \\ 0 & 0 & -1 \end{pmatrix} \begin{pmatrix} u_1^\sigma(t) \\ u_2^\sigma(t) \\ u_3^\sigma(t) \end{pmatrix} + \begin{pmatrix} 0 & -t(t+2) & \frac{2t+1}{2} \\ 0 & -\frac{(2t+3)(t+1)^2-2}{2(t+2)} & \frac{2t^2+7t+7}{2(t+2)} \\ 0 & -t(t+1) & t \end{pmatrix}$$

$$\times \begin{pmatrix} u_1(t) \\ u_2(t) \\ u_3(t) \end{pmatrix} + \begin{pmatrix} -1 & \frac{2t+1}{2} & \frac{1}{2} \\ 0 & \frac{2t+3}{2(t+2)} & \frac{1}{2(t+2)} \\ 0 & t & 0 \end{pmatrix} \begin{pmatrix} f_1(t) \\ f_2(t) \\ f_3(t) \end{pmatrix},$$

$$\begin{pmatrix} v_1(t) \\ v_2(t) \\ v_3(t) \end{pmatrix} = - \begin{pmatrix} 0 & 0 & 0 \\ 0 & t+1 & -1 \\ 0 & (t+1)^2 & -(t+1) \end{pmatrix} \begin{pmatrix} u_1(t) \\ u_2(t) \\ u_3(t) \end{pmatrix} - \begin{pmatrix} 0 & 0 & 0 \\ 0 & -1 & 0 \\ 0 & -(t+1) & 0 \end{pmatrix}$$

$$\times \begin{pmatrix} f_1(t) \\ f_2(t) \\ f_3(t) \end{pmatrix}, \quad t \in \mathbb{T},$$

or

$$\begin{pmatrix} u_1^\Delta(t) \\ u_2^\Delta(t) \\ u_3^\Delta(t) \end{pmatrix} = \begin{pmatrix} 0 \\ 0 \\ -u_3^\sigma(t) \end{pmatrix} + \begin{pmatrix} -t(t+2)u_2(t) + \frac{2t+1}{2}u_3(t) \\ -\frac{(2t+3)(t+1)^2-2}{2(t+2)}u_2(t) + \frac{2t^2+7t+7}{2(t+2)}u_3(t) \\ -t(t+1)u_2(t) + tu_3(t) \end{pmatrix}$$

$$+ \begin{pmatrix} -f_1(t) + \frac{2t+1}{2}f_2(t) + \frac{1}{2}f_2(t) \\ \frac{2t+3}{2(t+2)}f_2(t) + \frac{1}{2(t+2)}f_3(t) \\ tf_2(t) \end{pmatrix},$$

$$\begin{pmatrix} v_1(t) \\ v_2(t) \\ v_3(t) \end{pmatrix} = - \begin{pmatrix} 0 \\ (t+1)u_2(t) - u_3(t) \\ (t+1)^2u_2(t) - (t+1)u_3(t) \end{pmatrix} - \begin{pmatrix} 0 \\ -f_2(t) \\ -(t+1)f_3(t) \end{pmatrix}, \quad t \in \mathbb{T},$$

or

$$\begin{aligned} u_1^\Delta(t) &= -t(t+2)u_2(t) + \frac{2t+1}{2}u_3(t) - f_1(t) + \frac{2t+1}{2}f_2(t) + \frac{1}{2}f_2(t), \\ u_2^\Delta(t) &= -\frac{(2t+3)(t+1)^2-2}{2(t+2)}u_2(t) + \frac{2t^2+7t+7}{2(t+2)}u_3(t) \\ &\quad + \frac{2t+3}{2(t+2)}f_2(t) + \frac{1}{2(t+2)}f_3(t), \\ u_3^\Delta(t) &= -u_3^\sigma(t) - t(t+1)u_2(t) + tu_3(t) + tf_2(t), \\ v_1(t) &= 0, \\ v_2(t) &= -(t+1)u_2(t) + u_3(t) + f_2(t), \\ v_3(t) &= -(t+1)^2u_2(t) + (t+1)u_3(t) + (t+1)f_3(t), \quad t \in \mathbb{T}. \end{aligned}$$

1.1.3 Linear time-varying dynamic-algebraic equations of the third kind

In this section, we will investigate the following linear time-varying dynamic-algebraic equation:

$$A(t)(Bx)^{\Delta}(t) = C(t)x^{\sigma}(t) + f(t), \quad t \in I, \tag{1.23}$$

where $A, B, C : I \to \mathscr{M}_{m\times m}$ and $f : I \to \mathbb{R}^m$ are given.

Definition 1.4. Equation (1.23) will be said to be third kind linear time-varying dynamic-algebraic equation.

We will consider the solutions of (1.23) within the space $\mathscr{C}^1_B(I)$. Without loss of generality, we remove the explicit dependence on t.

1.1.3.1 A particular case

In this section, we will investigate the equation

$$Ax^{\Delta} = Cx^{\sigma} + f. \tag{1.24}$$

We will show that it can be reduced to equation (1.23). Let P be a $\mathscr{C}^1$-projector along $\ker A$. Then

$$AP = A$$

and from here,

$$\begin{aligned} Ax^{\Delta} &= APx^{\Delta} \\ &= A(Px^{\Delta}) \\ &= A((Px)^{\Delta} - P^{\Delta}x^{\sigma}) \\ &= A(Px)^{\Delta} - AP^{\Delta}x^{\sigma}. \end{aligned}$$

Then, equation (1.24) can be rewritten in the following manner:

$$A(Px)^{\Delta} - AP^{\Delta}x^{\sigma} = Cx^{\sigma} + f,$$

or

$$A(Px)^{\Delta} = (AP^{\Delta} + C)x^{\sigma} + f.$$

Denoting

$$C_1 = AP^{\Delta} + C,$$

we find

$$A(Px)^{\Delta} = C_1 x^{\sigma} + f, \tag{1.25}$$

and therefore we can consider equation (1.24) as a particular case of equation (1.23).

Example. Let $\mathbb{T} = 2^{\mathbb{N}_0}$ and the matrices A and C be as in (1.4). Then

$$\sigma(t) = 2t, \quad t \in \mathbb{T},$$

and

$$P(t) = \begin{pmatrix} 1 & 0 & 0 \\ 0 & 1 & -1 \\ 0 & 0 & 1-t \end{pmatrix}, \quad t \in \mathbb{T},$$

is a projector along $\ker A$. Also, we have

$$P^{\Delta}(t) = \begin{pmatrix} 0 & 0 & 0 \\ 0 & 0 & 0 \\ 0 & 0 & -1 \end{pmatrix}.$$

Hence,

$$\begin{aligned} C_1(t) &= A(t)P^{\Delta}(t) + C(t) \\ &= \begin{pmatrix} 1 & 0 & 0 \\ 0 & -t & 1 \\ 0 & 0 & 0 \end{pmatrix} \begin{pmatrix} 0 & 0 & 0 \\ 0 & 0 & 0 \\ 0 & 0 & -1 \end{pmatrix} + \begin{pmatrix} -t & 1 & t \\ 0 & 1 & 2t \\ t & 0 & 1 \end{pmatrix} \\ &= \begin{pmatrix} 0 & 0 & 0 \\ 0 & 0 & -1 \\ 0 & 0 & 0 \end{pmatrix} + \begin{pmatrix} -t & 1 & t \\ 0 & 1 & 2t \\ t & 0 & 1 \end{pmatrix} \\ &= \begin{pmatrix} -t & 1 & t \\ 0 & 1 & 2t-1 \\ t & 0 & 1 \end{pmatrix}. \end{aligned}$$

Equation (1.24) can be written as follows:

$$\begin{pmatrix} 1 & 0 & 0 \\ 0 & -t & 1 \\ 0 & 0 & 0 \end{pmatrix} \begin{pmatrix} x_1^{\Delta}(t) \\ x_2^{\Delta}(t) \\ x_3^{\Delta}(t) \end{pmatrix} = \begin{pmatrix} -t & 1 & t \\ 0 & 1 & 2t \\ t & 0 & 1 \end{pmatrix} \begin{pmatrix} x_1(t) \\ x_2(t) \\ x_3(t) \end{pmatrix} + \begin{pmatrix} f_1(t) \\ f_2(t) \\ f_3(t) \end{pmatrix}, \quad t \in \mathbb{T},$$

or

$$\begin{aligned} x_1^{\Delta}(t) &= -tx_1^{\sigma}(t) + x_2^{\sigma}(t) + tx_3^{\sigma}(t) + f_1(t), \\ -tx_2^{\Delta}(t) + x_3^{\Delta}(t) &= x_2^{\sigma}(t) + 2tx_3^{\sigma}(t) + f_2(t), \\ 0 &= tx_1^{\sigma}(t) + x_3^{\sigma}(t) + f_3(t), \quad t \in \mathbb{T}. \end{aligned}$$

This system, using (1.25), can be rewritten in the form

$$\begin{aligned} &\begin{pmatrix} 1 & 0 & 0 \\ 0 & -t & 1 \\ 0 & 0 & 0 \end{pmatrix} \left(\begin{pmatrix} 1 & 0 & 0 \\ 0 & 1 & -1 \\ 0 & 0 & 1-t \end{pmatrix} \begin{pmatrix} x_1(t) \\ x_2(t) \\ x_3(t) \end{pmatrix} \right)^{\Delta} \\ &= \begin{pmatrix} -t & 1 & t \\ 0 & 1 & 2t-1 \\ t & 0 & 1 \end{pmatrix} \begin{pmatrix} x_1^{\sigma}(t) \\ x_2^{\sigma}(t) \\ x_3^{\sigma}(t) \end{pmatrix} + \begin{pmatrix} f_1(t) \\ f_2(t) \\ f_3(t) \end{pmatrix}, \quad t \in \mathbb{T}, \end{aligned}$$

or

$$\begin{pmatrix} 1 & 0 & 0 \\ 0 & -t & 1 \\ 0 & 0 & 0 \end{pmatrix} \begin{pmatrix} x_1^{\Delta}(t) \\ x_2^{\Delta}(t) \\ x_3^{\Delta}(t) \end{pmatrix} = \begin{pmatrix} -tx_1^{\sigma}(t) + x_2^{\sigma}(t) + tx_3^{\sigma}(t) + f_1(t) \\ x_2^{\sigma}(t) + (2t-1)x_3^{\sigma}(t) + f_2(t) \\ tx_1^{\sigma}(t) + x_3^{\sigma}(t) + f_3(t) \end{pmatrix}, \quad t \in \mathbb{T},$$

or

$$\begin{aligned} x_1^{\Delta}(t) &= -tx_1^{\sigma}(t) + x_2^{\sigma}(t) + tx_3^{\sigma}(t) + f_1(t), \\ -t(x_2^{\Delta}(t) - x_3^{\Delta}(t)) + (1-2t)x_3^{\Delta}(t) - x_3(t) &= x_2^{\sigma}(t) + (2t-1)x_3^{\sigma}(t) + f_2(t), \\ 0 &= tx_1^{\sigma}(t) + x_3^{\sigma}(t), \quad t \in \mathbb{T}, \end{aligned}$$

or

$$\begin{aligned} x_1^{\Delta}(t) &= -tx_1^{\sigma}(t) + x_2^{\sigma}(t) + tx_3^{\sigma}(t) + f_1(t), \\ -tx_2^{\Delta}(t) + (1-t)x_3^{\Delta}(t) &= x_2^{\sigma}(t) + (2t-1)x_3^{\sigma}(t) + x_3(t) + f_2(t), \\ 0 &= tx_1^{\sigma}(t) + x_2^{\sigma}(t) + f_3(t), \quad t \in \mathbb{T}. \end{aligned}$$

1.1.3.2 Standard form index 1 problems

Consider the equation

$$A(Px)^{\Delta} = Cx^{\sigma} + f, \tag{1.26}$$

where $\ker A$ is a C^1-space, $C \in \mathscr{C}(I)$, P is a projector so that P^{σ} is a C^1-projector along $\ker A$. Denote

$$Q = I - P.$$

Then

$$AP^{\sigma} = A, \quad AQ^{\sigma} = 0.$$

Assume that

(C1) the matrix

$$A_1 = A + CQ^{\sigma}$$

is invertible.

Definition 1.5. Equation (1.26) is called regular with tractability index 1.

We have the following result.

Lemma 1.3. *Let* (C1) *hold. Then we have the following equations:*

$$A_1^{-1}A = P^{\sigma}, \quad A_1^{-1}CQ^{\sigma} = Q^{\sigma}.$$

Proof. By the definition of A_1, we get

$$\begin{aligned} A_1P^{\sigma} &= (A + CQ^{\sigma})P^{\sigma} \\ &= AP^{\sigma} + CQ^{\sigma}P^{\sigma} \\ &= AP^{\sigma} \\ &= A \end{aligned}$$

and

$$\begin{aligned} A_1Q^{\sigma} &= (A + CQ^{\sigma})Q^{\sigma} \\ &= AQ^{\sigma} + CQ^{\sigma}Q^{\sigma} \\ &= AQ^{\sigma}, \end{aligned}$$

which completes the proof. □

We multiply equation (1.26) by A_1^{-1} and, using the first equation and then the second equation of Lemma 1.3, we arrive at

$$A_1^{-1}A(Px)^{\Delta} = A_1^{-1}Cx^{\sigma} + A_1^{-1}f,$$

or

$$\begin{aligned} P^{\sigma}(Px)^{\Delta} &= A_1^{-1}Cx^{\sigma} + A_1^{-1}f \\ &= A_1^{-1}C(P^{\sigma} + Q^{\sigma})x^{\sigma} + A_1^{-1}f \\ &= A_1^{-1}CP^{\sigma}x^{\sigma} + Q^{\sigma}x^{\sigma} + A_1^{-1}f, \end{aligned}$$

i. e.,

$$P^\sigma(Px)^\Delta = A_1^{-1}CP^\sigma x^\sigma + Q^\sigma x^\sigma + A_1^{-1}f. \tag{1.27}$$

We multiply both sides of equation (1.27) by P^σ and, using

$$P^\sigma P^\sigma = P^\sigma \quad \text{and} \quad P^\sigma Q^\sigma = 0,$$

we obtain

$$P^\sigma P^\sigma(Px)^\Delta = P^\sigma A_1^{-1}CP^\sigma x^\sigma + P^\sigma Q^\sigma x^\sigma + P^\sigma A_1^{-1}f,$$

or

$$P^\sigma(Px)^\Delta = P^\sigma A_1^{-1}CP^\sigma x^\sigma + P^\sigma A_1^{-1}f. \tag{1.28}$$

Note that

$$P^\sigma(Px)^\Delta = (PPx)^\Delta - P^\Delta Px = (Px)^\Delta - P^\Delta Px.$$

Then, equation (1.28) can be rewritten in the form

$$(Px)^\Delta = P^\sigma A_1^{-1}CP^\sigma x^\sigma + P^\Delta Px + P^\sigma A_1^{-1}f.$$

Setting

$$Px = u,$$

we find

$$u^\Delta = P^\sigma A_1^{-1}Cu^\sigma + P^\Delta u + P^\sigma A_1^{-1}f. \tag{1.29}$$

Now, we multiply equation (1.27) by Q^σ and find

$$Q^\sigma P^\sigma(Px)^\Delta = Q^\sigma A_1^{-1}CP^\sigma x^\sigma + Q^\sigma Q^\sigma x^\sigma + Q^\sigma A_1^{-1}f,$$

or

$$0 = Q^\sigma A_1^{-1}CP^\sigma x^\sigma + Q^\sigma x^\sigma + Q^\sigma A_1^{-1}f.$$

Denoting

$$v = Qx,$$

from the latter equation we find

$$0 = Q^\sigma A_1^{-1} C u^\sigma + v^\sigma + Q^\sigma A_1^{-1} f,$$

or

$$v^\sigma = -Q^\sigma A_1^{-1} C u^\sigma - Q^\sigma A_1^{-1} f.$$

Using the latter equation and (1.29), we obtain the system

$$\begin{aligned} u^\Delta &= P^\sigma A_1^{-1} C u^\sigma + P^\Delta u + P^\sigma A_1^{-1} f, \\ v^\sigma &= -Q^\sigma A_1^{-1} C u^\sigma - Q^\sigma A_1^{-1} f. \end{aligned} \tag{1.30}$$

From the first equation of (1.30) we find u, and then we find v from the second equation of (1.30).

Example. Let $\mathbb{T} = \mathbb{N}$ and A, C, P be as in (1.12). Then

$$P^\Delta(t) = \begin{pmatrix} 0 & 0 & 0 \\ 0 & 0 & 0 \\ 0 & 0 & -1 \end{pmatrix}, \quad P^\sigma(t) = \begin{pmatrix} 1 & 0 & 0 \\ 0 & 1 & -1 \\ 0 & 0 & -t-1 \end{pmatrix},$$

$$Q^\sigma(t) = \begin{pmatrix} 0 & 0 & 0 \\ 0 & 0 & 1 \\ 0 & 0 & t+2 \end{pmatrix},$$

$$\begin{aligned} P^\sigma(t) A_1^{-1}(t) &= \begin{pmatrix} 1 & 0 & 0 \\ 0 & 1 & -1 \\ 0 & 0 & -t-1 \end{pmatrix} \begin{pmatrix} -1 & \frac{2t-1}{2} & \frac{1}{2} \\ 0 & -\frac{1}{2(t+1)} & \frac{1}{2(t+1)} \\ 0 & -1 & 0 \end{pmatrix} \\ &= \begin{pmatrix} -1 & \frac{2t-1}{2} & \frac{1}{2} \\ 0 & \frac{2t+1}{2(t+1)} & \frac{1}{2(t+1)} \\ 0 & t+1 & 0 \end{pmatrix}, \\ P^\sigma(t) A_1^{-1}(t) C(t) &= \begin{pmatrix} -1 & \frac{2t-1}{2} & \frac{1}{2} \\ 0 & \frac{2t+1}{2(t+1)} & \frac{1}{2(t+1)} \\ 0 & t+1 & 0 \end{pmatrix} \begin{pmatrix} 0 & 0 & 1 \\ 0 & -t & 1 \\ 0 & 2 & 1 \end{pmatrix} \\ &= \begin{pmatrix} 0 & \frac{2+t-2t^2}{2} & t \\ 0 & \frac{2-t-2t^2}{2(t+1)} & 1 \\ 0 & -t(t+1) & t+1 \end{pmatrix}, \end{aligned}$$

$$Q^\sigma(t) A_1^{-1}(t) = \begin{pmatrix} 0 & 0 & 0 \\ 0 & 0 & 1 \\ 0 & 0 & t+2 \end{pmatrix} \begin{pmatrix} -1 & \frac{2t-1}{2} & \frac{1}{2} \\ 0 & -\frac{1}{2(t+1)} & \frac{1}{2(t+1)} \\ 0 & -1 & 0 \end{pmatrix} = \begin{pmatrix} 0 & 0 & 0 \\ 0 & -1 & 0 \\ 0 & -t-2 & 0 \end{pmatrix},$$

$$Q^\sigma(t)A_1^{-1}(t)C(t) = \begin{pmatrix} 0 & 0 & 0 \\ 0 & -1 & 0 \\ 0 & -t-1 & 0 \end{pmatrix}\begin{pmatrix} 0 & 0 & 1 \\ 0 & -t & 1 \\ 0 & 2 & 1 \end{pmatrix}$$

$$= \begin{pmatrix} 0 & 0 & 0 \\ 0 & t & -1 \\ 0 & t(t+2) & -t-2 \end{pmatrix}, \quad t \in \mathbb{T}.$$

Hence, the system (1.30) takes the form

$$\begin{pmatrix} u_1^\Delta(t) \\ u_2^\Delta(t) \\ u_3^\Delta(t) \end{pmatrix} = \begin{pmatrix} 0 & \frac{2+t-2t^2}{2} & t \\ 0 & \frac{2-t-2t^2}{2(t+1)} & 1 \\ 0 & -t(t+1) & t+1 \end{pmatrix}\begin{pmatrix} u_1^\sigma(t) \\ u_2^\sigma(t) \\ u_3^\sigma(t) \end{pmatrix} + \begin{pmatrix} 0 & 0 & 0 \\ 0 & 0 & 0 \\ 0 & 0 & -1 \end{pmatrix}\begin{pmatrix} u_1(t) \\ u_2(t) \\ u_3(t) \end{pmatrix}$$

$$+ \begin{pmatrix} -1 & \frac{2t-1}{2} & \frac{1}{2} \\ 0 & \frac{2t+1}{2(t+1)} & \frac{1}{2(t+1)} \\ 0 & t+1 & 0 \end{pmatrix}\begin{pmatrix} f_1(t) \\ f_2(t) \\ f_3(t) \end{pmatrix},$$

$$\begin{pmatrix} v_1^\sigma(t) \\ v_2^\sigma(t) \\ v_3^\sigma(t) \end{pmatrix} = \begin{pmatrix} 0 & 0 & 0 \\ 0 & t & -1 \\ 0 & t(t+2) & -t-2 \end{pmatrix}\begin{pmatrix} u_1^\sigma(t) \\ u_2^\sigma(t) \\ u_3^\sigma(t) \end{pmatrix}$$

$$- \begin{pmatrix} 0 & 0 & 0 \\ 0 & -1 & 0 \\ 0 & -t-2 & 0 \end{pmatrix}\begin{pmatrix} f_1(t) \\ f_2(t) \\ f_3(t) \end{pmatrix}, \quad t \in \mathbb{T},$$

or

$$\begin{pmatrix} u_1^\Delta(t) \\ u_2^\Delta(t) \\ u_3^\Delta(t) \end{pmatrix} = \begin{pmatrix} \frac{2+t-2t^2}{2}u_2^\sigma(t) + tu_3^\sigma(t) \\ \frac{2-t-2t^2}{2(t+1)}u_2^\sigma(t) + tu_3^\sigma(t) \\ -t(t+1)u_2^\sigma(t) + (t+1)u_3^\sigma(t) \end{pmatrix}\begin{pmatrix} 0 \\ 0 \\ -u_3(t) \end{pmatrix}$$

$$+ \begin{pmatrix} -f_1(t) + \frac{2t-1}{2}f_2(t) + \frac{1}{2}f_3(t) \\ \frac{2t+1}{2(t+1)}f_2(t) + \frac{1}{2(t+1)}f_3(t) \\ (t+1)f_2(t) \end{pmatrix},$$

$$\begin{pmatrix} v_1^\sigma(t) \\ v_2^\sigma(t) \\ v_3^\sigma(t) \end{pmatrix} = -\begin{pmatrix} 0 \\ tu_2^\sigma(t) - u_3^\sigma(t) \\ t(t+2)u_2^\sigma(t) - (t+2)u_3^\sigma(t) \end{pmatrix} - \begin{pmatrix} 0 \\ -f_2(t) \\ -(t+2)f_3(t) \end{pmatrix}, \quad t \in \mathbb{T},$$

or

$$u_1^\Delta(t) = \frac{2+t-2t^2}{2}u_2^\sigma(t) + tu_3^\sigma(t) - f_1(t) + \frac{2t-1}{2}f_2(t) + \frac{1}{2}f_3(t),$$

$$u_2^\Delta(t) = \frac{2-t-2t^2}{2(t+1)}u_2^\sigma(t) + tu_3^\sigma(t) + \frac{2t+1}{2(t+1)}f_2(t) + \frac{1}{2(t+1)}f_3(t),$$

$$u_3^\Delta(t) = -t(t+1)u_2^\sigma(t) + (t+1)u_3^\sigma(t) + (t+1)f_2(t),$$
$$v_1^\sigma(t) = 0,$$
$$v_2^\sigma(t) = -tu_2^\sigma(t) + u_3^\sigma(t) + f_2(t),$$
$$v_3^\sigma(t) = -t(t+2)u_2^\sigma(t) + (t+2)u_3^\sigma(t) + (t+2)f_3(t), \quad t \in \mathbb{T}.$$

1.1.4 Linear time-varying dynamic-algebraic equations of the fourth kind

In this section, we will investigate the following linear time-varying dynamic-algebraic equation:

$$A(t)(Bx)^\Delta(t) = C(t)x(t) + f(t), \quad t \in I, \tag{1.31}$$

where $A : I \to \mathscr{M}_{n\times m}$, $B : I \to \mathscr{M}_{m\times n}$, $C : I \to \mathscr{M}_{n\times n}$, and $f : I \to \mathbb{R}^n$ are given.

Definition 1.6. Equation (1.31) will be said to be a linear time-varying dynamic-algebraic equation of the fourth kind.

We will consider the solutions of (1.31) within the space $\mathscr{C}_B^1(I)$. Below, we remove the explicit dependence on t for the sake of notational simplicity.

1.1.4.1 A particular case

In this section, we will consider the equation

$$Ax^\Delta = Cx + f. \tag{1.32}$$

Suppose that P is a $\mathscr{C}^1$-projector so that P^σ is a $\mathscr{C}^1$-projector along $\ker A$. Then

$$AP^\sigma = A$$

and, from equation (1.32), we find

$$Cx + f = AP^\sigma x^\Delta = A(Px)^\Delta - AP^\Delta x,$$

whereupon

$$A(Px)^\Delta = (C + AP^\Delta)x + f.$$

We set

$$A_1 = AP^\Delta + C.$$

Then, we get the equation

$$A(Px)^{\Delta} = A_1 x + f. \tag{1.33}$$

Thus, equation (1.32) can be reduced to equation (1.31).

Example. Let $\mathbb{T} = 2^{\mathbb{N}_0}$ and A, C be as in (1.4). We will search for a vector

$$y(t) = \begin{pmatrix} y_1(t) \\ y_2(t) \\ y_3(t) \end{pmatrix} \in \mathbb{R}^3, \quad t \in \mathbb{T},$$

so that

$$A(t)y(t) = 0, \quad t \in \mathbb{T}.$$

We have

$$\begin{pmatrix} 0 \\ 0 \\ 0 \end{pmatrix} = \begin{pmatrix} 1 & 0 & 0 \\ 0 & -t & 1 \\ 0 & 0 & 0 \end{pmatrix} \begin{pmatrix} y_1(t) \\ y_2(t) \\ y_3(t) \end{pmatrix} = \begin{pmatrix} y_1(t) \\ -ty_2(t) + y_3(t) \\ 0 \end{pmatrix}, \quad t \in \mathbb{T},$$

whereupon

$$\begin{aligned} y_1(t) &= 0, \\ y_3(t) &= ty_2(t), \quad t \in \mathbb{T}, \end{aligned}$$

and

$$Q^{\sigma}(t) = \begin{pmatrix} 0 & 0 & 0 \\ 0 & 0 & 2 \\ 0 & 0 & 2t \end{pmatrix}, \quad t \in \mathbb{T},$$

is such that

$$A(t)Q^{\sigma}(t) = 0, \quad t \in \mathbb{T}.$$

We have

$$Q(t) = \begin{pmatrix} 0 & 0 & 0 \\ 0 & 0 & 2 \\ 0 & 0 & t \end{pmatrix}, \quad t \in \mathbb{T},$$

and

$$P(t) = I - Q(t) = \begin{pmatrix} 1 & 0 & 0 \\ 0 & 1 & 0 \\ 0 & 0 & 1 \end{pmatrix} - \begin{pmatrix} 0 & 0 & 0 \\ 0 & 0 & 2 \\ 0 & 0 & t \end{pmatrix} = \begin{pmatrix} 1 & 0 & 0 \\ 0 & 1 & -2 \\ 0 & 0 & 1-t \end{pmatrix}, \quad t \in \mathbb{T},$$

and

$$P^\sigma(t) = \begin{pmatrix} 1 & 0 & 0 \\ 0 & 1 & -2 \\ 0 & 0 & 1-2t \end{pmatrix}, \quad t \in \mathbb{T}.$$

Next,

$$A(t)P^\sigma(t) = \begin{pmatrix} 1 & 0 & 0 \\ 0 & -t & 1 \\ 0 & 0 & 0 \end{pmatrix} \begin{pmatrix} 1 & 0 & 0 \\ 0 & 1 & -2 \\ 0 & 0 & 1-2t \end{pmatrix} = \begin{pmatrix} 1 & 0 & 0 \\ 0 & -t & 1 \\ 0 & 0 & 0 \end{pmatrix} = A(t), \quad t \in \mathbb{T}.$$

Therefore P^σ is a $\mathscr{C}^1$-projector along $\ker A$. Moreover,

$$P^\Delta(t) = \begin{pmatrix} 0 & 0 & 0 \\ 0 & 0 & 0 \\ 0 & 0 & -1 \end{pmatrix}, \quad t \in \mathbb{T},$$

and

$$\begin{aligned} A_1(t) &= A(t)P^\Delta(t) + C(t) \\ &= \begin{pmatrix} 1 & 0 & 0 \\ 0 & -t & 1 \\ 0 & 0 & 0 \end{pmatrix} \begin{pmatrix} 0 & 0 & 0 \\ 0 & 0 & 0 \\ 0 & 0 & -1 \end{pmatrix} + \begin{pmatrix} -t & 1 & t \\ 0 & 1 & 2t \\ t & 0 & 1 \end{pmatrix} \\ &= \begin{pmatrix} 0 & 0 & 0 \\ 0 & 0 & -1 \\ 0 & 0 & 0 \end{pmatrix} + \begin{pmatrix} -t & 1 & t \\ 0 & 1 & 2t \\ t & 0 & 1 \end{pmatrix} \\ &= \begin{pmatrix} -t & 1 & t \\ 0 & 1 & -1+2t \\ t & 0 & 1 \end{pmatrix}, \quad t \in \mathbb{T}. \end{aligned}$$

Equation (1.31) takes the form

$$\begin{pmatrix} 1 & 0 & 0 \\ 0 & -t & 1 \\ 0 & 0 & 0 \end{pmatrix} \begin{pmatrix} x_1^\Delta(t) \\ x_2^\Delta(t) \\ x_3^\Delta(t) \end{pmatrix} = \begin{pmatrix} -t & 1 & t \\ 0 & 1 & 2t \\ t & 0 & 1 \end{pmatrix} \begin{pmatrix} x_1(t) \\ x_2(t) \\ x_3(t) \end{pmatrix} + \begin{pmatrix} f_1(t) \\ f_2(t) \\ f_3(t) \end{pmatrix}, \quad t \in \mathbb{T},$$

or

$$\begin{pmatrix} x_1^\Delta(t) \\ -tx_2^\Delta(t) + x_3^\Delta(t) \\ 0 \end{pmatrix} = \begin{pmatrix} -tx_1(t) + x_2(t) + tx_3(t) \\ x_2(t) + 2tx_3(t) \\ tx_1(t) + x_3(t) \end{pmatrix} + \begin{pmatrix} f_1(t) \\ f_2(t) \\ f_3(t) \end{pmatrix}, \quad t \in \mathbb{T},$$

or

$$\begin{aligned} x_1^\Delta(t) &= -tx_1(t) + x_2(t) + tx_3(t) + f_1(t), \\ -x_2^\Delta(t) + x_3^\Delta(t) &= x_2(t) + 2tx_3(t) + f_2(t), \\ 0 &= tx_1(t) + x_3(t) + f_3(t), \quad t \in \mathbb{T}. \end{aligned}$$

Equation (1.33) can be rewritten in the form

$$\begin{aligned} &\begin{pmatrix} 1 & 0 & 0 \\ 0 & -t & 1 \\ 0 & 0 & 0 \end{pmatrix} \left(\begin{pmatrix} 1 & 0 & 0 \\ 0 & 1 & -2 \\ 0 & 0 & 1-t \end{pmatrix} \begin{pmatrix} x_1(t) \\ x_2(t) \\ x_3(t) \end{pmatrix} \right)^\Delta \\ &= \begin{pmatrix} -t & 1 & t \\ 0 & 1 & -1+2t \\ t & 0 & 1 \end{pmatrix} \begin{pmatrix} x_1(t) \\ x_2(t) \\ x_3(t) \end{pmatrix} + \begin{pmatrix} f_1(t) \\ f_2(t) \\ f_3(t) \end{pmatrix}, \quad t \in \mathbb{T}, \end{aligned}$$

or

$$\begin{aligned} \begin{pmatrix} 1 & 0 & 0 \\ 0 & -t & 1 \\ 0 & 0 & 0 \end{pmatrix} \begin{pmatrix} x_1(t) \\ x_2(t) - 2x_3(t) \\ (1-t)x_3(t) \end{pmatrix}^\Delta &= \begin{pmatrix} -tx_1(t) + x_2(t) + tx_3(t) \\ x_2(t) + (-1+2t)x_3(t) \\ tx_1(t) + x_3(t) \end{pmatrix} \\ &\quad + \begin{pmatrix} f_1(t) \\ f_2(t) \\ f_3(t) \end{pmatrix}, \quad t \in \mathbb{T}, \end{aligned}$$

or

$$\begin{pmatrix} 1 & 0 & 0 \\ 0 & -t & 1 \\ 0 & 0 & 0 \end{pmatrix} \begin{pmatrix} x_1^\Delta(t) \\ x_2^\Delta(t) - 2x_3^\Delta(t) \\ -x_3(t) + (1-2t)x_3^\Delta(t) \end{pmatrix} = \begin{pmatrix} -tx_1(t) + x_2(t) + tx_3(t) + f_1(t) \\ x_2(t) + (-1+2t)x_3(t) + f_2(t) \\ tx_1(t) + x_3(t) + f_3(t) \end{pmatrix},$$
$$t \in \mathbb{T},$$

or

$$\begin{aligned} x_1^\Delta(t) &= -tx_1(t) + x_2(t) + tx_3(t) + f_1(t), \\ -t\big(x_2^\Delta(t) - 2x_3^\Delta(t)\big) - x_3(t) + (1-2t)x_3^\Delta(t) &= x_2(t) + (-1+2t)x_3(t) + f_2(t), \\ 0 &= tx_2(t) + x_3(t) + f_3(t), \quad t \in \mathbb{T}, \end{aligned}$$

or

$$
\begin{aligned}
x_1^\Delta(t) &= -tx_1(t) + x_2(t) + tx_3(t) + f_1(t),\\
-tx_2^\Delta(t) + x_3^\Delta(t) &= x_2(t) + 2tx_3(t) + f_2(t),\\
0 &= tx_2(t) + x_3(t) + f_3(t), \quad t \in \mathbb{T}.
\end{aligned}
$$

1.1.5 Standard form index 1 problems

In this section, we will investigate the equation

$$A(Px)^\Delta = Cx + f, \tag{1.34}$$

where P is a $\mathscr{C}^1$-projector along $\ker A$. Set

$$Q = I - P.$$

Suppose that
(D1) $A_1 = A + CQ$ is invertible.

Definition 1.7. Equation (1.34) is called regular with tractability index 1.

By Lemma 1.1, it follows that

$$A_1^{-1}A = P, \quad A_1^{-1}CQ = Q.$$

We multiply (1.34) by A_1^{-1} and find

$$A_1^{-1}A(Px)^\Delta = A_1^{-1}Cx + A_1^{-1}f,$$

or

$$
\begin{aligned}
P(Px)^\Delta &= A_1^{-1}Cx + A_1^{-1}f\\
&= A_1^{-1}CPx + A_1^{-1}CQx + A_1^{-1}f\\
&= A_1^{-1}CPx + Qx + A_1^{-1}f,
\end{aligned}
$$

i. e.,

$$P(Px)^\Delta = A_1^{-1}CPx + Qx + A_1^{-1}f. \tag{1.35}$$

We multiply equation (1.35) by P, then, using

$$PP = P, \quad P(Px)^\Delta = (Px)^\Delta - P^\Delta P^\sigma x^\sigma,$$

and setting

$$Px = u,$$

we arrive at the equation

$$(Px)^{\Delta} = P^{\Delta}P^{\sigma}x^{\sigma} + PA_1^{-1}CPx + PA_1^{-1}f,$$

or

$$u^{\Delta} = P^{\Delta}u^{\sigma} + PA_1^{-1}Cu + PA_1^{-1}f. \tag{1.36}$$

Now, we multiply (1.35) by Q and, setting

$$Qx = v,$$

we find

$$\begin{aligned} 0 &= QA_1^{-1}CPx + Qx + QA_1^{-1}f \\ &= QA_1^{-1}Cu + v + QA_1^{-1}f. \end{aligned}$$

From the latter equation and (1.37), we obtain

$$\begin{aligned} u^{\Delta} &= P^{\Delta}u^{\sigma} + PA_1^{-1}Cu + PA_1^{-1}f, \\ v &= -QA_1^{-1}Cu - QA_1^{-1}f. \end{aligned} \tag{1.37}$$

Example. Let $\mathbb{T} = \mathbb{N}$ and A, C be as in (1.12). Let also

$$P(t) = \begin{pmatrix} 1 & 0 & 0 \\ 0 & 1 & -\frac{1}{t+1} \\ 0 & 0 & 0 \end{pmatrix}, \quad Q(t) = \begin{pmatrix} 0 & 0 & 0 \\ 0 & 0 & \frac{1}{t+1} \\ 0 & 0 & 1 \end{pmatrix}, \quad t \in \mathbb{T}.$$

We have

$$P(t) + Q(t) = \begin{pmatrix} 1 & 0 & 0 \\ 0 & 1 & -\frac{1}{t+1} \\ 0 & 0 & 0 \end{pmatrix} + \begin{pmatrix} 0 & 0 & 0 \\ 0 & 0 & \frac{1}{t+1} \\ 0 & 0 & 1 \end{pmatrix} = \begin{pmatrix} 1 & 0 & 0 \\ 0 & 1 & 0 \\ 0 & 0 & 1 \end{pmatrix}, \quad t \in \mathbb{T},$$

and

$$\begin{aligned} A(t)P(t) &= \begin{pmatrix} -1 & t+1 & -1 \\ 0 & 0 & 0 \\ 0 & 2(t+1) & -2 \end{pmatrix} \begin{pmatrix} 1 & 0 & 0 \\ 0 & 1 & -\frac{1}{t+1} \\ 0 & 0 & 0 \end{pmatrix} = \begin{pmatrix} -1 & t+1 & -1 \\ 0 & 0 & 0 \\ 0 & 2(t+1) & -2 \end{pmatrix} \\ &= A(t), \quad t \in \mathbb{T}, \end{aligned}$$

and

$$P(t)P(t) = \begin{pmatrix} 1 & 0 & 0 \\ 0 & 1 & -\frac{1}{t+1} \\ 0 & 0 & 0 \end{pmatrix}\begin{pmatrix} 1 & 0 & 0 \\ 0 & 1 & -\frac{1}{t+1} \\ 0 & 0 & 0 \end{pmatrix} = \begin{pmatrix} 1 & 0 & 0 \\ 0 & 1 & -\frac{1}{t+1} \\ 0 & 0 & 0 \end{pmatrix}$$
$$= P(t), \quad t \in \mathbb{T}.$$

Thus, P is a projector along $\ker A$. Next,

$$\begin{aligned} A_1(t) &= A(t) + C(t)Q(t) \\ &= \begin{pmatrix} -1 & t+1 & -1 \\ 0 & 0 & 0 \\ 0 & 2(t+1) & -2 \end{pmatrix} + \begin{pmatrix} 0 & 0 & 1 \\ 0 & -t & 1 \\ 0 & 2 & 1 \end{pmatrix}\begin{pmatrix} 0 & 0 & 0 \\ 0 & 0 & \frac{1}{t+1} \\ 0 & 0 & 1 \end{pmatrix} \\ &= \begin{pmatrix} -1 & t+1 & -1 \\ 0 & 0 & 0 \\ 0 & 2(t+1) & -2 \end{pmatrix} + \begin{pmatrix} 0 & 0 & 1 \\ 0 & 0 & \frac{1}{t+1} \\ 0 & 0 & \frac{t+3}{t+1} \end{pmatrix} \\ &= \begin{pmatrix} -1 & t+1 & 0 \\ 0 & 0 & \frac{1}{t+1} \\ 0 & 2(t+1) & -\frac{t-1}{t+1} \end{pmatrix}, \quad t \in \mathbb{T}. \end{aligned}$$

Hence,

$$\det A(t) = 2 \neq 0, \quad t \in \mathbb{T}.$$

Therefore A_1 is invertible. We will find the cofactors of A_1. We have

$$a_{11}(t) = \begin{vmatrix} 0 & \frac{1}{t+1} \\ 2(t+1) & -\frac{t-1}{t+1} \end{vmatrix} = -2, \qquad a_{12}(t) = -\begin{vmatrix} 0 & \frac{1}{t+1} \\ 0 & -\frac{t-1}{t+1} \end{vmatrix} = 0,$$
$$a_{13}(t) = \begin{vmatrix} 0 & 0 \\ 0 & 2(t+1) \end{vmatrix} = 0,$$
$$a_{21}(t) = -\begin{vmatrix} t+1 & 0 \\ 2(t+1) & -\frac{t-1}{t+1} \end{vmatrix} = t-1, \quad a_{22}(t) = \begin{vmatrix} -1 & 0 \\ 0 & -\frac{t-1}{t+1} \end{vmatrix} = \frac{t-1}{t+1},$$
$$a_{23}(t) = -\begin{vmatrix} -1 & t+1 \\ 0 & 2(t+1) \end{vmatrix} = 2(t+1),$$
$$a_{31}(t) = \begin{vmatrix} t+1 & -1 \\ 0 & 0 \end{vmatrix} = 0, \qquad a_{32}(t) = -\begin{vmatrix} -1 & 0 \\ 0 & \frac{1}{t+1} \end{vmatrix} = \frac{1}{t+1},$$
$$a_{33}(t) = \begin{vmatrix} -1 & t+1 \\ 0 & 0 \end{vmatrix} = 0, \quad t \in \mathbb{T}.$$

Hence,

$$A_1^{-1}(t) = \frac{1}{2}\begin{pmatrix} -2 & t-1 & 0 \\ 0 & \frac{t-1}{t+1} & \frac{1}{t+1} \\ 0 & 2(t+1) & 0 \end{pmatrix} = \begin{pmatrix} -1 & \frac{t-1}{2} & 0 \\ 0 & \frac{t-1}{2(t+1)} & \frac{1}{2(t+1)} \\ 0 & t+1 & 0 \end{pmatrix}, \quad t \in \mathbb{T},$$

and

$$\begin{aligned}
P(t)A_1^{-1}(t) &= \begin{pmatrix} 1 & 0 & 0 \\ 0 & 1 & -\frac{1}{t+1} \\ 0 & 0 & 0 \end{pmatrix}\begin{pmatrix} -1 & \frac{t-1}{2} & 0 \\ 0 & \frac{t-1}{2(t+1)} & \frac{1}{2(t+1)} \\ 0 & t+1 & 0 \end{pmatrix} \\
&= \begin{pmatrix} -1 & \frac{t-1}{2} & 0 \\ 0 & -\frac{t+3}{2(t+1)} & \frac{1}{2(t+1)} \\ 0 & 0 & 0 \end{pmatrix}, \\
P(t)A_1^{-1}(t)C(t) &= \begin{pmatrix} -1 & \frac{t-1}{2} & 0 \\ 0 & -\frac{t+3}{2(t+1)} & \frac{1}{2(t+1)} \\ 0 & 0 & 0 \end{pmatrix}\begin{pmatrix} 0 & 0 & 1 \\ 0 & -t & 1 \\ 0 & 2 & 1 \end{pmatrix} \\
&= \begin{pmatrix} 0 & -\frac{t(t-1)}{2} & \frac{t-3}{2} \\ 0 & \frac{t+2}{2} & -\frac{t+2}{2(t+1)} \\ 0 & 0 & 0 \end{pmatrix}, \\
Q(t)A_1^{-1}(t) &= \begin{pmatrix} 0 & 0 & 0 \\ 0 & 0 & \frac{1}{t+1} \\ 0 & 0 & 1 \end{pmatrix}\begin{pmatrix} -1 & \frac{t-1}{2} & 0 \\ 0 & \frac{t-1}{2(t+1)} & \frac{1}{2(t+1)} \\ 0 & t+1 & 0 \end{pmatrix} = \begin{pmatrix} 0 & 0 & 0 \\ 0 & 1 & 0 \\ 0 & t+1 & 0 \end{pmatrix}, \\
Q(t)A_1^{-1}(t)C(t) &= \begin{pmatrix} 0 & 0 & 0 \\ 0 & 1 & 0 \\ 0 & t+1 & 0 \end{pmatrix}\begin{pmatrix} 0 & 0 & 1 \\ 0 & -t & 1 \\ 0 & 2 & 1 \end{pmatrix} = \begin{pmatrix} 0 & 0 & 0 \\ 0 & -t & 1 \\ 0 & -t(t+1) & t+1 \end{pmatrix}, \\
P^{\Delta}(t) &= \begin{pmatrix} 0 & 0 & 0 \\ 0 & 0 & \frac{1}{(t+1)(t+2)} \\ 0 & 0 & 0 \end{pmatrix}, \quad t \in \mathbb{T}.
\end{aligned}$$

Then, the system (1.37) takes the form

$$\begin{aligned}
\begin{pmatrix} u_1^{\Delta}(t) \\ u_2^{\Delta}(t) \\ u_3^{\Delta}(t) \end{pmatrix} &= \begin{pmatrix} 0 & 0 & 0 \\ 0 & 0 & \frac{1}{(t+1)(t+2)} \\ 0 & 0 & 0 \end{pmatrix}\begin{pmatrix} u_1^{\sigma}(t) \\ u_2^{\sigma}(t) \\ u_3^{\sigma}(t) \end{pmatrix} + \begin{pmatrix} 0 & -\frac{t(t-1)}{2} & \frac{t-3}{2} \\ 0 & \frac{t+2}{2} & -\frac{t+2}{2(t+1)} \\ 0 & 0 & 0 \end{pmatrix} \\
&\quad \times \begin{pmatrix} u_1(t) \\ u_2(t) \\ u_3(t) \end{pmatrix} + \begin{pmatrix} -1 & \frac{t-1}{2} & 0 \\ 0 & -\frac{t+3}{2(t+1)} & \frac{1}{2(t+1)} \\ 0 & 0 & 0 \end{pmatrix}\begin{pmatrix} f_1(t) \\ f_2(t) \\ f_3(t) \end{pmatrix},
\end{aligned}$$

$$\begin{pmatrix} 0 \\ v_0(t) \\ v_1(t) \end{pmatrix} = -\begin{pmatrix} 0 & 0 & 0 \\ 0 & -t & 1 \\ 0 & -t(t+1) & t+1 \end{pmatrix}\begin{pmatrix} u_1(t) \\ u_2(t) \\ u_3(t) \end{pmatrix} - \begin{pmatrix} 0 & 0 & 0 \\ 0 & 1 & 0 \\ 0 & t+1 & 0 \end{pmatrix} \times \begin{pmatrix} f_1(t) \\ f_2(t) \\ f_3(t) \end{pmatrix}, \quad t \in \mathbb{T},$$

or

$$\begin{pmatrix} u_1^\Delta(t) \\ u_2^\Delta(t) \\ u_3^\Delta(t) \end{pmatrix} = \begin{pmatrix} 0 \\ \frac{1}{(t+1)(t+2)} u_3^\sigma(t) \\ 0 \end{pmatrix} + \begin{pmatrix} -\frac{t(t-1)}{2} u_2(t) + \frac{t-3}{2} u_3(t) \\ \frac{t+2}{2} u_2(t) - \frac{t+2}{2(t+1)} u_3(t) \\ 0 \end{pmatrix} + \begin{pmatrix} -f_1(t) + \frac{t-1}{2} f_2(t) \\ -\frac{t+3}{2(t+1)} f_2(t) + \frac{1}{2(t+1)} f_3(t) \\ 0 \end{pmatrix},$$

$$\begin{pmatrix} 0 \\ v_0(t) \\ v_1(t) \end{pmatrix} = -\begin{pmatrix} 0 \\ -tu_2(t) + u_3(t) \\ -t(t+1)u_2(t) + (t+1)u_3(t) \end{pmatrix} - \begin{pmatrix} 0 \\ f_2(t) \\ (t+1)f_2(t) \end{pmatrix}, \quad t \in \mathbb{T},$$

or

$$\begin{aligned} u_1^\Delta(t) &= -\frac{t(t-1)}{2} u_2(t) + \frac{t-3}{2} u_3(t) - f_1(t) + \frac{t-1}{2} f_2(t), \\ u_2^\Delta(t) &= \frac{1}{(t+1)(t+2)} u_3^\sigma(t) + \frac{t+2}{2} u_2(t) - \frac{t+2}{2(t+1)} u_3(t) \\ &\quad - \frac{t+3}{2(t+1)} f_2(t) + \frac{1}{2(t+1)} f_3(t), \\ u_3^\Delta(t) &= 0, \\ v_0(t) &= tu_2(t) - u_3(t) + f_2(t), \\ v_1(t) &= t(t+1)u_2(t) - (t+1)u_3(t) - (t+1)f_2(t), \quad t \in \mathbb{T}. \end{aligned}$$

1.2 Jets and jet spaces

In this chapter, we introduce jets of one independent time scale variable, as well as jets of n independent real variables and one independent time scale variable. We deduct some of their properties and then we define jet spaces and total derivative in jet variables.

Suppose that $\mathbb{T}$ is a time scale with forward jump operator and delta differentiation operator σ and Δ, respectively.

1.2.1 Taylor's formula for a function of one independent time scale variable

We introduce the generalized monomials $h_k : \mathbb{T} \times \mathbb{T} \to \mathbb{R}$, $k \in \mathbb{N}_0$, defined recursively by

$$h_0(t,s) = 1,$$
$$h_k(t,s) = \int_s^t h_{k-1}(\tau,s)\Delta\tau, \quad k \in \mathbb{N}, \quad t,s \in \mathbb{T}.$$

Then

$$h_1(t,s) = \int_s^t \Delta\tau = t - s,$$
$$h_k^{\Delta_t}(t,s) = h_{k-1}(t,s), \quad k \in \mathbb{N}, \quad t,s \in \mathbb{T}.$$

Example. Let $\mathbb{T} = \mathbb{R}$. Then

$$h_k(t,s) = \frac{(t-s)^k}{k!}, \quad k \in \mathbb{N}, \quad t,s \in \mathbb{T}.$$

Example. Let $\mathbb{T} = \mathbb{Z}$. We define

$$t^{(0)} = 1,$$
$$t^{(k)} = \prod_{j=0}^{k-1}(t-j), \quad k \in \mathbb{N}, \quad t \in \mathbb{T}.$$

Then

$$h_0(t,s) = (t-s)^{(0)},$$
$$h_1(t,s) = \int_s^t h_0(\tau,s)\Delta\tau = \int_s^t \Delta\tau = t - s = \frac{(t-s)^{(1)}}{1!}.$$

Assume that

$$h_k(t,s) = \frac{(t-s)^{(k)}}{k!}, \quad t,s \in \mathbb{T}, \tag{1.38}$$

for some $k \in \mathbb{N}$. We will prove that

$$h_{k+1}(t,s) = \frac{(t-s)^{(k+1)}}{(k+1)!}.$$

Indeed, we have

$$\begin{aligned}\left(\frac{(t-s)^{(k+1)}}{(k+1)!}\right)^{\Delta_t} &= (\sigma(t)-s)^{(k+1)} - (t-s)^{(k+1)}\\ &= \frac{1}{(k+_1)!}((\sigma(t)-s)(\sigma(t)-s-1)\cdots(\sigma(t)-s-k)\\ &\quad -(t-s)(t-s-1)\cdots(t-s-k))\\ &= \frac{1}{(k+1)!}((t+1-s)(t+1-s-1)\cdots(t+1-s-k)\\ &\quad -(t-s)(t-s-1)\cdots(t-s-k))\\ &= \frac{1}{(k+1)!}((t+1-s)(t-s)\cdots(t-s-k+1)\\ &\quad -(t-s)(t-s-1)\cdots(t-s-k))\\ &= \frac{1}{(k+1)!}(t-s)\cdots(t-s-k+1)(t+1-s-t+s+k)\\ &= \frac{1}{(k+1)!}(t-s)\cdots(t-s-k+1)(k+1)\\ &= \frac{1}{k!}(t-s)(t-s-1)\cdots(t-s-k+1)\\ &= \frac{(t-s)^{(k)}}{k!}\\ &= h_k(t,s), \quad t,s\in\mathbb{T}.\end{aligned}$$

Therefore (1.38) holds for any $k \in \mathbb{N}$.

Theorem 1.1. *We have*

$$h_{k+m+1}(t,t_0) = \int_{t_0}^{t} h_k(t,\sigma(s))h_m(s,t_0)\Delta s, \quad t,t_0\in\mathbb{T}, \quad k,m\in\mathbb{N}_0.$$

Proof. Let

$$g(t) = \int_{t_0}^{t} h_k(t,\sigma(s))h_m(s,t_0)\Delta s.$$

Then, using the chain rule, we get

$$g^{\Delta}(t) = h_k(\sigma(t),\sigma(t))h_m(\sigma(t),t_0) + \int_{t_0}^{t} h_{k-1}(t,\sigma(s)),h_m(s,t_0)\Delta s$$

$$
\begin{aligned}
&= \int_{t_0}^{t} h_{k-1}(t,\sigma(s))h_m(s,t_0)\Delta s,\\
g^{\Delta^2}(t) &= h_{k-1}(\sigma(t),\sigma(t))h_m(t,t_0) + \int_{t_0}^{t} h_{k-2}(t,\sigma(s))h_m(s,t_0)\Delta s\\
&= \int_{t_0}^{t} h_{k-2}(t,\sigma(s))h_m(s)\Delta s,\\
&\vdots\\
g^{\Delta^k}(t) &= \int_{t_0}^{t} h_m(s,t_0)\Delta s = h_{m+1}(t,t_0),\\
g^{\Delta^{k+1}}(t) &= h_{m+1}(t,t_0),\\
&\vdots\\
g^{\Delta^{k+m}}(t) &= h_{k+m+1}(t,t_0).
\end{aligned}
$$

This completes the proof. □

We define

$$
g_0(t,s) = 1, \quad g_{k+1}(t,s) = \int_{s}^{t} g_k(\sigma(\tau),s)\Delta\tau, \quad k \in \mathbb{N}_0, \quad t,s \in \mathbb{T}.
$$

Lemma 1.4. *Let $n \in \mathbb{N}$. If f is n times differentiable and p_k, $0 \le k \le n-1$, are differentiable at some $t \in \mathbb{T}$ with*

$$
p^{\Delta}_{k+1}(t) = p^{\sigma}_{k}(t) \quad \text{for all } 0 \le k \le n-2, n \ge 2,
$$

then we have

$$
\left(\sum_{k=0}^{n-1} (-1)^k f^{\Delta^k}(t)p_k(t)\right)^{\Delta} = (-1)^{n-1} f^{\Delta^n}(t)p^{\sigma}_{n-1}(t) + f(t)p^{\Delta}_0(t).
$$

Proof. We have

$$
\begin{aligned}
\left(\sum_{k=0}^{n-1} (-1)^k f^{\Delta^k}(t)p_k(t)\right)^{\Delta} &= \sum_{k=0}^{n-1} (-1)^k (f^{\Delta^k}(t)p_k(t))^{\Delta}\\
&= \sum_{k=0}^{n-1} (-1)^k (f^{\Delta^{k+1}}(t)p^{\sigma}_k(t) + f^{\Delta^k}(t)p^{\Delta}_k(t))
\end{aligned}
$$

$$\begin{aligned}
&= \sum_{k=0}^{n-1}(-1)^k f^{\Delta^{k+1}}(t)p_k^\sigma(t) + \sum_{k=0}^{n-1}(-1)^k f^{\Delta^k}(t)p_k^\Delta(t) \\
&= \sum_{k=0}^{n-2}(-1)^k f^{\Delta^{k+1}}(t)p_k^\sigma(t) + (-1)^{n-1}f^{\Delta^n}(t)p_{n-1}^\sigma(t) \\
&\quad + \sum_{k=0}^{n-1}(-1)^k f^{\Delta^k}(t)p_k^\Delta(t) + f^{\Delta^0}(t)p_0^\Delta(t) \\
&= \sum_{k=0}^{n-2}(-1)^k f^{\Delta^{k+1}}(t)p_{k+1}^\Delta(t) + (-1)^{n-1}f^{\Delta^n}(t)p_{n-1}^\sigma(t) \\
&\quad + \sum_{k=0}^{n-2}(-1)^{k+1} f^{\Delta^{k+1}}(t)p_{k+1}^\Delta(t) + f(t)p_0^\Delta(t) \\
&= (-1)^{n-1}f^{\Delta^n}(t)p_{n-1}^\sigma(t) + f(t)p_0^\Delta(t).
\end{aligned}$$

This completes the proof. □

Lemma 1.5. *The functions $g_n(t,s)$ satisfy, for all $t \in \mathbb{T}$, the relationship*

$$g_n(\rho^k(t), t) = 0 \quad \textit{for all } n \in \mathbb{N} \textit{ and all } 0 \le k \le n-1.$$

Proof. Let $n \in \mathbb{N}$ be arbitrarily chosen. Then

$$g_n(\rho^0(t), t) = g_n(t,t) = \int_t^t g_{n-1}(\sigma(\tau), t)\Delta\tau = 0.$$

Assume that

$$g_{n-1}(\rho^k(t), t) = 0 \quad \text{and} \quad g_n(\rho^k(t), t) = 0$$

for some $0 \le k < n-1$.

We will prove that

$$g_n(\rho^{k+1}(t), t) = 0.$$

Case 1. $\rho^k(t)$ is left-dense. Then

$$\rho^{k+1}(t) = \rho(\rho^k(t)) = \rho^k(t).$$

Consequently, using the induction assumption, we have

$$g_n(\rho^{k+1}(t), t) = g_n(\rho^k(t), t) = 0.$$

Case 2. $\rho^k(t)$ is left-scattered. Then

$$\rho(\rho^k(t)) < \rho^k(t)$$

and there is no $s \in \mathbb{T}$ such that $\rho^{k+1}(t) < s < \rho^k(t)$. Hence,

$$\sigma(\rho^{k+1}(t)) = \rho^k(t).$$

Therefore

$$g_n(\sigma(\rho^{k+1}(t)), t) = g_n(\rho^{k+1}(t), t) + \mu(\rho^{k+1}(t)) g_n^{\Delta}(\rho^{k+1}(t), t),$$

or

$$g_n(\rho^k(t), t) = g_n(\rho^{k+1}(t), t) + \mu(\rho^{k+1}(t)) g_n^{\Delta}(\rho^{k+1}(t), t),$$

whereupon

$$\begin{aligned} g_n(\rho^{k+1}(t), t) &= g_n(\rho^k(t), t) - \mu(\rho^{k+1}(t)) g_n^{\Delta}(\rho^{k+1}(t), t) \\ &= g_n(\rho^k(t), t) - \mu(\rho^{k+1}(t)) g_{n-1}(\sigma(\rho^{k+1}(t)), t) \\ &= g_n(\rho^k(t), t) - \mu(\rho^{k+1}(t)) g_{n-1}(\rho^k(t), t) \\ &= 0. \end{aligned}$$

This completes the proof. □

Lemma 1.6. *Let $n \in \mathbb{N}$ and suppose that f is $(n-1)$ times differentiable at $\rho^{n-1}(t)$. Then*

$$f(t) = \sum_{k=0}^{n-1} (-1)^k f^{\Delta^k}(\rho^{n-1}(t)) g_k(\rho^{n-1}(t), t).$$

Proof.

I. Let $n = 1$. Then

$$\sum_{k=0}^{0} (-1)^k f^{\Delta^k}(\rho^0(t)) g_k(\rho^0(t), t) = (-1)^0 f^{\Delta^0}(t) g_0(t, t) = f(t).$$

II. Assume that

$$f(t) = \sum_{k=0}^{m-1} (-1)^k f^{\Delta^k}(\rho^{m-1}(t)) g_k(\rho^{m-1}(t), t)$$

for some $m \in \mathbb{N}$. We will prove that

$$f(t) = \sum_{k=0}^{m} (-1)^k f^{\Delta^k}(\rho^m(t)) g_k(\rho^m(t), t).$$

Case 1. $\rho^{m-1}(t)$ is left-dense. Then

$$\rho^m(t) = \rho(\rho^{m-1}(t)) = \rho^{m-1}(t).$$

Hence by the induction assumption, we obtain

$$\begin{aligned}
&\sum_{k=0}^{m}(-1)^k f^{\Delta^k}(\rho^m(t))g_k(\rho^m(t),t)\\
&\quad=\sum_{k=0}^{m-1}(-1)^k f^{\Delta^k}(\rho^m(t))g_k(\rho^m(t),t)+(-1)^m f^{\Delta^m}(\rho^m(t))g_m(\rho^m(t),t)\\
&\quad=\sum_{k=0}^{m-1}(-1)^k f^{\Delta^k}(\rho^{m-1}(t))g_k(\rho^{m-1}(t),t)\\
&\qquad+(-1)^m f^{\Delta^m}(\rho^{m-1}(t))g_m(\rho^{m-1}(t),t)\\
&\qquad\text{(now we apply Lemma 1.5 } (g_m(\rho^{m-1}(t),t)=0))\\
&\quad=\sum_{k=0}^{m-1}(-1)^k f^{\Delta^k}(\rho^{m-1}(t))g_k(\rho^{m-1}(t),t)\\
&\qquad\text{(now we apply the induction assumption)}\\
&\quad=f(t).
\end{aligned}$$

Case 2. $\rho^{m-1}(t)$ is left-scattered. Then

$$\rho^m(t) = \rho(\rho^{m-1}(t)) < \rho^{m-1}(t)$$

and there is no $s \in \mathbb{T}$ such that

$$\rho^m(t) < s < \rho^{m-1}(t).$$

Also,

$$\sigma(\rho^m(t)) = \rho^{m-1}(t).$$

Hence,

$$g_k(\sigma(\rho^m(t)),t) = g_k(\rho^{m-1}(t),t).$$

Therefore

$$\begin{aligned}
g_k(\rho^{m-1}(t),t) &= g_k(\sigma(\rho^m(t)),t)\\
&= g_k(\rho^m(t),t) + \mu(\rho^m(t))g_k^{\Delta}(\rho^m(t),t)
\end{aligned}$$

$$
\begin{aligned}
&= g_k(\rho^m(t), t) + \mu(\rho^m(t)) g_{k-1}(\sigma(\rho^m(t)), t) \\
&= g_k(\rho^m(t), t) + \mu(\rho^m(t)) g_{k-1}(\rho^{m-1}(t), t),
\end{aligned}
$$

whereupon

$$
g_k(\rho^m(t), t) = g_k(\rho^{m-1}(t), t) - \mu(\rho^m(t)) g_{k-1}(\rho^{m-1}(t), t).
$$

Consequently,

$$
\begin{aligned}
&\sum_{k=0}^{m} (-1)^k f^{\Delta^k}(\rho^m(t)) g_k(\rho^m(t), t) \\
&\quad = f(\rho^m(t)) + \sum_{k=1}^{m} (-1)^k f^{\Delta^k}(\rho^m(t)) g_k(\rho^m(t), t) \\
&\quad = f(\rho^m(t)) + \sum_{k=1}^{m} (-1)^k f^{\Delta^k}(\rho^m(t)) g_k(\rho^{m-1}(t), t) \\
&\qquad + \sum_{k=1}^{m} (-1)^{k-1} f^{\Delta^k}(\rho^m(t)) \mu(\rho^m(t)) g_{k-1}(\rho^{m-1}(t), t) \\
&\quad = f(\rho^m(t)) + \sum_{k=1}^{m-1} (-1)^k f^{\Delta^k}(\rho^m(t)) g_k(\rho^{m-1}(t), t) \\
&\qquad + (-1)^m f^{\Delta^m}(\rho^m(t)) g_m(\rho^{m-1}(t), t) \\
&\qquad + \sum_{k=0}^{m-1} (-1)^k f^{\Delta^{k-1}}(\rho^m(t)) \mu(\rho^m(t)) g_k(\rho^{m-1}(t), t) \\
&\quad = \sum_{k=0}^{m-1} (-1)^k f^{\Delta^k}(\rho^m(t)) g_k(\rho^{m-1}(t), t) \\
&\qquad + \sum_{k=0}^{m-1} (-1)^k \mu(\rho^m(t)) f^{\Delta^{k+1}}(\rho^m(t)) g_k(\rho^{m-1}(t), t) \\
&\quad = \sum_{k=0}^{m-1} (-1)^k (f^{\Delta^k}(\rho^m(t)) + \mu(\rho^m(t)) (f^{\Delta^k})^{\Delta}(\rho^m(t))) g_k(\rho^{m-1}(t), t) \\
&\quad = \sum_{k=0}^{m-1} (-1)^k f^{\Delta^k}(\sigma(\rho^m(t))) g_k(\rho^{m-1}(t), t) \\
&\quad = \sum_{k=0}^{m-1} (-1)^k f^{\Delta^k}(\rho^{m-1}(t)) g_k(\rho^{m-1}(t), t) \\
&\quad = f(t).
\end{aligned}
$$

This completes the proof. □

Theorem 1.2 (Taylor's formula). *Let $n \in \mathbb{N}$. Suppose f is n times differentiable on $\mathbb{T}^{\kappa^n}$. Let $\alpha \in \mathbb{T}^{\kappa^{n-1}}$, $t \in \mathbb{T}$. Then*

$$f(t) = \sum_{k=0}^{n-1} (-1)^k g_k(\alpha, t) f^{\Delta^k}(\alpha) + \int_{\alpha}^{\rho^{n-1}(t)} (-1)^{n-1} g_{n-1}(\sigma(\tau), t) f^{\Delta^n}(\tau) \Delta\tau.$$

Proof. We note that, applying Lemma 1.4 with $p_k = g_k$, we have

$$\begin{aligned}\left(\sum_{k=0}^{n-1} (-1)^k g_k(\tau, t) f^{\Delta^k}(\tau)\right)_{\tau}^{\Delta} &= (-1)^{n-1} f^{\Delta^n}(\tau) g_{n-1}(\sigma(\tau), t) + f(\tau) g_0^{\Delta}(\tau, t) \\ &= (-1)^{n-1} f^{\Delta^n}(\tau) g_{n-1}(\sigma(\tau), t) \quad \text{for all } \tau \in \mathbb{T}^{\kappa^n}.\end{aligned}$$

We integrate the latter relation from α to $\rho^{n-1}(t)$ and get

$$\begin{aligned}&\int_{\alpha}^{\rho^{n-1}(t)} \left(\sum_{k=0}^{n-1} (-1)^k g_k(\tau, t) f^{\Delta^k}(\tau)\right)_{\tau}^{\Delta} \Delta\tau \\ &\quad = \int_{\alpha}^{\rho^{n-1}(t)} (-1)^{n-1} f^{\Delta^n}(\tau) g_{n-1}(\sigma(\tau), t) \Delta\tau,\end{aligned}$$

or

$$\begin{aligned}&\sum_{k=0}^{n-1} (-1)^k g_k(\rho^{n-1}(t), t) f^{\Delta^k}(\rho^{n-1}(t)) - \sum_{k=0}^{n-1} (-1)^k g_k(\alpha, t) f^{\Delta^k}(\alpha) \\ &\quad = \int_{\alpha}^{\rho^{n-1}(t)} (-1)^{n-1} f^{\Delta^n}(\tau) g_{n-1}(\sigma(\tau), t) \Delta\tau.\end{aligned}$$

Hence, applying Lemma 1.6,

$$f(t) - \sum_{k=0}^{n-1} (-1)^k g_k(\alpha, t) f^{\Delta^k}(\alpha) = \int_{\alpha}^{\rho^{n-1}(t)} (-1)^{n-1} f^{\Delta^n}(\tau) g_{n-1}(\sigma(\tau), t) \Delta\tau.$$

This completes the proof. □

Theorem 1.3. *The functions g_n and h_n satisfy the relationship*

$$h_n(t, s) = (-1)^n g_n(s, t)$$

for all $t \in \mathbb{T}$ and all $s \in \mathbb{T}^{\kappa^n}$.

Proof. Let $t \in \mathbb{T}$ and $s \in \mathbb{T}^{\kappa^n}$ be arbitrarily chosen. We apply Theorem 1.2 for $\alpha = s$ and $f(\tau) = h_n(\tau, s)$. We observe that

$$f^{\Delta^k}(\tau) = h_{n-k}(\tau, s), \quad 0 \le k \le n.$$

Hence,

$$f^{\Delta^k}(s) = h_{n-k}(s, s) = 0, \quad 0 \le k \le n-1,$$
$$f^{\Delta^n}(s) = h_0(s, s) = 1, \quad f^{\Delta^{n+1}}(\tau) = 0.$$

From here, using Taylor's formula, we get

$$\begin{aligned} f(t) &= h_n(t,s) \\ &= \sum_{k=0}^{n}(-1)^k g_k(\alpha,t) f^{\Delta^k}(\alpha) + \int_{\alpha}^{\rho^n(t)} (-1)^n g_n(\sigma(\tau),t) f^{\Delta^{n+1}}(\tau)\Delta\tau \\ &= \sum_{k=0}^{n}(-1)^k g_k(s,t) f^{\Delta^k}(s) + \int_{s}^{\rho^n(t)} (-1)^n g_n(\sigma(\tau),t) f^{\Delta^{n+1}}(\tau)\Delta\tau \\ &= \sum_{k=0}^{n-1}(-1)^k g_k(s,t) f^{\Delta^k}(s) + (-1)^n g_n(s,t) f^{\Delta^n}(s) \\ &= (-1)^n g_n(s,t) f^{\Delta^n}(s) \\ &= (-1)^n g_n(s,t), \end{aligned}$$

i. e.,

$$h_n(t,s) = (-1)^n g_n(s,t).$$

This completes the proof. □

From Theorems 1.2 and 1.3, the following theorem, known as the Taylor formula of order n around α, follows.

Theorem 1.4 (Taylor's formula). *Let $n \in \mathbb{N}$. Suppose f is n times differentiable on $\mathbb{T}^{\kappa^n}$. Let also, $\alpha \in \mathbb{T}^{\kappa^{n-1}}$, $t \in \mathbb{T}$. Then*

$$f(t) = \sum_{k=0}^{n-1} h_k(t,\alpha) f^{\Delta^k}(\alpha) + \int_{\alpha}^{\rho^{n-1}(t)} h_{n-1}(t,\sigma(\tau)) f^{\Delta^n}(\tau)\Delta\tau.$$

Now we will formulate and prove another variant of Taylor's formula.

Theorem 1.5 (Taylor's formula). *Let $n \in \mathbb{N}$. Suppose that the function f is $(n+1)$ times differentiable on $\mathbb{T}^{\kappa^{n+1}}$. Let $\alpha \in \mathbb{T}^{\kappa^{n+1}}$, $t \in \mathbb{T}$, and $t > \alpha$. Then*

$$f(t) = \sum_{k=0}^{n} h_k(t,a) f^{\Delta^k}(a) + \int_a^t h_n(t,\sigma(\tau)) f^{\Delta^{n+1}}(\tau)\Delta\tau. \tag{1.39}$$

Proof. Let

$$g(t) = f^{\Delta^{n+1}}(t).$$

Then f solves the problem

$$x^{\Delta^{n+1}} = g(t), \quad x^{\Delta^k}(a) = f^{\Delta^k}(a), \quad k \in \{0,\dots,n\}.$$

We have that

$$y(t,s) = h_n(t,\sigma(s))$$

is the Cauchy function for $y^{\Delta^{n+1}} = 0$. Hence, it follows that

$$\begin{aligned} f(t) &= u(t) + \int_a^t y(t,\sigma(\tau)) g(\tau)\Delta\tau \\ &= u(t) + \int_a^t h_n(t,\sigma(s)) g(s)\Delta s, \end{aligned} \tag{1.40}$$

where, u solves the initial value problem

$$u^{\Delta^{n+1}} = 0, \quad u^{\Delta^m}(a) = f^{\Delta^m}(a), \quad m \in \{0,\dots,n\}.$$

We set

$$w(t) = \sum_{k=0}^{n} h_k(t,a) f^{\Delta^k}(a). \tag{1.41}$$

We have

$$w^{\Delta^m}(t) = \sum_{k=0}^{n} h_{k-m}(t,a) f^{\Delta^k}(a), \quad m \in \{0,\dots,n\},$$

and hence

$$w^{\Delta^m}(a) = \sum_{k=0}^{n} h_{k-m}(a,a) f^{\Delta^k}(a) = f^{\Delta^m}(a), \quad m \in \{0,\dots,n\},$$

i. e., w solves (1.40). Consequently, $w = u$. Hence using (1.41), we obtain (1.39). This completes the proof. □

Example. Let $\mathbb{T} = 2^{\mathbb{N}_0}$ and

$$f(t) = t^4 + t, \quad t \in \mathbb{T}. \tag{1.42}$$

We will apply Taylor's formula of order 3 for f around $\alpha = 1$. Here

$$\sigma(t) = 2t, \quad t \in \mathbb{T},$$

and

$$\begin{aligned}
h_0(t,1) &= 1, \quad h_1(t,1) = \int_1^t \Delta s = t - 1, \\
h_2(t,1) &= \int_1^t h_1(\tau,1)\Delta\tau = \int_1^t (\tau - 1)\Delta\tau = \int_1^t \tau\Delta\tau - \int_1^t \Delta\tau \\
&= \frac{\tau^2}{3}\Big|_{\tau=1}^{\tau=t} - t + 1 = \frac{t^2}{3} - \frac{1}{3} - t + 1 = \frac{t^2}{3} - t + \frac{2}{3}, \\
h_2(t,\sigma(\tau)) = h_2(t,2\tau) &= \int_{2\tau}^t h_1(s,2\tau)\Delta s = \int_{2\tau}^t (s - 2\tau)\Delta s \\
&= \int_{2\tau}^t s\Delta s - 2\tau \int_{2\tau}^t \Delta s = \frac{s^2}{3}\Big|_{s=2\tau}^{s=t} - 2\tau(t - 2\tau) \\
&= \frac{t^2}{3} - \frac{4}{3}\tau^2 - 2\tau t + 4\tau^2 = \frac{t^2}{3} - 2\tau t + \frac{8}{3}\tau^2, \quad t, \tau \in \mathbb{T}, \quad \tau < t.
\end{aligned}$$

Next,

$$\begin{aligned}
&f(1) = 2, \\
&f^{\Delta}(t) = (2^3 + 2^2 + 2 + 1)t^3 + 1 = 15t^3 + 1, \\
&f^{\Delta}(1) = 16, \\
&f^{\Delta^2}(t) = 15(2^2 + 2 + 1)t^2 = 105t^2, \\
&f^{\Delta^2}(1) = 105, \\
&f^{\Delta^3}(t) = 105(2 + 1)t = 315t, \quad t \in \mathbb{T}.
\end{aligned}$$

Now, applying Taylor's formula of order 3 around $t = 1$, we find

$$f(1) + h_1(t,1)f^{\Delta}(1) + h_2(t,1)f^{\Delta^2}(1) + \int_1^t h_2(t,\sigma(\tau))f^{\Delta^3}(\tau)\Delta\tau$$

$$
\begin{aligned}
&= 2 + 16(t-1) + 105\left(\frac{t^2}{3} - t + \frac{2}{3}\right) + 315\int_1^t \left(\frac{t^2}{3} - 2\tau t + \frac{8}{3}\tau^2\right)\tau\Delta\tau \\
&= 2 + 16t - 16 + 35t^2 - 105t + 70 + 105t^2\int_1^t \tau\Delta\tau - 630t\int_1^t \tau^2\Delta\tau + 840\int_1^t \tau^3\Delta\tau \\
&= 35t^2 - 89t + 56 + 105t^2\frac{\tau^2}{3}\Big|_{\tau=1}^{\tau=t} - 630t\frac{\tau^3}{2^2+2+1}\Big|_{\tau=1}^{\tau=t} + 840\frac{\tau^4}{2^3+2^2+2+1}\Big|_{\tau=1}^{\tau=t} \\
&= 35t^2 - 89t + 56 + 35t^4 - 35t^2 - 90t^4 + 90t + 56t^4 - 56 \\
&= t^4 + t \\
&= f(t), \quad t \in \mathbb{T}.
\end{aligned}
$$

Theorem 1.6. *For any $k \in \mathbb{N}_0$, we have*

$$
0 \le h_k(t,s) \le \frac{(t-s)^k}{k!}, \quad t \ge s. \tag{1.43}
$$

Proof. Let

$$
g(t) = (t-s)^{k+1}, \quad t, s \in \mathbb{T}, \quad k \in \mathbb{N}.
$$

Then

$$
\begin{aligned}
g^\Delta(t) &= \lim_{y\to t}\frac{g(\sigma(t)) - g(y)}{\sigma(t) - y} \\
&= \lim_{y\to t}\frac{(\sigma(t)-s)^{k+1} - (y-s)^{k+1}}{\sigma(t) - y} \\
&= \lim_{y\to t}\frac{(\sigma(t)-y)\sum_{\nu=0}^{k}(\sigma(t)-s)^\nu(y-s)^{k-\nu}}{\sigma(t)-y} \\
&= \lim_{y\to t}\sum_{\nu=0}^{k}(\sigma(t)-s)(y-s)^{k-\nu} \\
&= \sum_{\nu=0}^{k}(\sigma(t)-s)^\nu(t-s)^{k-\nu}, \quad t, s \in \mathbb{T}, \quad k \in \mathbb{N}.
\end{aligned}
$$

Note that the inequalities (1.43) are true for $k = 0$. Assume that they are true for some $k \in \mathbb{N}$. We will prove the inequalities (1.43) for $k+1$. We have

$$
\begin{aligned}
0 &\le h_{k+1}(t,s) \\
&= \int_s^t h_k(\tau.s)\Delta\tau
\end{aligned}
$$

$$\begin{aligned}
&\le \frac{1}{k!}\int_s^t (\tau-s)^k \Delta\tau \\
&= \frac{1}{(k+1)!}\int_s^t \sum_{\nu=0}^{k} (\tau-s)^k \Delta\tau \\
&= \frac{1}{(k+1)!}\int_s^t \sum_{\nu=0}^{k} (\tau-s)^\nu (\tau-s)^{k-\nu} \Delta\tau \\
&\le \frac{1}{(k+1)!}\int_s^t \sum_{\nu=0}^{k} (\sigma(\tau)-s)^\nu (\tau-s)^{k-\nu} \Delta\tau \\
&= \frac{1}{(k+1)!}\int_s^t g^\Delta(\tau)\Delta\tau \\
&= \frac{1}{(k+1)!} g(\tau)\Big|_{\tau=s}^{\tau=t} \\
&= \frac{1}{(k+1)!} (\tau-s)^{k+1}\Big|_{\tau=s}^{\tau=t} \\
&= \frac{(t-s)^{k+1}}{(k+1)!}, \quad t,s \in \mathbb{T}, \quad t \ge s.
\end{aligned}$$

By the principle of mathematical induction, it follows that (1.43) is true for any $k \in \mathbb{N}$. This completes the proof. □

Let

$$R_n(t,a) = \int_a^{\rho^{n-1}(t)} h_{n-1}(t,\sigma(\tau)) f^{\Delta^n}(\tau)\Delta\tau.$$

Theorem 1.7. *Let $t \in \mathbb{T}$, $t \ge a$ and*

$$M_n(t) = \sup\{|f^{\Delta^n}(\tau)| : \tau \in [a,t]\}.$$

Then

$$|R_n(t,a)| \le M_n(t)\frac{(t-a)^n}{(n-1)!}.$$

Proof. Let $\tau \in [a,t)$. Then $a \le \sigma(\tau) \le t$ and, applying (1.43), we get

$$0 \le h_{n-1}(t,\sigma(\tau)) \le \frac{(t-\sigma(\tau))^{n-1}}{(n-1)!} \le \frac{(t-\tau)^{n-1}}{(n-1)!} \le \frac{(t-a)^{n-1}}{(n-1)!}.$$

Hence,

$$
\begin{aligned}
|R_n(t,a)| &= \left| \int_a^{\rho^{n-1}(t)} h_{n-1}(t,\sigma(\tau)) f^{\Delta^n}(\tau)\Delta\tau \right| \\
&\le \int_a^t h_{n-1}(t,\sigma(\tau)) |f^{\Delta^n}(\tau)|\Delta\tau \\
&\le M_n(t) \int_a^t \frac{(t-a)^{n-1}}{(n-1)!}\Delta\tau \\
&= M_n(t)\frac{(t-a)^n}{(n-1)!}.
\end{aligned}
$$

This completes the proof. □

1.2.2 Taylor's formula for a function of *n* independent real variables and one independent time scale variable

Suppose that S is an open and convex set in $\mathbb{R}^n$, $I \subseteq \mathbb{T}$, $t_0 \in I$, $f : S\times I \to \mathbb{R}$, $f \in \mathscr{C}^{k+1}(S\times I)$, $(x_1^0,\ldots,x_n^0), (x_1,\ldots,x_n) \in S$, $t \in \mathbb{T}$, $t \ge t_0$. Then, applying the classical Taylor's formula for a function of n independent real variables and then the Taylor's formula for a function of one independent time scale variable, we obtain

$$
\begin{aligned}
& f(x_1,\ldots,x_n,t) - f(x_1^0,\ldots,x_n^0,t_0) \\
&\quad = f(x_1,\ldots,x_n,t) - f(x_1^0,\ldots,x_n^0,t) + f(x_1^0,\ldots,x_n^0,t) - f(x_1^0,\ldots,x_n^0,t_0) \\
&\quad = \sum_{|\alpha|\le k} \frac{(x_1-x_1^0)^{\alpha_1}\cdots(x_n-x_n^0)^{\alpha_n}}{\alpha_1!\cdots\alpha_n!} \frac{\partial^{\alpha_1}\cdots\partial^{\alpha_n}}{\partial x_1^{\alpha_1}\cdots\partial x_n^{\alpha_n}} f(x_1^0,\ldots,x_n^0,t) \\
&\qquad + R_{n,k}(x_1-x_1^0,\ldots,x_n-x_n^0,t) \\
&\qquad + \sum_{l=0}^k h_l(t,t_0)\frac{\partial^l}{\Delta t^l} f(x_1^0,\ldots,x_n^0,t_0) + R_{1k}(x_1^0,\ldots,x_n^0,t_0) \\
&\quad = \sum_{|\alpha|\le k} \frac{(x_1-x_1^0)^{\alpha_1}\cdots(x_n-x_n^0)^{\alpha_n}}{\alpha_1!\cdots\alpha_n!} \frac{\partial^{\alpha_1}\cdots\partial^{\alpha_n}}{\partial x_1^{\alpha_1}\cdots\partial x_n^{\alpha_n}} \sum_{l=0}^k h_l(t,t_0)\frac{\partial^l}{\Delta t^l} f(x_1^0,\ldots,x_n^0,t_0) \\
&\qquad + R_{n,k}(x_1-x_1^0,\ldots,x_n-x_n^0,t) + R_{1k}(x_1^0,\ldots,x_n^0,t,t_0) \\
&\qquad + \sum_{l=0}^k h_l(t,t_0)\frac{\partial^l}{\Delta t^l} f(x_1^0,\ldots,x_n^0,t_0) \\
&\quad = \sum_{|\alpha|\le k}\sum_{l=0}^k \frac{(x_1-x_1^0)^{\alpha_1}\cdots(x_n-x_n^0)^{\alpha_n}}{\alpha_1!\cdots\alpha_n!} h_l(t,t_0) \frac{\partial^{\alpha_1}\cdots\partial^{\alpha_n}}{\partial x_1^{\alpha_1}\cdots\partial x_n^{\alpha_n}} \frac{\partial^l}{\Delta t^l} f(x_1^0,\ldots,x_n^0,t_0) \\
&\qquad + \sum_{l=0}^k h_l(t,t_0)\frac{\partial^l}{\Delta t^l} f(x_1^0,\ldots,x_n^0,t_0) \\
&\qquad + R_{n,k}(x_1-x_1^0,\ldots,x_n-x_n^0,t) + R_{1k}(x_1^0,\ldots,x_n^0,t,t_0),
\end{aligned}
$$

where $R_{nk}(\cdot,\dots,\cdot,\cdot)$ is the remainder in the Taylor's formula for a function of n independent real variables, $R_{1k}(\cdot,\dots,\cdot,\cdot)$ is the remainder in the Taylor's formula for a function of one independent time scale variable, $\alpha = (\alpha_1,\dots,\alpha_n) \in \mathbb{N}_0^n$, $|\alpha| = \alpha_1+\dots+\alpha_n$. Therefore

$$\begin{aligned}
f(x_1,\dots,x_n,t) \\
&= f(x_1^0,\dots,x_n^0,t_0) \\
&\quad + \sum_{|\alpha|\le k}\sum_{l=0}^{k} \frac{(x_1-x_1^0)^{\alpha_1}\cdots(x_n-x_n^0)^{\alpha_n}}{\alpha_1!\cdots\alpha_n!} h_l(t,t_0)\frac{\partial^{\alpha_1}\cdots\partial^{\alpha_n}}{\partial x_1^{\alpha_1}\cdots\partial x_n^{\alpha_n}}\frac{\partial^l}{\Delta t^l} f(x_1^0,\dots,x_n^0,t_0) \\
&\quad + \sum_{l=0}^{k} h_l(t,t_0)\frac{\partial^l}{\Delta t^l} f(x_1^0,\dots,x_n^0,t_0) \\
&\quad + R_{n,k}(x_1-x_1^0,\dots,x_n-x_n^0,t) + R_{1k}(x_1^0,\dots,x_n^0,t,t_0). \qquad (1.44)
\end{aligned}$$

Definition 1.8. Equation (1.44) is said to be the Taylor's formula of order (k,k) for a function of n independent real variables and one independent time scale variable around $(x_1^0,\dots,x_n^0,t_0)$.

Let now $f \in \mathscr{C}^k(S,\mathscr{C}^m(I))$. As above, one can deduct the following formula:

$$\begin{aligned}
f(x_1,\dots,x_n,t) \\
&= f(x_1^0,\dots,x_n^0,t_0) \\
&\quad + \sum_{|\alpha|\le k}\sum_{l=0}^{m} \frac{(x_1-x_1^0)^{\alpha_1}\cdots(x_n-x_n^0)^{\alpha_n}}{\alpha_1!\cdots\alpha_n!} h_l(t,t_0)\frac{\partial^{\alpha_1}\cdots\partial^{\alpha_n}}{\partial x_1^{\alpha_1}\cdots\partial x_n^{\alpha_n}}\frac{\partial^l}{\Delta t^l} f(x_1^0,\dots,x_n^0,t_0) \\
&\quad + \sum_{l=0}^{m} h_l(t,t_0)\frac{\partial^l}{\Delta t^l} f(x_1^0,\dots,x_n^0,t_0) \\
&\quad + R_{n,k}(x_1-x_1^0,\dots,x_n-x_n^0,t) + R_{1m}(x_1^0,\dots,x_n^0,t,t_0). \qquad (1.45)
\end{aligned}$$

Definition 1.9. Equation (1.45) is said to be the Taylor's formula of order (k,m) for a function of n independent real variables and one independent time scale variable around $(x_1^0,\dots,x_n^0,t_0)$.

1.2.3 Jets of a function of one independent time scale variable

Let $t_0 \in \mathbb{T}$.

Definition 1.10. Let f be $(k+1)$ times delta differentiable in a neighborhood U of t_0. Then the k-jet of f at t_0 is defined to be the function

$$(J_{t_0}^k f)(x) = f(t_0) + h_1(z,t_0)f^{\Delta}(t_0) + \cdots + h_k(z,t_0)f^{\Delta^k}(t_0).$$

Example. Let $\mathbb{T} = 2^{\mathbb{N}_0}$ and f be as in (1.42). We will find $(J_1^2 f)(z)$. We have

$$h_1(z,1) = z - 1,$$
$$h_2(z,1) = \frac{z^2}{3} - z + \frac{2}{3}.$$

Then

$$\begin{aligned}(J_1^2 f)(z) &= f(1) + h_1(z,1)f^{\Delta}(1) + h_2(z,1)f^{\Delta^2}(1)\\ &= 2 + 16(z-1) + 105\left(\frac{z^2}{3} - z + \frac{2}{3}\right)\\ &= 2 + 16z - 16 + 35z^2 - 105z + 70\\ &= 35z^2 + 89z + 56.\end{aligned}$$

Theorem 1.8. *Let f and g be $(k+1)$ times delta differentiable in a neighborhood U of t_0. Then*

$$(J_{t_0}^k(af + bg))(z) = a(J_{t_0}^k f)(z) + b(J_{t_0}^k g)(z)$$

for any $a, b \in \mathbb{R}$.

Proof. We have

$$\begin{aligned}&(J_{t_0}^k(af+bg))(z)\\ &= (af+bg)(t_0) + h_1(z,t_0)(af+bg)^{\Delta}(t_0) + \cdots + h_k(z,t_0)(af+bg)^{\Delta^k}(t_0)\\ &= af(t_0) + ah_1(z,t_0)f^{\Delta}(t_0) + \cdots + ah_k(z,t_0)f^{\Delta^k}(t_0)\\ &\quad + bg(t_0) + bh_1(z,t_0)g^{\Delta}(t_0) + \cdots + bh_k(z,t_0)g^{\Delta^k}(t_0)\\ &= a(J_{t_0}^k f)(z) + b(J_{t_0}^k g)(z).\end{aligned}$$

This completes the proof. □

Remark 1.1. Let f and g be $(k+1)$ times delta differentiable in a neighborhood U of t_0. If f^{Λ} exists for any $\Lambda \in S_l^{(k)}$, where $S_l^{(k)}$ is the set consisting of all possible strings of length k containing σ exactly l times and Δ exactly $k - l$ times, then

$$(fg)^{\Delta^k} = \sum_{j=0}^{k}\left(\sum_{\Lambda \in S_j^{(k)}} f^{\Lambda}\right) g^{\Delta^l}.$$

Therefore it is impossible to deduce in the general case a representation for a k-jet $(J_{t_0}^k(fg))(z)$ containing jets of f and jets of g. The same situation is for a representation in the general case for a jet of the composition of f and g via jets of f and jets of g.

1.2.4 Jets of a function of n independent real variables and one independent time scale variable

Definition 1.11. Let S be open and connected set in $\mathbb{R}^n$ that contains $x^0 = (x_1^0, \dots, x_n^0) \in \mathbb{R}^n$, $t_0 \in \mathbb{T}$, and U be a neighborhood of t_0. Suppose that $f : S \times U \to \mathbb{R}$, $f \in \mathscr{C}^{k+1}(S, \mathscr{C}^{m+1}(U))$. Then the (k, m)-jet of f at (x^0, t_0) is defined to be

$$
\begin{aligned}
&(J^{k,m}_{(x^0,t_0)} f)(z_1, \dots, z_n, z_{n+1}) \\
&\quad = f(x_1^0, \dots, x_n^0, t_0) \\
&\qquad + \sum_{|\alpha| \le k} \sum_{l=0}^{m} \frac{(z_1 - x_1^0)^{\alpha_1} \cdots (z_n - x_n^0)^{\alpha_n}}{\alpha_1! \cdots \alpha_n!} h_l(z_{n+1}, t_0) \frac{\partial^{\alpha_1} \cdots \partial^{\alpha_n}}{\partial x_1^{\alpha_1} \cdots \partial x_n^{\alpha_n}} \frac{\partial^l}{\Delta t^l} f(x_1^0, \dots, x_n^0, t_0) \\
&\qquad + \sum_{l=0}^{m} h_l(z_{n+1}, t_0) \frac{\partial^l}{\Delta t^l} f(x_1^0, \dots, x_n^0, t_0).
\end{aligned}
$$

When $k = m$, we will write $(J^{k}_{(x^0,t_0)} f)(z_1, \dots, z_n, z_{n+1})$.

Let

$$
f(x_1, \dots, x_n, t) = f_1(x_1, \dots, x_n) f_2(t),
$$

where $f_1 \in \mathscr{C}^k(S)$ and $f_2 \in \mathscr{C}^m(U)$. Then

$$
\begin{aligned}
&(J^{k,m}_{(x^0,t_0)} f)(z_1, \dots, z_n, z_{n+1}) \\
&\quad = f(x_1^0, \dots, x_n^0, t_0) \\
&\qquad + \sum_{|\alpha| \le k} \sum_{l=0}^{m} \frac{(z_1 - x_1^0)^{\alpha_1} \cdots (z_n - x_n^0)^{\alpha_n}}{\alpha_1! \cdots \alpha_n!} h_l(z_{n+1}, t_0) \frac{\partial^{\alpha_1} \cdots \partial^{\alpha_n}}{\partial x_1^{\alpha_1} \cdots \partial x_n^{\alpha_n}} \frac{\partial^l}{\Delta t^l} f(x_1^0, \dots, x_n^0, t_0) \\
&\qquad + \sum_{l=0}^{m} h_l(z_{n+1}, t_0) \frac{\partial^l}{\Delta t^l} f(x_1^0, \dots, x_n^0, t_0) \\
&\quad = f_1(x_1^0, \dots, x_n^0) f_2(t_0) \\
&\qquad + \sum_{|\alpha| \le k} \sum_{l=0}^{m} \frac{(z_1 - x_1^0)^{\alpha_1} \cdots (z_n - x_n^0)^{\alpha_n}}{\alpha_1! \cdots \alpha_n!} h_l(z_{n+1}, t_0) \\
&\qquad \times \frac{\partial^{\alpha_1} \cdots \partial^{\alpha_n}}{\partial x_1^{\alpha_1} \cdots \partial x_n^{\alpha_n}} f_1(x_1^0, \dots, x_n^0) \frac{\partial^l}{\Delta t^l} f_2(t_0) \\
&\qquad + f_1(x_1^0, \dots, x_n^0) \sum_{l=0}^{m} h_l(z_{n+1}, t_0) \frac{\partial^l}{\Delta t^l} f_2(t_0) \\
&\quad = f_1(x_1^0, \dots, x_n^0) f_2(t_0) \\
&\qquad + \left(\sum_{|\alpha| \le k} \frac{(z_1 - x_1^0)^{\alpha_1} \cdots (z_n - x_n^0)^{\alpha_n}}{\alpha_1! \cdots \alpha_n!} \frac{\partial^{\alpha_1} \cdots \partial^{\alpha_n}}{\partial x_1^{\alpha_1} \cdots \partial x_n^{\alpha_n}} f_1(x_1^0, \dots, l x_n^0) \right)
\end{aligned}
$$

$$\times\left(\sum_{l=0}^{m} h_l(z_{n+1},t_0)\frac{\partial^l}{\Delta t^l}f_2(t_0)\right)+f_1(x_1^0,\dots,x_n^0)(J_{t_0}^m f))(z_{n+1})$$
$$=f(x_1^0,\dots,x_n^0,t_0)+(J_{x^0}^k f_1)(z_1,\dots,z_n)(J_{t_0}^m f_2)(z_{n+1})$$
$$+f_1(x_1^0,\dots,x_n^0)(J_{t_0}^m f))(z_{n+1}).$$

Example. Let $\mathbb{T}=2^{\mathbb{N}_0}$ and

$$f_1(x)=x^4+x^2+x,\quad x\in\mathbb{R},$$

as well as

$$g(x,t)=f_1(x)f(t),\quad x\in\mathbb{R},\quad t\in\mathbb{T},$$

where f is the function in (1.42). We will find

$$(J_{(1,1)}^{3,2}f)(z_1,z_2).$$

We have

$$g(x,t)=(x^4+x^2+x)(t^4+t),\quad (x,t)\in\mathbb{R}\times\mathbb{T},$$
$$g(1,1)=3\cdot 2=6,$$

and

$$\begin{aligned} & & f_1(1)&=3,\\ f_1'(x)&=4x^3+2x+1, & f_1'(1)&=7,\\ f_1''(x)&=12x^2+2, & f_1''(1)&=14,\\ f_1'''(x)&=24x, & f_1'''(1)&=24.\end{aligned}$$

Then

$$\begin{aligned}(J_1^3 f_1)(z_1)&=f_1(1)+f_1'(1)z_1+\frac{1}{2}z_1^2f_1''(1)+\frac{1}{6}z_1^3f_1'''(1)\\ &=3+7z_1+7z_1^2+4z_1^3.\end{aligned}$$

Now, we find

$$\begin{aligned}(J_{(1,1)}^{3,2}g)(z_1,z_2)&=g(1,1)+(J_1^3f_1)(z_1)(J_1^2f)(z_2)+f_1(1)(J_1^2f_2)(z_2)\\ &=6+(3+7z_1+7z_1^2+4z_1^3)(35z_2^2+89z_2+56)\\ &\quad+3(35z_2^2+89z_2+56)\\ &=6+(6+7z_1+7z_1^2+4z_1^3)(35z_2^2+89z_2+56).\end{aligned}$$

1.2.5 Jet spaces

Suppose that $f : \mathbb{R}^{p+1} \times \mathbb{T} \to \mathbb{R}$,

$$f = f(x_1, \dots, x_{p-1}, t)$$

depends on $p-1$ independent real variables and one independent time scale variable t. For convenience, we introduce the notation $x_p = t$ and will write ∂x_p instead of Δt. The function f has

$$p_k = \begin{pmatrix} p+k-1 \\ k \end{pmatrix}$$

different kth partial derivatives

$$\partial_J f = \frac{\partial^k f}{\partial x_{j_1} \cdots \partial x_{j_k}}$$

indexed by unordered multiindices

$$J = (j_1, \dots, j_k), \quad 1 \le j_k \le p,$$

of order $k = \#J$. Thus, if we have q dependent variables $(u^1, \dots, u^q)$, we will require

$$q_k = q p_k$$

different coordinates

$$u_J^\alpha = \partial_J f^\alpha(x)$$

of a function

$$u = f(x).$$

For the total space

$$E = X \times U \simeq \mathbb{R}^{p-1} \times \mathbb{T} \times \mathbb{R}^q,$$

the nth jet space

$$J^n = J^n E = X \times U^{(n)}$$

is the Euclidean space of dimension

$$p + q^{(n)} = p + q \begin{pmatrix} p+n \\ n \end{pmatrix},$$

whose coordinates consist of the p independent variables x_i, the q dependent variables u^α, and the derivative coordinates u_J^α, $\alpha = 1, \dots, q$, of order

$$1 \le \#J \le n.$$

Definition 1.12. The space $U^{(n)}$ will be called the vertical space or the fiber.

The points of the vertical space $U^{(n)}$ are denoted by $u^{(n)}$ and consist of all the dependent variables and their derivatives up to order n. Thus, the coordinates of a typical point $z \in J^n$ are denoted by

$$(x, u^{(n)}).$$

Because the derivative coordinates $u^{(n)}$ form a subset of the derivative coordinates $u^{(n+k)}$, there is a projector

$$\pi_n^{n+k} : J^{n_k} \to J^n$$

on the jet space with

$$\pi_n^{n+k}(x, u^{(n+k)}) = (x, u^{(n)}).$$

In particular, we have that

$$\pi_0^n(x, u^{(n)}) = (x, u)$$

is the projector from J^n to $E = J^0$. If $M \subset E$ is an open subset, then

$$J^n M = (\pi_0^n)^{-1} M \subset J^n E$$

is the open subset of the nth jet space which projects back down to M.

1.2.6 Total derivatives

Definition 1.13. A smooth real-valued function $F : J^n \to \mathbb{R}$, defined on an open subset of the nth jet space, is called a differentiable function.

Definition 1.14. Let $F(x, u^{(n)})$ be a differentiable function of order n. The total derivative of F with respect to x_i is the $(n+1)$th order differential function $D_i F$ that satisfies

$$D_i F(x, f^{(n+1)}(x)) = \frac{\partial}{\partial x_i} F(x, f^{(n)}(x))$$

for any smooth function $u = f(x)$.

Example. In the case of one independent time scale variable t and one dependent variable, the total derivative of a given function

$$F(t, u, u^{\Delta}(t), \ldots, u^{\Delta^n}(t))$$

is given by

$$\begin{aligned} &D_t F(t, u(t), u^{\Delta}(*t), \ldots, u^{\Delta^n}(t)) \\ &= \frac{\partial}{\Delta t} F(c_1, u(c_1), u^{\Delta}(c_1), \ldots, u^{\Delta^n}(c_1)) + u^{\Delta} \frac{\partial}{\partial u} F(\sigma(c_2), u(c_2), u^{\Delta}(c_2), \ldots, u^{\Delta^n}(c_2)) \\ &\quad + \cdots + u^{\Delta^{n+1}}(t) \frac{\partial}{\partial u^{\Delta^n}} F(\sigma(c_{n+1}), u(\sigma(c_{n+1})), \ldots, u^{\Delta^{n-1}}(\sigma(c_{n+1})), u^{\Delta^n}(c_{n+1})) \end{aligned}$$

for some $c_j \in [t, \sigma(t)]$, $j \in \{1, \ldots, n+1\}$.

1.3 Nonlinear dynamic systems

In this chapter, we will investigate the following nonlinear dynamic system:

$$f((g(x(t), t))^{\Delta}, x(t), t) = 0, \tag{1.46}$$

where $f : \mathbb{R}^n \times D_f \times I_f \to \mathbb{R}^k$, $D_f \times I_f \subseteq \mathbb{R}^m \times \mathbb{T}$ is continuous and has continuous classical derivatives f_y, f_x, as well as a continuous delta derivative f_t^{Δ}, $g : D_f \times I_f \to \mathbb{R}^n$ is continuous and has a continuous classical derivative g_x and a continuous delta derivative g_t^{Δ}.

A solution x of the nonlinear dynamic system (1.46) is a function defined on an interval $I_* \subseteq I_f$ so that $x(t) \in D_f$ for any $t \in I_*$, the function $g(x(\cdot), \cdot)$ is continuously delta differentiable on I_* and x satisfies equation (1.46) pointwise on I_*.

1.3.1 Properly involved derivatives

Define

$$F(y^1(t), y^2(t), x(t), t) = \int_0^1 \frac{\partial}{\partial y} f(hy^1(t) + (1-h)y^2(t), x(t), t)dh,$$

$$G(x^1(t), x^2(t), t) = \int_0^1 \frac{\partial}{\partial x} g(hx^1(t) + (1-h)x^2(t), t)dh,$$

where

$$\frac{\partial}{\partial y}f(hy^1(t)+(1-h)y^2(t),x(t),t)$$
$$=\left(\frac{\partial}{\partial y_j}f(y_1^1(t),\dots,y_{j-1}^1(t),hy_j^1(t)+(1-h)y_j^2(t),y_{j+1}^2(t),\dots,y_n^2(t),x(t),t)\right)_{i,j=1}^{k,n}$$

and

$$\frac{\partial}{\partial x}g(hx^1(t)+(1-h)x^2(t),t)$$
$$=\left(\frac{\partial}{\partial x_j}g(x_1^1(t),\dots,x_{j-1}^1(t),hx_j^1(t)+(1-h)x_j^2(t),x_{j+1}^2(t),\dots,x_m^2(t),\sigma(t))\right)_{i,j=1}^{n,m},$$

and $y^1(t),y^2(t)\in\mathbb{R}^n$, $x(t),x^1(t),x^2(t)\in D_f$, $t\in I_f$.

Definition 1.15. The nonlinear dynamic equation (1.46) has a properly involved derivative, also called a properly stated leading term, if $\operatorname{im} G$ and $\ker F$ are $\mathscr{C}^1$-subspaces and the transversality condition

$$\begin{aligned}&\ker F(y^1(t),y^2(t),x(t),t)\oplus\operatorname{im} G(x^1(t),x^2(t),t)=\mathbb{R}^n,\\&y^1(t),y^2(t)\in\mathbb{R}^n,\quad x(t),x^1(t),x^2(t)\in D_f,\quad t\in I_f,\end{aligned}\tag{1.47}$$

holds.

Example. Let $\mathbb{T}=2^{\mathbb{N}_0}$. Consider the following nonlinear dynamic system:

$$\begin{aligned}(x_1(t)-2x_2(t)x_3(t))^\Delta+x_2(t)-q_1(t)&=0,\\x_1(t)+x_2(t)-q_2(t)&=0,\\x_2(t)-2x_3(t)-q_3(t)&=0,\quad t\in\mathbb{T}.\end{aligned}$$

Here

$$n=1,\quad k=m=3.$$

We have

$$f(y^1,y^2,x,t)=\begin{pmatrix}1\\0\\0\end{pmatrix}y+\begin{pmatrix}x_2\\x_1+x_2\\x_2-2x_3\end{pmatrix}-q(t),\quad x\in\mathbb{R}^3,\quad y^1,y^2\in\mathbb{R},\quad t\in\mathbb{T},$$

and

$$g(x^1,x^2,t)=x_1(t)-2x_2(t)x_3(t),\quad x^1,x^2\in\mathbb{R}^3,\quad t\in\mathbb{T}.$$

Then

$$F(y^1(t), y^2(t), x(t), t) = \int_0^1 \begin{pmatrix} 1 \\ 0 \\ 0 \end{pmatrix} dh = \begin{pmatrix} 1 \\ 0 \\ 0 \end{pmatrix}, \quad x \in \mathbb{R}^3, \quad y^1, y^2 \in \mathbb{R}, \quad t \in \mathbb{T}.$$

Hence,

$$\ker F(y^1(t), y^2(t), x(t), t) = \{0\}, \quad y^1(t), y^2(t)\mathbb{R}, \quad x(t) \in \mathbb{R}^3, \quad t \in \mathbb{T}.$$

Next,

$$\begin{aligned} G(x^1(t), x^2(t), t) &= \int_0^1 (1, -2x_3^2(t), -2x_2^1(t))dh \\ &= (1, -2x_3^2(t), -2x_2^1(t)), \quad x^1(t), x^2(t) \in \mathbb{R}^3, \quad t \in \mathbb{T}. \end{aligned}$$

Then

$$\ker G(x^1(t), x^2(t), t) = \{0\}, \quad x^1(t), x^2(t) \in \mathbb{R}^3, \quad t \in \mathbb{T},$$

and

$$\operatorname{im} G(x^1(t), x^2(t), t) = \mathbb{R}, \quad x^1(t), x^2(t) \in \mathbb{R}^3, \quad t \in \mathbb{T}.$$

Consequently,

$$\begin{aligned} \ker F(y^1(t), y^2(t), x(t), t) \oplus \operatorname{im} G(x^1(t), x^2(t), t) &= \mathbb{R}, \\ y^1(t), y^2(t) \in \mathbb{T}, \quad x(t), x^1(t), x^2(t) \in \mathbb{R}^3, \quad t \in \mathbb{T}. \end{aligned}$$

Thus, the considered nonlinear dynamic system has a properly involved derivative.

Note that the nonlinear dynamic system (1.46) covers equation (1.31) with

$$\begin{aligned} f(y, x, t) &= A(t)y - C(t)x - f(t), \\ g(x, t) &= B(t)x. \end{aligned}$$

If equation (1.46) has a properly involved derivative, then F and G have constant rank on their definition domain.

1.3.2 Constraints and consistent initial values

Consider equation (1.46) subject to the initial condition

$$x(t_0) = x_0, \tag{1.48}$$

where $t_0 \in I_f$ and $x_0 \in D_f$.

Definition 1.16. For a given nonlinear dynamic system (1.46) and a given $t_0 \in I_f$, the value $x_0 \in D_f$ is said to be consistent if the IVP (1.46), (1.48) possesses a solution.

Definition 1.17. For a nonlinear dynamic system (1.46) with a properly involved derivative, the set

$$\begin{aligned}\mathscr{M}_0(t) = \{x(t) \in D_f : \exists y(t) \in \mathbb{R}^n \text{ such that} \\ y(t) - g_t^\Delta(x(\sigma(t)), t) \in \operatorname{im} G(x^\sigma(t), x(t), t), \\ f(y(t), x(t), t) = 0\}\end{aligned}$$

is called obvious restriction set or obvious constraint of the equation (1.46) at $t \in I_f$.

Example. Let $\mathbb{T} = \mathbb{Z}$. Consider the nonlinear dynamic equation

$$\begin{aligned}x_1^\Delta(t) &= 2x_1(t), \\ 3(x_1(t))^4 + (x_2(t))^2 &= 1 + q(t), \quad t \in \mathbb{T},\end{aligned} \tag{1.49}$$

with two equations and two unknown functions on

$$D_f = \{x \in \mathbb{R}^2 : x_2 > 0\},$$

with $I_f = \mathbb{T}$, $q \in \mathscr{C}(I_f)$, $q > -1$ on I_f. This equation can be rewritten in the form (1.46) using

$$\begin{aligned}f(y, x, t) &= \begin{pmatrix} y - 2x_1 \\ 3x_1^4 + x_2^2 - 1 - q(t) \end{pmatrix}, \\ g(x, t) &= x_1, \quad x \in D_f, \quad y \in \mathbb{R}, \quad t \in I_f.\end{aligned}$$

We have

$$\begin{aligned}&F(y^1(t), y^2(t), x(t), t) = \begin{pmatrix} 1 \\ 0 \end{pmatrix}, \\ &\quad G(x^1(t), x^2(t), t) = (1, 0), \quad y^1(t), y^2(t) \in \mathbb{R}, \quad x(t), x^1(t), x^2(t) \in D_f, \quad t \in I_f.\end{aligned}$$

From here,

$$\begin{aligned}&\ker F(y^1(t), y^2(t), x(t), t) = \ker G(x^1(t), x^2(t), t) = \{0\}, \\ &y^1(t), y^2(t) \in \mathbb{R}, \quad x^1(t), x^2(t), x(t) \in D_f, \quad t \in I_f.\end{aligned}$$

Therefore

$$\operatorname{im} G(x^1(t), x^2(t), t) = \mathbb{R}, \quad x^1(t), x^2(t) \in D_f, \quad t \in I_f,$$

and

$$\ker F(y^1(t), y^2(t), x(t), t) \oplus \operatorname{im} G(x^1(t), x^2(t), t) = \mathbb{R},$$
$$y^1(t), y^2(t) \in \mathbb{R}, \quad x(t), x^1(t), x^2(t) \in D_f, \quad t \in I_f.$$

Thus, equation (1.49) has a properly involved derivative. From the second equation of (1.49), we conclude that the solutions of (1.49) must lie in

$$\mathscr{M}_0(t) = \{x \in D_f : 3x_1^4 + x_2^2 - 1 - q(t) = 0\}.$$

Let

$$x_1(0) = x_{10}.$$

Then

$$x_{20} = \pm\sqrt{1 + q(0) - 3x_{10}^4}.$$

By the first equation of (1.49), we find

$$x_1(t) = e_2(t, 0)x_{10} = x_{10}e^{\int_0^2 \log(1+2)\Delta\tau} = x_{10}e^{t\log 3} = x_{10}3^t, \quad t \in \mathbb{T},$$

and then

$$x_2(t) = \sqrt{1 + q(t) - 3x_{10}^4 3^{4t}}, \quad t \in \mathbb{T}.$$

It is clear that through each point of $\mathscr{M}_0(0)$ passes exactly one solution.

Now, we consider equation (1.46). Suppose that $x \in \mathscr{C}^1(I_f)$ and set

$$u(t) = g(x(t), t), \quad t \in I_f.$$

Using the generalized Pötzsche chain rule, we find

$$\begin{aligned} u^\Delta(t) &= \left(\int_0^1 g(x(t) + h\mu(t)x^\Delta(t), t)dh\right)x^\Delta(t) + g_t^\Delta(x(\sigma(t)), t) \\ &= G(x^\sigma(t), x(t), t)x^\Delta(t) + g_t^\Delta(x(\sigma(t)), t), \quad t \in I_f, \end{aligned}$$

or

$$u^\Delta(t) - g_t^\Delta(x(\sigma(t)), t) = G(x^\sigma(t), x(t), t)x^\Delta(t), \quad t \in I_f.$$

From the inclusion

$$u^\Delta(t) - g_t^\Delta(x(\sigma(t)), t) \in \operatorname{im} G(x^\sigma(t), x(t), t), \quad t \in I_f, \tag{1.50}$$

it follows that there is a $w(t) \in \operatorname{im} G(x^\sigma(t), x(t), t)$, $t \in I_f$, such that

$$u^{\Delta}(t) = G(x^{\sigma}(t), x(t), t)w(t) + g_t^{\Delta}(x^{\sigma}(t), t), \quad t \in I_f.$$

Note that the inclusion (1.50) holds trivially in the case when $G(x(t), t)$ has full row rank for any $t \in I_f$. For any solution of equation (1.46), we have the following valid identities:

$$f(u^{\Delta}(t), x(t), t) = 0, \quad t \in I_f,$$

and

$$f(G(x^{\sigma}(t), x(t), t)w(t) + g_t^{\Delta}(x^{\sigma}(t), t)) = 0, \quad t \in I_f,$$

and then the values $x(t)$ belong to the set

$$\widetilde{\mathscr{M}_0}(t) = \{x \in D_f : \exists y \in \mathbb{R}^n : f(y, x, t) = 0\}.$$

The sets $\mathscr{M}_0(t)$ and $\widetilde{\mathscr{M}_0}(t)$ are defined for any $t \in I_f$. We have

$$\mathscr{M}_0(t) \subseteq \widetilde{\mathscr{M}_0}(t), \quad t \in I_f.$$

If $G(x^{\sigma}(t), x(t), t)$ has full row rank, we have

$$\mathscr{M}_0(t) = \widetilde{\mathscr{M}_0}(t), \quad t \in I_f.$$

Theorem 1.9. *Let equation* (1.46) *have a properly involved derivative. Then, for each* $t \in I_f$ *and each* $x(t) \in \mathscr{M}_0(t)$, *there is a unique* $y(t) \in \mathbb{R}^n$ *such that*

$$y(t) - g_t^{\Delta}(x(\sigma(t)), t) \in \operatorname{im} G(x^{\sigma}(t), x(t), t)$$

and

$$f(y(t), x(t), t) = 0.$$

Proof. Let $t_1 \in I_f$ and $x^1(t_1) \in \mathscr{M}_0(t_1)$ be arbitrarily chosen. Suppose that there are $y^1(t_1), y^2(t_1) \in \mathbb{R}^n$ so that

$$\begin{aligned} y^1(t_1) - g_t^{\Delta}(x^1(\sigma(t_1)), x^1(t_1), t_1) &\in G(x^1(\sigma(t_1)), x^1(t_1), t_1), \\ y^2(t_1) - g_t^{\Delta}(x^1(\sigma(t_1)), x^1(t_1), t_1) &\in G(x^1(\sigma(t_1)), x^1(t_1), t_1). \end{aligned} \tag{1.51}$$

Let

$$N = \ker G(x^1(\sigma(t_1)), x^1(t_1), t_1)$$

and

$$w^1(t_1) = G(x^1(\sigma(t_1)), x^1(t_1), t_1)^+ (y^1(t_1) - g_t^{\Delta}(x^1(t_1), t_1)),$$

$$w^2(t_1) = G(x^1(\sigma(t_1)), x^1(t_1), t_1)^+ (y^2(t_1) - g_t^\Delta(x^1(t_1), t_1)).$$

Here $G(x^1(\sigma(t_1)), x^1(t_1), t_1)^+$ is the Moore–Penrose inverse of $G(x^1(\sigma(t_1)), x^1(t_1), t_1)$. Then $w^1(t_1), w^2(t_1) \in N^\perp$ and

$$\begin{aligned} &G(x^1(\sigma(t_1)), x^1(t_1), t_1)w^1(t_1) \\ &\quad = G(x^1(\sigma(t_1)), x^1(t_1), t_1)G(x^1(\sigma(t_1)), x^1(t_1), t_1)^+ (y^1(t_1) - g_t^\Delta(x^1(t_1), t_1)), \\ &G(x^1(\sigma(t_1)), x^1(t_1), t_1)w^2(t_1) \\ &\quad = G(x^1(\sigma(t_1)), x^1(t_1), t_1)G(x^1(\sigma(t_1)), x^1(t_1), t_1)^+ (y^2(t_1) - g_t^\Delta(x^1(t_1), t_1)). \end{aligned}$$

By (1.51), it follows that there are $q_1(t_1)$, $q_2(t_1)$ such that

$$\begin{aligned} y^1(t_1) - g_t^\Delta(x^1(\sigma(t_1)), x^1(t_1), t_1) &= G(x^1(\sigma(t_1)), x^1(t_1), t_1)q_1(t_1), \\ y^2(t_1) - g_t^\Delta(x^1(\sigma(t_1)), x^1(t_1), t_1) &= G(x^1(\sigma(t_1)), x^1(t_1), t_1)q_2(t_1). \end{aligned}$$

Then

$$\begin{aligned} &G(x^1(\sigma(t_1)), x^1(t_1), t_1)w^1(t_1) \\ &\quad = G(x^1(\sigma(t_1)), x^1(t_1), t_1)G(x^1(\sigma(t_1)), x^1(t_1), t_1)^+ G(x^1(\sigma(t_1)), x^1(t_1), t_1)q_1(t_1) \\ &\quad = G(x^1(\sigma(t_1)), x^1(t_1), t_1)q_1(t_1) \\ &\quad = y^1(t_1) - g_t^\Delta(x^1(\sigma(t_1)), x^1(t_1), t_1). \end{aligned}$$

As above,

$$G(x^1(\sigma(t_1)), x^1(t_1), t_1)w^2(t_1) = y^2(t_1) - g_t^\Delta(x^1(\sigma(t_1)), x^1(t_1), t_1).$$

Thus,

$$y^1(t_1) - y^2(t_1) = G(x^1(\sigma(t_1)), x^1(t_1), t_1)(w^1(t_1) - w^2(t_1))$$

and

$$\begin{aligned} f(g_t^\Delta(x^1(\sigma(t_1)), t_1) + G(x^1(\sigma(t_1)), x^1(t_1), t_1)w^1(t_1), x^1(t_1), t_1) &= 0, \\ f(g_t^\Delta(x^1(\sigma(t_1)), t_1) + G(x^1(\sigma(t_1)), x^1(t_1), t_1)w^2(t_1), x^1(t_1), t_1) &= 0. \end{aligned}$$

Hence,

$$\begin{aligned} 0 &= f(g_t^\Delta(x^1(\sigma(t_1)), t_1) + G(x^1(\sigma(t_1)), x^1(t_1), t_1)w^1(t_1), x^1(t_1), t_1) \\ &\quad - f(g_t^\Delta(x^1(\sigma(t_1)), t_1) + G(x^1(\sigma(t_1)), x^1(t_1), t_1)w^2(t_1), x^1(t_1), t_1) \\ &= \left(\int_0^1 \frac{\partial}{\partial y} f(s(g_t^\Delta(x^1(\sigma(t_1)), t_1) + G(x^1(\sigma(t_1)), x^1(t_1), t_1)w^1(t_1)) \right. \end{aligned}$$

$$+ (1-s)(g_t^{\Delta}(x^1(\sigma(t_1)), t_1) + G(x^1(\sigma(t_1)), x^1(t_1), t_1)w^2(t_1))x^1(t_1), t_1)ds\Big)$$

$$\times G(x^1(\sigma(t_1)), x(t_1), t_1)(w^1(t_1) - w^2(t_1)).$$

Since equation (1.46) has a properly involved derivative, we have

$$\begin{aligned}\ker\Bigg(\Bigg(\int_0^1 \frac{\partial}{\partial y} f(s(g_t^{\Delta}(x^1(\sigma(t_1)), t_1) + G(x^1(\sigma(t_1)), x^1(t_1), t_1)w^1(t_1))\\ + (1-s)(g_t^{\Delta}(x^1(\sigma(t_1)), t_1) + G(x^1(\sigma(t_1)), x^1(t_1), t_1)w^2(t_1))x^1(t_1), t_1)ds\Bigg)\\ \times G(x^1(\sigma(t_1)), x^1(t_1), t_1)\Bigg)\\ = \ker G(x^1(\sigma(t_1)), x^1(t_1), t_1).\end{aligned}$$

Therefore

$$w^1(t_1) - w^2(t_1) \in N.$$

Since $w^1(t_1), w^2(t_2) \in N^{\perp}$, the latter relation is possible if and only if

$$w^1(t_1) = w^2(t_1).$$

Then

$$y^1(t_1) = y^2(t_1).$$

This completes the proof. □

Theorem 1.10. *Let equation* (1.46) *have a properly involved derivative and let*

$$\ker F(y^1(t), y^2(t), x(t), t), \quad y^1(t), y^2(t) \in \mathbb{R}^n, x(t) \in D_f, \quad t \in I_f,$$

be independent of the choice of y^1 *and* y^2. *Suppose that there exists a projector* $R(x(t), t)$, $x(t) \in D_f$, $t \in I_f$, *such that*

$$\begin{aligned}\operatorname{im} R(x(t), t) &= \operatorname{im} G(x^1(t), x^2(t), t),\\ \ker R(x(t), t) &= \ker F(y^1(t), y^2(t), x(t), t),\\ y^1(t), y^2(t) \in \mathbb{R}^n, &\quad x(t), x^1(t), x^2(t) \in D_f, \quad t \in I_f.\end{aligned}$$

Then, we have the following:

1.

$$f(y(t), x(t), t) = f(R(x(t), t)y(t), x(t), t), \quad y(t) \in \mathbb{R}^n, \quad x(t) \in D_d, \quad t \in I_f.$$

2. *Projector R is continuously-differentiable on $D_f \times I_f$.*
3. $\mathscr{M}_0(t) = \widetilde{\mathscr{M}_0}(t)$ *for any* $t \in I_f$.

Proof.
1. Let $t \in I_f$, $x(t) \in D_f$, $y(t) \in \mathbb{R}^n$ be arbitrarily chosen. Set

$$\eta(t) = (I - R(x(t), t))y(t).$$

Then

$$\begin{aligned} &f(y(t), x(t), t) - f(R(x(t), t)y(t), x(t), t) \\ &\quad = \int_0^1 \frac{\partial}{\partial y} f(sy(t) + (1 - s)R(x(t), t)y(t), x(t), t) ds (I - R(x(t), t))y(t) \\ &\quad = \int_0^1 \frac{\partial}{\partial y} f(sy(t) + (1 - s)R(x(t), t)y(t), x(t), t) ds \eta(t) \\ &\quad = F(y(t), R(x(t), t)y(t), x(t), t)\eta(t). \end{aligned} \tag{1.52}$$

Note that

$$\eta(t) \in \operatorname{im}(I - R(x(t), t)) = \ker F(y(t), R(x(t), t)y(t), x(t), t).$$

Therefore

$$F(y(t), R(x(t), t)y(t), x(t), t) = 0$$

and

$$f(y(t), x(t), t) = f(R(x(t), t)y(t), x(t), t).$$

2. The function R is continuously differentiable because it is a projector defined on $\mathscr{C}^1$-subspaces.
3. Let $t \in I_f$, $x(t) \in \widetilde{M}_0(t)$ be arbitrarily chosen and fixed. Let also, $y^1(t) \in \mathbb{R}^n$ be such that

$$0 = f(y^1(t), x(t), t) = f(R(x(t), t)y^1(t), x(t), t).$$

Define

$$y(t) = R(x(t),t)y^1(t) + (I - R(x(t),t))g_t^{\Delta}(x(\sigma(t)),t).$$

Then

$$\begin{aligned} y(t) - g_t^{\Delta}(x(\sigma(t)),t) &= R(x(t),t)(y^1(t) - g_t^{\Delta}(x(\sigma(t)),t)) \\ &\in \operatorname{im} R(x(t),t) = \operatorname{im} G(x^1(t),x^2(t),t). \end{aligned}$$

By the definition of $y(t)$, we find

$$\begin{aligned} R(x(t),t)y(t) &= R(x(t),t)R(x(t),t)y^1(t) + R(x(t),t)(I - R(x(t),t))g_t^{\Delta}(x(\sigma(t)),t) \\ &= R(x(t),t)y^1(t). \end{aligned}$$

From here and using claim 1, we arrive at

$$\begin{aligned} f(y(t),x(t),t) &= f(R(x(t),t)y(t),x(t),t) = f(R(x(t),t)y^1(t),x(t),t) \\ &= f(y^1(t),x(t),t) = 0. \end{aligned}$$

Consequently, $x(t) \in \mathscr{M}_0(t)$. Because $x(t) \in \widetilde{\mathscr{M}_0}(t)$, and since it is an element of $\mathscr{M}_0(t)$, we get the inclusion

$$\widetilde{\mathscr{M}_0}(t) \subseteq \mathscr{M}_0(t).$$

Hence, using

$$\mathscr{M}_0(t) \subseteq \widetilde{\mathscr{M}_0}(t),$$

we get

$$\mathscr{M}_0(t) = \widetilde{\mathscr{M}_0}(t).$$

This completes the proof. □

1.3.3 Linearization

Define

$$H(y(t),x^1(t),x^2(t),t) = \int_0^1 \frac{\partial}{\partial y} f(y(\sigma(t)),sx^1(t) + (1-s)x^2(t),t)dh,$$

where

$$\frac{\partial}{\partial y} f(y(t), hx^1(t) + (1-h)x^2(t), t)$$
$$= \left(\frac{\partial}{\partial x_j} f\Big(y(\sigma(t), x_1^1(t), \ldots, x_{j-1}^1(t), hx_j^1(t) + (1-h)x_j^2(t),\right.$$
$$\left. x_{j+1}^2(t), \ldots, x_m^2(t), t)\Big)\right)_{i,j=1}^{k,m},$$

with $y(t) \in \mathbb{R}^n$, $x^1(t), x^2(t) \in D_f$, $t \in I_f$. Let $I_* \subseteq I$ and $x_* \in \mathscr{C}^1(I_*)$ be such that $x_*(t) \in D_f$, $t \in I_f$, and

$$g(x_*(\cdot), \cdot) \in \mathscr{C}^1(I_*).$$

Set

$$\begin{aligned} A(t) &= F((g(x_*(t), t))^{\Delta\sigma}, (g(x_*(t), t))^{\Delta}, x_*(t), t), \\ B(t) &= G(x_*^\sigma(t), x_*(t), t), \\ C(t) &= H((g(x_*(t), t))^{\Delta}, x_*^\sigma(t), x_*(t), t), \quad t \in I_*. \end{aligned}$$

Consider the equation

$$A(t)(B(t)x(t))^{\Delta} + C(t)x(t) = q(t), \quad t \in I_*. \tag{1.53}$$

Definition 1.18. Equation (1.53) is said to be a linearization of equation (1.46) along the reference function x_*.

Note that the reference function x_* is not necessarily a solution of equation (1.46).

Example. Let $\mathbb{T} = \mathbb{Z}$. Consider the following nonlinear dynamic system:

$$\begin{aligned} 3((x_1(t))^2 + x_2(t)x_3(t))^{\Delta} &= q_1(t), \\ 2x_1(t) + x_3(t) &= q_2(t), \\ x_2(t) + x_3(t) &= q_3(t), \quad t \in \mathbb{T}. \end{aligned}$$

Here, $n = 1$, $k = m = 3$, and

$$\sigma(t) = t + 1, \quad t \in \mathbb{T}.$$

We have

$$\begin{aligned} f(y, x, t) &= \begin{pmatrix} 3 \\ 0 \\ 0 \end{pmatrix} y + \begin{pmatrix} 0 \\ 2x_1 + x_3 \\ x_2 + x_3 \end{pmatrix} - q(t), \\ g(x, t) &= x_1^2 + x_2 x_3. \end{aligned}$$

Then

$$F(y^1(t),y^2(t),x(t),t) = \int_0^1 \begin{pmatrix} 3 \\ 0 \\ 0 \end{pmatrix} dh = \begin{pmatrix} 3 \\ 0 \\ 0 \end{pmatrix},$$

$$y^1(t),y^2(t) \in \mathbb{R}, \quad x(t) \in \mathbb{R}^3, \quad t \in I_f,$$

and

$$\begin{aligned} G(x^1(t),x^2(t),t) &= \int_0^1 (2hx_1^1(t) + 2(1-h)x_1^2(t), x_3^2(t), x_2^1(t))dh \\ &= (x_1^1(t) - x_1^2(t), x_3^2(t), x_2^1(t)), \quad x^1(t),x^2(t) \in \mathbb{R}^3, \quad t \in I_f, \end{aligned}$$

as well as

$$H(y(t),x^1(t),x^2(t),t) = \int_0^1 \begin{pmatrix} 0 & 0 & 0 \\ 2 & 0 & 1 \\ 0 & 1 & 1 \end{pmatrix} dh = \begin{pmatrix} 0 & 0 & 0 \\ 2 & 0 & 1 \\ 0 & 1 & 1 \end{pmatrix},$$

$$y(t) \in \mathbb{R}, \quad x^1(t),x^2(t) \in \mathbb{R}^3, \quad t \in \mathbb{T}.$$

Let $x_* \in \mathscr{C}^1(\mathbb{T})$ be arbitrarily chosen. Then

$$\begin{aligned} F(y^\sigma(t),y(t),x_*(t),t) &= \begin{pmatrix} 3 \\ 0 \\ 0 \end{pmatrix}, \\ G(x_*^\sigma(*t),x_*(t),t) &= (x_{*1}^\sigma(t) - x_{*1}(t), x_{*3}(t), x_{*1}^\sigma(t)) \\ &= (x_{*1}(t+1) - x_{*1}(t), x_{*3}(t), x_{*1}(t+1)), \\ H(y(t),x_*^\sigma(t),x_*(t),t) &= \begin{pmatrix} 0 & 0 & 0 \\ 2 & 0 & 1 \\ 0 & 1 & 1 \end{pmatrix}, \quad t \in \mathbb{T}, \quad y(t) \in \mathbb{R}. \end{aligned}$$

Next,

$$\begin{aligned} (g(x_*(t),t))^\Delta &= G(x_*^\sigma(t),x_*(t),t)x_*^\Delta(t) \\ &= (x_{*1}(t+1) - x_{*1}(t), x_{*3}(t), x_{*1}(t+1)) \begin{pmatrix} x_{*1}^\Delta(t) \\ x_{*2}^\Delta(t) \\ x_{*3}^\Delta(t) \end{pmatrix} \\ &= (x_{*1}(t+1) - x_{*1}(t))x_{*1}^\Delta(t) + x_{*3}(t)x_{*2}^\Delta(t) + x_{*1}(t+1)x_{*3}^\Delta(t), \\ A(t) = F((g(x_*(t),t))^{\Delta\sigma}, &(g(x_*(t),t))^\Delta, x_*(t),t) = \begin{pmatrix} 3 \\ 0 \\ 0 \end{pmatrix}, \end{aligned}$$

$$B(t) = G(x_*^\sigma(t), x_*(t), t) = (x_{*1}(t+1) - x_{*1}(t), x_{*3}(t), x_{*1}(t+1)),$$
$$C(t) = H((g(x_*(t), t))^\Delta, x_*^\sigma(t), x_*(t)) = \begin{pmatrix} 0 & 0 & 0 \\ 2 & 0 & 1 \\ 0 & 1 & 1 \end{pmatrix}, \quad t \in \mathbb{T}.$$

Therefore

$$\begin{pmatrix} 3 \\ 0 \\ 0 \end{pmatrix} \left((x_{*1}(t+1) - x_{*1}(t), x_{*3}(t), x_{*1}(t+1)) \begin{pmatrix} x_1(t) \\ x_2(t) \\ x_3(t) \end{pmatrix} \right)^\Delta$$
$$+ \begin{pmatrix} 0 & 0 & 0 \\ 2 & 0 & 1 \\ 0 & 1 & 1 \end{pmatrix} \begin{pmatrix} x_1(t) \\ x_2(t) \\ x_3(t) \end{pmatrix} = q(t), \quad t \in \mathbb{T},$$

or

$$\begin{pmatrix} 3 \\ 0 \\ 0 \end{pmatrix} ((x_{*1}(t+1) - x_{*1}(t))x_1(t) + x_{*3}(t)x_2(t) + x_{*1}(t+1)x_3(t))^\Delta$$
$$+ \begin{pmatrix} 0 \\ 2x_1(t) + x_3(t) \\ x_2(t) + x_3(t) \end{pmatrix} = q(t), \quad t \in \mathbb{T},$$

or

$$\begin{aligned} ((x_{*1}(t+1) - x_{*1}(t))x_1(t) + x_{*3}(t)x_2(t) + x_{*1}(t+1)x_3(t))^\Delta &= q_1(t), \\ 2x_1(t) + x_3(t) &= q_2(t), \\ x_2(t) + x_3(t) &= q_3(t), \quad t \in \mathbb{T}. \end{aligned}$$

If equation (1.46) has a properly involved derivative, then the decomposition

$$\ker A(t) \oplus \operatorname{im} B(t) = \mathbb{R}^n, \quad t \in I_*,$$

holds if $\ker A(t)$, $t \in I_*$, and $\operatorname{im} B(t)$, $t \in I_*$, are $\mathscr{C}$-subspaces. If the subspace $\ker F(y^1(t), y^2(t), x(t), t)$, $y^1(t), y^2(t) \in \mathbb{R}^n$, $x(t) \in D_f$, $t \in I_*$, does not depend on y^1 and y^2 and $x_* \in \mathscr{C}^1(I_*)$, by Theorem 1.10 it follows that $\ker A(t)$, $t \in I_*$, and $\operatorname{im} B(t)$, $t \in I_*$, are $\mathscr{C}^1$-subspaces.

We set

$$\begin{aligned} B_*(x(t), t) &= G(x^\sigma(t), x(t), t), \\ A_*(x^1(t), x(t), t) &= F((B_*(x(t), t)x^1(t) + g_t^\Delta(x(\sigma(t)), t))^\sigma, \\ &\qquad B_*(x(t), t)x^1(t) + g_t^\Delta(x(\sigma(t)), t), x(t), t), \\ C_*(x^1(t), x(t), t) &= H(B_*(x(t), t)x^1(t) = g_t^\Delta(x(\sigma(t)), t), x^\sigma(t), x(t), t), \end{aligned}$$

for $x^1(t), x(t) \in D_f$, $t \in I_f$. Then, we have the following:

$$\begin{aligned}
A(t) &= F((g(x_*(t),t))^{\Delta\sigma}, (g(x_*(t),t))^{\Delta}, x_*(t), t) \\
&= F((G(x_*^\sigma(t), x_*(t), t)x_*^\Delta(t) + g_t^\Delta(x_*(\sigma(t)), t))^\sigma, \\
&\qquad G(x_*^\sigma(t), x_*(t), t)x_*^\Delta(t) + g_t^\Delta(x_*(\sigma(t)), t), x_*(t), t) \\
&= F((B_*(x_*(t), t)x_*^\Delta(t) + g_t^\Delta(x_*(\sigma(t)), t))^\sigma, \\
&\qquad B_*(x_*(t), t)x_*^\Delta(t) + g_t^\Delta(x_*(\sigma(t)), t), x_*(t), t) \\
&= A_*(x_*^\Delta(t), x_*(t), t), \\
B(t) &= G(x_*^\sigma(t), x_*(t), t) = B_*(x_*(t), t), \\
C(t) &= H((g(x_*(t),t))^\Delta, x_*^\sigma(t), x_*(t), t) \\
&= H(G(x_*^\sigma(t), x_*(t), t)x_*^\Delta(t) + g_t^\Delta(x_*(\sigma(t)), t), x_*^\sigma(t), x_*(t), t) \\
&= H(B_*(x_*(t), t)x_*^\Delta(t) + g_t^\Delta(x_*(\sigma(t)), t), x_*^\sigma(t), x_*(t), t) \\
&= C_*(x_*^\Delta(t), x_*(t), t), \quad t \in I_*.
\end{aligned}$$

Theorem 1.11. *Let equation* (1.46) *have a properly involved derivative. Then the decomposition*

$$\ker A_*(x^1(t), x(t), t) \oplus \operatorname{im} B_*(x(t), t) = \mathbb{R}^n \tag{1.54}$$

holds for any $x^1(t) \in \mathbb{R}^m$, $x(t) \in D_f$, $t \in I_f$, *and* $\ker A_*$ *and* $\operatorname{im} B_*$ *are* $\mathscr{C}$*-subspaces.*

Proof. Since equation (1.46) has a properly involved derivative, we have

$$\ker F(y^1(t) y^2(t), x(t), t) \oplus \operatorname{im} G(x^1(t), x^2(t), t)$$

for any $y^1(t), y^2(t) \in \mathbb{R}^n$, $x^1(t), x^2(t) \in \mathbb{R}^m$, $x(t) \in D_f$, $t \in I_f$. For each triple $(\overline{\overline{x}}(t), \overline{x}(t), \overline{t}) \in \mathbb{R}^m \times D_f \times I_f$, we set

$$\begin{aligned}
y^1(\overline{t}) &= (B_*(\overline{x}(\overline{t}), \overline{t})\overline{\overline{x}}(t) + g_t^\Delta(\overline{x}(\sigma(\overline{t})), \overline{t}))^\sigma, \\
y^2(t) &= B_*(\overline{x}(\overline{t}), \overline{t})\overline{\overline{x}}(t) + g_t^\Delta(\overline{x}(\sigma(\overline{t})), \overline{t}), \\
x^1(\overline{t}) &= \overline{x}^\sigma(\overline{t}), \\
x^2(\overline{t}) &= \overline{x}(\overline{t}).
\end{aligned}$$

Then

$$\begin{aligned}
F(y^1(\overline{t}), y^2(\overline{t}), x(\overline{t}), \overline{t}) &= A_*(\overline{\overline{x}}(\overline{t}), \overline{x}(\overline{t}), \overline{t}), \\
G(x^1(\overline{t}), x^2(\overline{t}), \overline{t}) &= B_*(\overline{x}(\overline{t}), \overline{t}),
\end{aligned}$$

and

$$\ker A_*(\overline{\overline{x}}(\overline{t}), \overline{x}(\overline{t}), \overline{t}) \oplus \operatorname{im} B_*(\overline{x}(\overline{t}), \overline{t}) = \mathbb{R}^n.$$

Because A_* and B_* are continuous matrix functions with constant rank, we have that $\ker A_*$ and $\operatorname{im} B_*$ are $\mathscr{C}$-subspaces. This completes the proof. □

Definition 1.19. Let equation (1.46) have a properly involved derivative. The projector-valued function R defined by

$$\begin{aligned}\operatorname{im} R(x^1(t), x(t), t) &= \operatorname{im} B_*(x(t), t),\\ \ker R(x^1(t), x(t), t) &= \ker A_*(x^1(t), x(t), t),\end{aligned}$$

for $x^1(t) \in \mathbb{R}^m$, $x(t) \in D_f$, $t \in I_f$, is said to be a border projector function or border projector of equation (1.46).

The basic assumption below is as follows:

(E1)
1. The function f is classically continuously differentiable with respect to its first and second arguments and delta continuously differentiable with respect to its third argument on $\mathbb{R}^n \times D_f \times I_f$. The functions F and H are continuous on $\mathbb{R}^n \times \mathbb{R}^n \times D_f \times I_f$. The function g is classically continuously differentiable with respect to its first argument and delta continuously differentiable with respect to its second argument on $D_f \times I_f$.
2. Equation (1.46) has a properly involved derivative.
3. If $\ker F(y^1, y^2, x, t)$ depends on y^1 and y^2, then suppose that g has a continuous classical second derivative with respect to its first argument and a continuous delta second derivative with respect to its second argument on $D_f \times I_f$.
4. The transversality conditions (1.47) and (1.54) are equivalent.

1.3.4 Regular linearized equations with tractability index 1

Suppose that (E1) holds and the matrices A_*, B_*, C_* are defined as in the previous section. Then A_*, B_*, C_* and the border projector are continuous matrix functions. Assume that the linearized equation (1.53) is regular with tractability index 1. Denote

$$N_0(x(t), t) = \ker B_*(x(t), t), \quad x(t) \in D_f, \quad t \in I_f,$$

and let Q_0 be a projector onto B_*,

$$P_0(x(t), t) = I - Q_0(x(t), t), \quad x(t) \in D_f, \quad t \in I_f.$$

We can choose P_0 and Q_0 to be continuous. Let B_*^{-1} be the $\{1, 2\}$-inverse of B_* defined by

$$
\begin{aligned}
B_*(x(t),t)B_*^-(x^1(t),x(t),t)B_*(x(t),t) &= B_*(x(t),t),\\
B_*^-(x^1(t),x(t),t)B_*(x(t),t)B_*^{-1}(x^1(t),x(t),t) &= B_*^-(x^1(t),x(t),t),\\
B_*(x(t),t)B_*^-(x^1(t),x(t),t) &= R(x^1(t),x(t),t),\\
B_*^-(x^1(t),x(t),t)B_*(x(t),t) &= P_0(x^1(t),x(t),t),\\
x^1(t)\in\mathbb{R}^m,\quad x(t)\in D_f,\quad t&\in I_f,
\end{aligned}
\tag{1.55}
$$

where $R(x^1(t),x(t),t)$, $x^1(t)\in\mathbb{R}^m$, $x(t)\in D_f$, $t\in I_f$, is a continuous projector along $\ker A_*(x^1(t),x(t),t)$, $x^1(t)\in\mathbb{R}^m$, $x(t)\in D_f$, $t\in I_f$. Note that B_*^- is uniquely determined by (1.55). Suppose that

$$
\begin{aligned}
G_0(x^1(t),x(t),t) &= A_*(x^1(t),x(t),t)B_*(x(t),t),\\
\Pi_0(x(t),t) &= P_0(x(t),t),\\
C_0(x^1(t),x(t),t) &= C_*(x^1(t),x(t),t),\\
G_1(x^1(t),x(t),t) &= G_0(x^1(t),x(t),t)+C_0(x^1(t),x(t),t)Q_0(x(t),t),\\
N_1(x^1(t),x(t),t) &= \ker G_1(x^1(t),x(t),t),\\
\Pi_1(x^1(t),x(t),t) &= \Pi_0(x(t),t)P_1(x^1(t),x(t),t),\quad x^1(t)\in\mathbb{R}^m,\quad x(t)\in D_f,\quad t\in I_f,
\end{aligned}
$$

where $Q_1(x^1(t),x(t),t)$, $x^1(t)\in\mathbb{R}^m$, $x(t)\in D_f$, $t\in I_f$, is a continuous projector onto $N_1(x^1(t),x(t),t)$, $x^1(t)\in\mathbb{R}^m$, $x(t)\in D_f$, $t\in I_f$, and

$$
P_1(x^1(t),x(t),t) = I - Q_1(x^1(t),x(t),t),\quad x^1(t)\in\mathbb{R}^m,\quad x(t)\in D_f,\quad t\in I_f.
$$

The total derivative of $B_*\Pi_0B_*^-$ in jet variables we will denote as follows:

$$
\begin{aligned}
\mathrm{Diff}_1(x^2(t),x^1(t),x(t),t) &= D_t(B_*\Pi_0B_*^-)(x^1(t),x(t),t),\\
x^1(t),x^2(t)\in\mathbb{R}^m,\quad x(t)&\in D_f,\quad t\in I_f.
\end{aligned}
$$

The new jet variable $x^2(t)\in\mathbb{R}^m$, $t\in I_f$, can be considered as a place holder for $x^{\Delta^2}(t)$, $t\in I_f$. We have that there are $c_1,c_2\in[t,\sigma(t)]$ such that

$$
\begin{aligned}
\mathrm{Diff}_1(x^2(t),x^\Delta(t),x(t),t) &= \frac{\partial}{\Delta t}(B_*\Pi_0B_*^-)(x^\Delta(t),x(t),t)\\
&\quad + x^\Delta(t)\frac{\partial}{\partial x}(B_*\Pi_0B_*^-)(x^\Delta(c_1),x(c_1),\sigma(c_1))\\
&\quad + x^{\Delta^2}(t)\frac{\partial}{\partial x^1}(B_*\Pi_0B_*^-)(x^\Delta(c_2),x(\sigma(c_2)),\sigma(c_2)).
\end{aligned}
$$

Example. Let $\mathbb{T}=\mathbb{Z}$. Consider the following nonlinear dynamic system:

$$
\begin{aligned}
x_1^\Delta(t)+2x_1(t) &= 0,\\
(x_1(t))^2+(x_2(t))^2-2 &= t^2
\end{aligned}
$$

on

$$D_f = \{x \in \mathbb{R}^2 : x_2 > 0\}, \quad I_f = \mathbb{T}.$$

Here $n = 1$, $m = k = 2$, and

$$f(y, x, t) = \begin{pmatrix} y + x_1 \\ x_1^2 + x_2^2 - 2 - t^2 \end{pmatrix},$$
$$g(x, t) = x_1, \quad y \in \mathbb{R}, \quad x \in \mathbb{R}^2, \quad t \in \mathbb{T}.$$

Then

$$f_y(y, x, t) = \begin{pmatrix} 1 \\ 0 \end{pmatrix}, \quad f_x(y, x, t) = \begin{pmatrix} 1 & 0 \\ 2x_1 & 2x_2 \end{pmatrix},$$
$$g_x(x, t) = (1, 0), \qquad g_t(x, t) = 0, \quad y \in \mathbb{R}, \quad x \in \mathbb{R}^2, \quad t \in \mathbb{T}.$$

Hence,

$$F(y^1(t), y^2(t), x(t), t) = \int_0^1 \begin{pmatrix} 1 \\ 0 \end{pmatrix} dh = \begin{pmatrix} 1 \\ 0 \end{pmatrix},$$
$$G(x^1(t), x^2(t), t) = \int_0^1 (1, 0) dh = (1, 0),$$
$$H(y(t), x^1(t), x^2(t), t)$$
$$= \int_0^1 \begin{pmatrix} 1 & 0 \\ 2sx_1^1(t) + 2(1-s)x_1^2(t) & 2sx_2^1(t) + 2(1-s)x_2^2(t) \end{pmatrix} ds$$
$$= \begin{pmatrix} 1 & 0 \\ x_1^1(t)s^2\big|_{s=0}^{s=1} - x_1^2(t)(1-s)^2\big|_{s=0}^{s=1} & x_2^1(t)s^2\big|_{s=0}^{s=1} - x_2^2(t)(1-s)^2\big|_{s=0}^{s=1} \end{pmatrix}$$
$$= \begin{pmatrix} 1 & 0 \\ x_1^1(t) + x_1^2(t) & x_2^1(t) + x_2^2(t) \end{pmatrix},$$
$$x^1(t), x^2(t) \in D_f, \quad y(t), y^1(t), y^2(t) \in \mathbb{R}, \quad t \in \mathbb{T}.$$

Therefore

$$B_*(x(t), t) = (1, 0),$$
$$A_*(x^1(t), x(t), t) = \begin{pmatrix} 1 \\ 0 \end{pmatrix},$$
$$C_*(x^1(t), x(t), t) = H((1, 0)x^1(t), x(t+1), x(t), t)$$
$$= \begin{pmatrix} 1 & 0 \\ x_1(t+1) + x_1(t) & x_2(t+1) + x_2(t) \end{pmatrix},$$
$$x^1(t) \in \mathbb{R}^2, \quad x(t) \in D_f, \quad t \in I_f.$$

Next,

$$\begin{aligned} G_0(x^1(t), x(t), t) &= A_*(x^1(t), x(t), t)B_*(x(t), t) \\ &= \begin{pmatrix} 1 \\ 0 \end{pmatrix} (1, 0) \\ &= \begin{pmatrix} 1 & 0 \\ 0 & 0 \end{pmatrix}, \quad x^1(t) \in \mathbb{R}^2, \quad x(t) \in D_f, \quad t \in I_f. \end{aligned}$$

Let

$$z(t) = \begin{pmatrix} z_1(t) \\ z_2(t) \end{pmatrix} \in \mathbb{R}^2, \quad t \in \mathbb{T},$$

be such that

$$G_0(x^1(t), x(t), t)z(t) = 0, \quad x^1(t) \in \mathbb{R}^2, \quad x(t) \in D_f, \quad t \in \mathbb{T}.$$

Then

$$\begin{pmatrix} 1 & 0 \\ 0 & 0 \end{pmatrix} \begin{pmatrix} z_1(t) \\ z_2(t) \end{pmatrix} = \begin{pmatrix} 0 \\ 0 \end{pmatrix}, \quad t \in \mathbb{T},$$

or

$$\begin{pmatrix} z_1(t) \\ 0 \end{pmatrix} = \begin{pmatrix} 0 \\ 0 \end{pmatrix}, \quad t \in \mathbb{T},$$

whereupon

$$z_1(t) = 0, \quad t \in \mathbb{T}.$$

Let

$$Q_0 = \begin{pmatrix} 0 & 0 \\ 0 & p \end{pmatrix},$$

where $p \in \mathbb{R}$, $p \neq 0$, is chosen such that

$$Q_0 = Q_0 Q_0.$$

We have

$$\begin{pmatrix} 0 & 0 \\ 0 & p \end{pmatrix} = \begin{pmatrix} 0 & 0 \\ 0 & p \end{pmatrix} \begin{pmatrix} 0 & 0 \\ 0 & p \end{pmatrix} = \begin{pmatrix} 0 & 0 \\ 0 & p^2 \end{pmatrix},$$

whereupon

$$p^2 = p$$

and

$$p = 1.$$

Consequently,

$$Q_0 = \begin{pmatrix} 0 & 0 \\ 0 & 1 \end{pmatrix}$$

and

$$\begin{aligned} G_1(x^1(t), x(t), t) &= G_0(x^1(t), x(t), t) + C_*(x^1(t), x(t), t)Q_0 \\ &= \begin{pmatrix} 1 & 0 \\ 0 & 0 \end{pmatrix} + \begin{pmatrix} 1 & 0 \\ x_1(t+1) + x_1(t) & x_2(t+1) + x_2(t) \end{pmatrix} \begin{pmatrix} 0 & 0 \\ 0 & 1 \end{pmatrix} \\ &= \begin{pmatrix} 1 & 0 \\ 0 & 0 \end{pmatrix} + \begin{pmatrix} 0 & 0 \\ 0 & x_1(t) + x_2(t) \end{pmatrix} \\ &= \begin{pmatrix} 1 & 0 \\ 0 & x_1(t) + x_2(t) \end{pmatrix}, \quad x^1(t) \in \mathbb{R}^2, \quad x(t) \in D_f, \quad t \in \mathbb{T}. \end{aligned}$$

Note that

$$\det G_0(x^1(t), x(t), t) = x_1(t) + x_2(t) \neq 0, \quad x^1(t) \in \mathbb{R}^2, \quad x(t) \in D_f, \quad t \in I_f.$$

Samir H. Saker

2 Muckenhoupt and Gehring weights on time scales

Abstract: This chapter deals with the fundamental properties of the Muckenhoupt and Gehring weights on time scales. We also present some results related to the self-improving properties of the Muckenhoupt and Gehring classes and some higher integrability results for nonincreasing functions on time scales. This chapter is organized as follows: In Section 2.1, we present some essential preliminaries on time scales that will act as prerequisites to the main results. Section 2.2 deals with the definitions and properties of the classical weights in the integral forms and the definitions and properties of the discrete weights, and then we write the definitions of Muckenhoupt and Gehring weights on time scales. Section 2.3 deals with the fundamental properties of the Muckenhoupt and Gehring classes on time scales. Section 2.4 deals with essential relations between the norms of these classes on time scales. Section 2.5 deals with the self-improving properties of Muckenhoupt and Gehring classes. In Section 2.6, we present some higher integrability results for nonincreasing functions on time scales. Our approach is based on proving some properties of integral operators with powers, Hölder inequality, chain rules, as well as some connecting relations between Muckenhoupt and Gehring classes on time scales.

> A science is said to be useful if its development tends to accentuate the existing inequalities in the distribution of wealth, or more directly promotes the destruction of human life.
> Godfrey Harold Hardy (1877–1947).

2.1 Preliminaries on time scales

We assume the reader has a good background in time scale calculus. For the reader who is not familiar with this calculus, we present to him, in this section, some preliminaries, definitions, concepts, and the basic dynamic inequalities on time scales that will be needed throughout the book. The results in this section will cover delta derivatives and integrals. For the notions used below, we refer the reader to the books [8, 9], and for more details about related inequalities on time scales, we refer the reader to the books [1, 2].

A time scale is an arbitrary nonempty closed subset of the real numbers. Throughout the chapter, we denote the time scale by the symbol $\mathbb{T}$. For example, the real numbers $\mathbb{R}$, the integers $\mathbb{Z}$, and the natural numbers $\mathbb{N}$ are time scales. For $t \in \mathbb{T}$, we define

Samir H. Saker, Department of Mathematics, Mansoura University, Mansoura 35516, Egypt, e-mail: shsaker@mans.edu.eg

https://doi.org/10.1515/9783111182971-002

the forward jump operator $\sigma : \mathbb{T} \to \mathbb{T}$ by $\sigma(t) := \inf\{s \in \mathbb{T} : s > t\}$. A time scale $\mathbb{T}$ equipped with the order topology is metrizable and is a K_σ-space, i. e., it is a union of at most countably many compact sets. The metric on $\mathbb{T}$ which generates the order topology is given by $d(r; s) := \mu(r; s)$, where $\mu(\cdot) = \mu(\cdot; \tau)$ for a fixed $\tau \in \mathbb{T}$ is defined as follows. The mapping $\mu : \mathbb{T} \to \mathbb{R}^+ = [0, \infty)$ such that $\mu(t) := \sigma(t) - t$ is called the graininess function.

When $\mathbb{T} = \mathbb{R}$, we see that $\sigma(t) = t$ and $\mu(t) \equiv 0$ for all $t \in \mathbb{T}$, and when $\mathbb{T} = \mathbb{N}$, we have that $\sigma(t) = t + 1$ and $\mu(t) \equiv 1$ for all $t \in \mathbb{T}$. The backward jump operator $\rho : \mathbb{T} \to \mathbb{T}$ is defined by $\rho(t) := \sup\{s \in \mathbb{T} : s < t\}$. The mapping $\nu : \mathbb{T} \to \mathbb{R}_0^+$ such that $\nu(t) = t - \rho(t)$ is called the backward graininess function. If $\sigma(t) > t$, we say that t is right-scattered, while if $\rho(t) < t$, we say that t is left-scattered. Also, if $t < \sup \mathbb{T}$ and $\sigma(t) = t$, then t is called right-dense, and if $t > \inf \mathbb{T}$ and $\rho(t) = t$, then t is called left-dense. If $\mathbb{T}$ has a left-scattered maximum m, then $\mathbb{T}^k = \mathbb{T} - \{m\}$. Otherwise $\mathbb{T}^k = \mathbb{T}$. In summary,

$$\mathbb{T}^k = \begin{cases} \mathbb{T}\backslash(\rho \sup \mathbb{T}, \sup \infty), & \text{if } \sup \mathbb{T} < \infty, \\ \mathbb{T}, & \text{if } \sup \mathbb{T} = \infty. \end{cases}$$

Likewise, $\mathbb{T}_k$ is defined as the set $\mathbb{T}_k = \mathbb{T}\backslash[\inf \mathbb{T}, \sigma(\inf \mathbb{T})]$ if $|\inf \mathbb{T}| < \infty$, and $\mathbb{T}_k = \mathbb{T}$ if $\inf \mathbb{T} = -\infty$. For a weight $f : \mathbb{T} \to \mathbb{R}$, we define the derivative f^Δ at $t \in \mathbb{T}$ as follows: If there exists a number $\alpha \in \mathbb{R}$ such that for all $\varepsilon > 0$ there exists a neighborhood U of t with

$$|f(\sigma(t)) - f(s) - \alpha(\sigma(t) - s)| \le \varepsilon|\sigma(t) - s|,$$

for all $s \in U$, then f is said to be differentiable at t, and we call α the delta derivative of f at t and denote it by $f^\Delta(t)$. For example, if $\mathbb{T} = \mathbb{R}$, then

$$f^\Delta(t) = f'(t) = \lim_{\Delta t \to 0} \frac{f(t + \Delta t) - f(t)}{\Delta t}, \quad \text{for all } t \in \mathbb{T}.$$

If $\mathbb{T} = \mathbb{N}$, then $f^\Delta(t) = f(t + 1) - f(t)$ for all $t \in \mathbb{T}$. For a weight $f : \mathbb{T} \to \mathbb{R}$, the (delta) derivative is defined by

$$f^\Delta(t) = \frac{f(\sigma(t)) - f(t)}{\sigma(t) - t},$$

if f is continuous at t and t is right-scattered. If t is not right-scattered then the derivative is defined by

$$f^\Delta(t) = \lim_{s \to t} \frac{f(\sigma(t)) - f(s)}{t - s} = \lim_{t \to \infty} \frac{f(t) - f(s)}{t - s},$$

provided this limit exists. A useful formula is

$$f^\sigma = f + \mu f^\Delta \quad \text{where } f^\sigma := f \circ \sigma.$$

A weight $f : [a, b] \to \mathbb{R}$ is said to be right-dense continuous (rd-continuous) if it is right continuous at each right-dense point and there exists a finite left limit at all left-dense points, and f is said to be differentiable if its derivative exists. The space of rd-continuous functions is denoted by $C_{rd}(\mathbb{T}, \mathbb{R})$. A time scale $\mathbb{T}$ is said to be regular if the following two conditions are satisfied simultaneously:

(a) For all $t \in \mathbb{T}$, $\sigma(\rho(t)) = t$,

(b) For all $t \in \mathbb{T}$, $\rho(\sigma(t)) = t$.

Remark 2.1. If $\mathbb{T}$ is a regular time scale, then both operators ρ and σ are invertible with $\sigma^{-1} = \rho$ and $\rho^{-1} = \sigma$.

The following theorem gives the product and quotient rules for the derivative of the product fg and the quotient f/g (where $gg^\sigma \neq 0$) of two delta differentiable functions f and g.

Theorem 2.1. *Assume $f; g : \mathbb{T} \to \mathbb{R}$ are delta differentiable at $t \in \mathbb{T}$. Then*

$$(fg)^\Delta = f^\Delta g + f^\sigma g^\Delta = fg^\Delta + f^\Delta g^\sigma, \tag{2.1}$$

$$\left(\frac{f}{g}\right)^\Delta = \frac{f^\Delta g - fg^\Delta}{gg^\sigma}. \tag{2.2}$$

By using the product rule, we see that the derivative of $f(t) = (t - \alpha)^m$ for $m \in \mathbb{N}$, and $\alpha \in \mathbb{T}$ can be calculated as

$$f^\Delta(t) = ((t-\alpha)^m)^\Delta = \sum_{\nu=0}^{m-1} (\sigma(t) - \alpha)^\nu (t - \alpha)^{m-\nu-1}. \tag{2.3}$$

As a special case when $\alpha = 0$, we see that the derivative of $f(t) = t^m$ for $m \in \mathbb{N}$ can be calculated as

$$(t^m)^\Delta = \sum_{\gamma=0}^{m-1} \sigma^\gamma(t) t^{m-\gamma-1}.$$

Note that when $\mathbb{T} = \mathbb{R}$, we have

$$\sigma(t) = t, \quad \mu(t) = 0, \quad f^\Delta(t) = f'(t).$$

When $\mathbb{T} = \mathbb{Z}$, we have

$$\sigma(t) = t + 1, \quad \mu(t) = 1, \quad f^\Delta(t) = \Delta f(t), \quad \int_a^b f(t)\Delta t = \sum_{t=a}^{b-1} f(t).$$

When $\mathbb{T} = h\mathbb{Z}$, $h > 0$, we have $\sigma(t) = t + h$, $\mu(t) = h$,

$$f^{\Delta}(t) = \Delta_h f(t) = \frac{f(t+h)-f(t)}{h}, \quad \int_a^b f(t)\Delta t = \sum_{k=0}^{\frac{b-a-h}{h}} f(a+kh)h.$$

When $\mathbb{T} = \{t : t = q^k, k \in \mathbb{N}_0, q > 1\}$, we have $\sigma(t) = qt$, $\mu(t) = (q-1)t$,

$$f^{\Delta}(t) = \Delta_q f(t) = \frac{(f(qt) - f(t))}{(q-1)\,t}, \quad \int_{t_0}^{\infty} f(t)\Delta t = \sum_{k=0}^{\infty} f(q^k)\mu(q^k).$$

When $\mathbb{T} = \mathbb{N}_0^2 = \{t^2 : t \in \mathbb{N}\}$, we have $\sigma(t) = (\sqrt{t}+1)^2$ and

$$\mu(t) = 1 + 2\sqrt{t}, \quad f^{\Delta}(t) = \Delta_0 f(t) = (f((\sqrt{t}+1)^2) - f(t))/1 + 2\sqrt{t}.$$

When $\mathbb{T} = \mathbb{T}_n = \{t_n : n \in \mathbb{N}\}$, where (t_n) are the harmonic numbers that are defined by $t_0 = 0$ and $t_n = \sum_{k=1}^{n} \frac{1}{k}$, $n \in \mathbb{N}_0$, we have

$$\sigma(t_n) = t_{n+1}, \quad \mu(t_n) = \frac{1}{n+1}, \quad f^{\Delta}(t) = \Delta_1 f(t_n) = (n+1)f(t_n).$$

When $\mathbb{T}_2$=$\{\sqrt{n} : n \in \mathbb{N}\}$, we have $\sigma(t) = \sqrt{t^2+1}$,

$$\mu(t) = \sqrt{t^2+1} - t, \quad f^{\Delta}(t) = \Delta_2 f(t) = \frac{(f(\sqrt{t^2+1}) - f(t))}{\sqrt{t^2+1} - t}.$$

When $\mathbb{T}_3$=$\{\sqrt[3]{n} : n \in \mathbb{N}\}$, we have $\sigma(t) = \sqrt[3]{t^3+1}$ and

$$\mu(t) = \sqrt[3]{t^3+1} - t, \quad f^{\Delta}(t) = \Delta_3 f(t) = \frac{(f(\sqrt[3]{t^3+1}) - f(t))}{\sqrt[3]{t^3+1} - t}.$$

For $a, b \in \mathbb{T}$, and a delta-differentiable weight f, the Cauchy integral of f^{Δ} is defined by

$$\int_a^b f^{\Delta}(t)\Delta t = f(b) - f(a).$$

Theorem 2.2. *Let $f, g \in C_{rd}([a,b], \mathbb{R})$ be rd-continuous functions, $a, b, c \in \mathbb{T}$, and $\alpha, \beta \in \mathbb{R}$. Then, the following are true:*
1. $\int_a^b [\alpha f(t) + \beta g(t)]\Delta t = \alpha \int_a^b f(t)\Delta t + \beta \int_a^b g(t)\Delta t,$
2. $\int_a^b f(t)\Delta t = -\int_b^a f(t)\Delta t,$
3. $\int_a^c f(t)\Delta t = \int_a^b f(t)\Delta t + \int_b^c f(t)\Delta t,$
4. $|\int_a^b f(t)\Delta t| \le \int_a^b |f(t)|\Delta t.$

An integration by parts formula reads

$$\int_a^b f(t)g^\Delta(t)\Delta t = f(t)g(t)|_a^b - \int_a^b f^\Delta(t)g^\sigma(t)\Delta t, \tag{2.4}$$

and improper integrals are defined as

$$\int_a^\infty f(t)\Delta t = \lim_{b\to\infty}\int_a^b f(t)\Delta t.$$

Note that when $\mathbb{T} = \mathbb{R}$, we have

$$\int_a^b f(t)\Delta t = \int_a^b f(t)dt.$$

When $\mathbb{T} = \mathbb{Z}$, we have

$$\int_a^b f(t)\Delta t = \sum_{t=a}^{b-1} f(t).$$

When $\mathbb{T} = h\mathbb{Z}$, $h > 0$, we have

$$\int_a^b f(t)\Delta t = \sum_{k=0}^{\frac{b-a-h}{h}} f(a+kh)h.$$

When $\mathbb{T} = \{t : t = q^k, k \in \mathbb{N}_0, q > 1\}$, we have

$$\int_{t_0}^\infty f(t)\Delta t = \sum_{k=0}^\infty f(q^k)\mu(q^k).$$

Note that the integration formula on a discrete time scale is defined by

$$\int_a^b f(t)\Delta t = \sum_{t\in(a,b)} f(t)\mu(t).$$

It is well known that rd-continuous functions possess antiderivatives. If f is rd-continuous and $F^\Delta = f$, then

$$\int_t^{\sigma(t)} f(s)\Delta s = F(\sigma(t)) - F(t) = \mu(t)F^\Delta(t) = \mu(t)f(t).$$

Theorem 2.3. *If $a, b \in \mathbb{T}$ and $f \in C_{rd}(\mathbb{T}, \mathbb{R})$ is such that $f(t) \geq 0$ for all $a \leq t < b$, then*

$$\int_a^b f(t)\Delta t \geq 0.$$

Lemma 2.1. *Let $v \in C^1_{rd}(\mathbb{T}, \mathbb{R})$ be strictly increasing and $\widetilde{\mathbb{T}} = v(\mathbb{T})$ be a time scale. If $f \in C_{rd}(\mathbb{T}, \mathbb{R})$, then for $a, b \in \mathbb{T}$, we have*

$$\int_a^b f(x)v^{\Delta}(x)\Delta x = \int_{v(a)}^{v(b)} f(v^{-1}(y))\widetilde{\Delta}y.$$

Throughout the chapter, we will use the following results:

$$\int_{t_0}^{\infty} \frac{\Delta s}{s^{\nu}} = \infty, \quad \text{if } 0 \leq \nu \leq 1, \quad \text{and} \quad \int_{t_0}^{\infty} \frac{\Delta s}{s^{\nu}} < \infty, \quad \text{if } \nu > 1,$$

and, without loss of generality, we assume that $\sup \mathbb{T} = \infty$, and define the time scale interval $[a, b]_{\mathbb{T}}$ by $[a, b]_{\mathbb{T}} := [a, b] \cap \mathbb{T}$. The two chain rules that we will use in this chapter are given in the next two lemmas.

Lemma 2.2. *Let $f : \mathbb{R} \to \mathbb{R}$ be continuously differentiable and suppose $g : \mathbb{T} \to \mathbb{R}$ is delta differentiable. Then $f \circ g : \mathbb{T} \to \mathbb{R}$ is delta differentiable and*

$$f^{\Delta}(g(t)) = f'(g(\zeta))g^{\Delta}(t), \quad \text{for } \zeta \in [t, \sigma(t)]. \tag{2.5}$$

Lemma 2.3. *Let $f : \mathbb{R} \to \mathbb{R}$ be continuously differentiable and suppose $g : \mathbb{T} \to \mathbb{R}$ is delta differentiable. Then $f \circ g : \mathbb{T} \to \mathbb{R}$ is delta differentiable and the formula*

$$(f \circ g)^{\Delta}(t) = \left\{ \int_0^1 f'(g(t) + h\mu(t)g^{\Delta}(t))dh \right\} g^{\Delta}(t) \tag{2.6}$$

holds.

In the following, we present the Jensen, Hölder, and Minkowski inequalities that will be used later in the proofs of the results in this book. For more details, we refer the reader to the book [1].

Theorem 2.4. *Let a, $b \in \mathbb{T}$ and c, $d \in \mathbb{R}$. Suppose that $g \in C_{rd}([a, b], (c, d))$ and $h \in C_{rd}([a, b]_{\mathbb{T}}, \mathbb{R})$ with*

$$\int_a^b |h(s)|\Delta s > 0.$$

If $F \in C((c,d),\mathbb{R})$ is convex, then

$$F\left(\frac{\int_a^b |h(s)|g(s)\Delta s}{\int_a^b |h(s)|\Delta s}\right) \le \frac{\int_a^b |h(s)|F(g(s))\Delta s}{\int_a^b |h(s)|\Delta s}. \tag{2.7}$$

If F is strictly convex, then the inequality $\le$ can be replaced by $<$.

A special case of Theorem 2.4, when $F(x) = x^{\alpha}$ is presented in the following result.

Corollary 2.1. *Let $\mathbb{T}$ be a time scale with $I \subset I_0$ and $a, b \in \mathbb{T}$ and consider $g \in C_{rd}([a,b]_{\mathbb{T}},(c,d))$. Then*

$$\left(\frac{1}{|I|}\int_I g(s)\Delta s\right)^{\alpha} \le \frac{1}{|I|}\int_I g^{\alpha}(s)\Delta s \tag{2.8}$$

holds for $\alpha < 0$ or $\alpha > 1$, and

$$\left(\frac{1}{|I|}\int_I g(s)\Delta s\right)^{\alpha} \ge \frac{1}{|I|}\int_I g^{\alpha}(s)\Delta s \tag{2.9}$$

holds for $\alpha \in (0,1)$.

Theorem 2.5. *Let $a, b \in \mathbb{T}$. For $f, g \in C_{rd}(\mathbb{I},\mathbb{R})$, the Hölder inequality is given by*

$$\int_a^b |u(t)v(t)|\Delta t \le \left[\int_a^b |v(t)|^p\Delta t\right]^{\frac{1}{p}}\left[\int_a^b |u(t)|^q\Delta t\right]^{\frac{1}{q}}, \tag{2.10}$$

where $p > 1$ and $\frac{1}{p}+\frac{1}{q} = 1$, a, $b \in \mathbb{T}$ and u, $v \in C_{rd}(\mathbb{T},\mathbb{R})$. This inequality is reversed if $0 < p < 1$ and if $p < 0$ or $q < 0$. For example, if $p = 1/\gamma < 1$, then

$$\int_a^b |u(t)v(t)|\Delta t \ge \left[\int_a^b |u(t)|^{1/\gamma}\Delta t\right]^{\gamma}\left[\int_a^b |v(t)|^{-1/(\gamma-1)}\Delta t\right]^{-(\gamma-1)}. \tag{2.11}$$

Theorem 2.6. *Let $h, f, g \in C_r([a,b]_{\mathbb{T}},[0,\infty))$. If $1/p + 1/q = 1$, with $p > 1$, then*

$$\int_a^b h(t)f(t)g(t)\Delta t \le \left(\int_a^b h(t)f^p(t)\Delta t\right)^{1/p}\left(\int_a^b h(t)g^q(t)\Delta t\right)^{1/q}. \tag{2.12}$$

The following theorems give the reverse Hölder-type inequality on time scales.

Theorem 2.7. *Let a, $b \in \mathbb{T}$ with $a < b$ and f and g be two positive functions defined on the interval $[a,b]_{\mathbb{T}}$ such that $0 < m \le f^p/g^q \le M < \infty$. Then for $p > 1$ and $q > 1$ with $1/p + 1/q = 1$, we have*

$$\left(\int_a^b f^p(t)\Delta t\right)^{1/p}\left(\int_a^b g^q(t)\Delta t\right)^{1/q} \leq \left(\frac{M}{m}\right)^{\frac{1}{pq}}\int_a^b f(t)g(t)\Delta t. \tag{2.13}$$

Next, we present the Hölder-type inequality in two dimensions on time scales.

Theorem 2.8. *Let $a, b \in \mathbb{T}$ with $a < b$ and f and g be two rd-continuous functions defined on the square $[a,b]_{\mathbb{T}} \times [a,b]_{\mathbb{T}}$. Then*

$$\begin{aligned}&\int_a^b\int_a^b |f(x,y)g(x,y)|\Delta x\Delta y\\ &\quad\leq \left(\int_a^b\int_a^b |f(x,y)|^p\Delta x\Delta y\right)^{1/p}\left(\int_a^b\int_a^b |g(x,y)|^q\Delta x\Delta y\right)^{1/q},\end{aligned} \tag{2.14}$$

where $p > 1$ and $q = p/(p-1)$.

In the following, we present the Minkowski inequality on time scales.

Theorem 2.9. *Let $f, g, h \in C_{rd}([a,b]_{\mathbb{T}}, \mathbb{R})$ and $p > 1$. Then*

$$\begin{aligned}&\left(\int_a^b |h(x)||f(x)+g(x)|^p\Delta x\right)^{\frac{1}{p}}\\ &\quad\leq \left(\int_a^b |h(x)||f(x)|^p\Delta x\right)^{\frac{1}{p}} + \left(\int_a^b |h(x)||g(x)|^p\Delta x\right)^{\frac{1}{p}}.\end{aligned} \tag{2.15}$$

Now, we present the definition of Δ-measurable functions on time scales to define the Lebesgue Δ-measure on $\mathbb{T}$. Let $\mathbb{T}$ be a time scale. Denote by S the family of all left-closed and right-open intervals of $\mathbb{T}$ of the form $[a,b) = \{t \in \mathbb{T} : a \leq t < b\}$ with $a, b \in \mathbb{T}$ and $a \leq b$. The interval $[a,a)$ is understood as the empty set. Obviously, the set weight $m : S \to [0,\infty)$, where S is a semiring of subsets of $\mathbb{T}$, defined by $m([a,b)) = b - a$ is a countably additive measure. An outer measure $m^* : P(\mathbb{T}) \to [0,\infty]$ generated by m is defined by

$$m^*(A) = \inf\left\{\sum_{n=1}^{\infty} m(A_n) : A_n \text{ is a sequence of } S \text{ with } A \subset \bigcup_{n=1}^{\infty} A_n\right\}.$$

If there is no sequence (A_n) of S such that $A \subset \bigcup_{n=1}^{\infty} A_n$, then we let $m^*(A) = \infty$. We define the family $S(m^*)$ of all m^*-measurable subsets of $\mathbb{T}$, i. e.,

$$S(m^*) = \{E \in \mathbb{T} : m^*(A) = m^*(A \cap E) + m^*(A \cap E^{\mathscr{C}}) \text{ for all } A \subset \mathbb{T}\}.$$

The collection $S(m^*)$ of all m^*-measurable sets is a σ-algebra and the restriction of m^* to $S(m^*)$ which we denote by m_Δ is a countably additive measure on $S(m^*)$. This measure

m_Δ is a Carathéodory extension of the set weight m associated with the family S, and called the Lebesgue Δ-measure on $\mathbb{T}$. We say that $f:\mathbb{T}\to\mathbb{R}$ is a measurable function if $f^{-1}(O)\in S(m^*)$ for every open subset of O of $\mathbb{R}$. We say that f belongs to $L^p_\Delta(\mathbb{T})$ provided that

$$\|f\|_{L^p_\Delta(\mathbb{T})}=\left(\int_{\mathbb{T}}|f|^p\Delta s\right)^{1/p}<\infty,\quad \text{if } 1\le p<\infty.$$

Throughout this chapter, we will assume that the functions are nonnegative rd-continuous functions, Δ-differentiable, locally delta integrable, and the left hand sides of the inequalities exits if the right hand sides exist. We also assume that all the constants and boundaries of the integrals that will appear in the inequalities are positive real numbers.

2.2 Background on weights

In this section, we introduce the necessary background on weights. We fix an interval $I_0\subset\mathbb{R}_+=[0,\infty)$ and consider subinterval I of I_0 of the form $[0,s]$, for $0<s<\infty$ and denote by $|I|$ the Lebesgue measure of I. A weight is a nonnegative locally integrable function defined on a bounded interval $I\subset I_0$ with values in $[0,+\infty)$. The classical Muckenhoupt class of weights A_p has been introduced by Muckenhoupt [37] in connection with the boundedness of the Hardy and Littlewood maximal operator in the space

$$L^p_w(\mathbb{R}_+)=\left\{f:\|f\|=\left(\int_0^\infty|f(x)|^p\omega(x)dx\right)^{\frac{1}{p}}<\infty\right\}.$$

A nonnegative weight ω is said to belong to the Muckenhoupt class $A_p(\mathscr{C})$ on the interval I_0 for $p>1$ and $\mathscr{C}>1$ (independent of p) if the inequality

$$\frac{1}{|I|}\int_I\omega(x)dx\le\mathscr{C}\left(\frac{1}{|I|}\int_I\omega^{\frac{1}{1-p}}(x)dx\right)^{1-p}\tag{2.16}$$

holds for every subinterval $I\subset I_0$. For $p>1$, we define the A_p-norm of the weight ω by

$$[A_p(\omega)]:=\sup_{I\subset I_0}\left(\frac{1}{|I|}\int_I\omega(x)dx\right)\left(\frac{1}{|I|}\int_I\omega^{\frac{-1}{p-1}}(x)dx\right)^{p-1}<\infty.$$

The weight ω is said to belong to the Muckenhoupt class $A_1(\mathscr{C})$ on the interval I_0 if the inequality

$$\frac{1}{|I|}\int_I\omega(x)dx\le\mathscr{C}\inf_{x\in I}\omega(x),\quad \text{for } \mathscr{C}>1,$$

holds for every subinterval $I \subset I_0$, and we define the A_1-norm by

$$[A_1(\omega)] := \sup_{I \subset I_0} \left(\frac{1}{|I|} \int_I \omega(x)dx \right) \left(\inf_{x \in I} \omega(x) \right)^{-1} < \infty.$$

The weight ω is said to belong to the Muckenhoupt class $A_\infty(\mathscr{C})$ on the interval I_0 if the inequality

$$\left(\frac{1}{|I|} \int_I \omega(x)dx \right) \left(\exp \frac{1}{|I|} \int_I \log \frac{1}{\omega(x)} dx \right) \leq \mathscr{C}, \quad \text{for } \mathscr{C} > 1,$$

holds for every subinterval $I \subset I_0$, and we define the A_∞-norm by

$$[A_\infty(\omega)] := \sup_{I \subset I_0} \left(\frac{1}{|I|} \int_I \omega(x)dx \right) \left(\exp \frac{1}{|I|} \int_I \log \frac{1}{\omega(x)} dx \right) < \infty.$$

In [37] Muckenhoupt proved the following result.

Theorem 2.10. *If $1 < p < \infty$ and ω satisfies the A_p-condition* (2.16) *on the interval I_0 with constant $\mathscr{C}$, then there exist constants q and $\mathscr{C}_1$ depending on p and $\mathscr{C}$ such that $1 < q < p$ and ω satisfies the A_q-condition*

$$\left(\frac{1}{|I|} \int_I \omega(t)dt \right) \left(\frac{1}{|I|} \int_I \omega^{-\frac{1}{q-1}}(t)dt \right)^{q-1} \leq \mathscr{C}_1, \tag{2.17}$$

for every subinterval $I \subset I_0$.

In other words, the Muckenhoupt result (see also Coifman and Fefferman [15]) for *self-improving* property states that: if $\omega \in A_p(\mathscr{C})$, then there exist a constant $\epsilon > 0$ and a positive constant $\mathscr{C}_1$ such that $\omega \in A_{p-\epsilon}(\mathscr{C}_1)$, and then

$$A_p(\mathscr{C}) \subset A_{p-\epsilon}(\mathscr{C}_1). \tag{2.18}$$

Further, Muckenhoupt proved the following result:

Theorem 2.11. *If $1 < p < \infty$ and $\omega(x) \in A_p(\mathscr{C})$ on the interval I with a constant $\mathscr{C}$, then there exist constants r and $\mathscr{C}_1$ depending only on p and $\mathscr{C}$ such that $\omega^r(x) \in A_p(\mathscr{C}_1)$ for $r > 1$.*

Despite a variety of ideas related to weighted inequalities that appeared with the birth of singular integrals, it was only in the 1970s that a better understanding of the subject was obtained and the full characterization of the weights w for which the Hardy–Littlewood maximal operator

$$\mathcal{M}f(x) := \sup_{x\in I} \frac{1}{|I|}\int_I f(y)dy, \tag{2.19}$$

is bounded on $L^p_w(\mathbb{R}_+)$ by means of the so-called A_p-*condition* was achieved by Muckenhoupt and published in 1972 (see [37]). Muckenhoupt's result became a landmark in the theory of weighted inequalities because most of the previously known results for classical operators had been obtained for special classes of weights (like power weights) and has been extended to cover several operators like Hardy operator, Hilbert operator, Calderón–Zygmund singular integral operators, fractional integral operators, etc.

A year later after the Muckenhoupt paper, a different class of weights satisfying the reverse Hölder inequality has been introduced and developed by Gehring [21, 22] in connection with the integrability properties of the gradient of quasiconformal mappings.

A weight ω is said to belong to the Gehring class $G_q(\mathcal{K}), 1 < q < \infty$ on the interval I_0, if there exists a constant $\mathcal{K} > 1$ such that the inequality

$$\left(\frac{1}{|I|}\int_I \omega^q(x)dx\right)^{\frac{1}{q}} \leq \mathcal{K}\left(\frac{1}{|I|}\int_I \omega(x)dx\right) \tag{2.20}$$

holds for every subinterval $I \subset I_0$, and we define the G_q-norm by

$$[G_q(\omega)] := \sup_{I\subset I_0}\left[\left(\frac{1}{|I|}\int_I \omega^q(x)dx\right)^{\frac{1}{q}}\left(\frac{1}{|I|}\int_I \omega(x)dx\right)^{-1}\right]^{\frac{q}{q-1}}.$$

The weight ω is said to belong to the Gehring class $G_\infty(\mathcal{K})$ if the inequality

$$\sup_I\left(\sup_{x\in I}\omega(x)\right)\left(\frac{1}{|I|}\int_I \omega(x)dx\right)^{-1} \leq \mathcal{K},$$

holds for every subinterval $I \subset I_0$. The weight ω is said to belong to the Gehring class $G_1(\mathcal{K})$ if the inequality

$$\exp\left(\frac{1}{|I|}\int_I \frac{\omega(x)}{\frac{1}{|I|}\int_I \omega(x)dx}\log\left(\frac{\omega(x)}{\frac{1}{|I|}\int_I \omega(x)dx}\right)dx\right) \leq \mathcal{K},$$

holds for every interval $I \subset I_0$. Gehring proved that if (2.20) holds, then there exist a $p > q$ and a positive constant $\mathcal{K}_1$ such that

$$\frac{1}{|I|}\int_I \omega^p(x)dx \leq \mathcal{K}_1\left(\frac{1}{|I|}\int_I \omega(x)dx\right)^p. \tag{2.21}$$

In other words, the Gehring result for *self-improving* property states that: if $\omega \in G_q(\mathcal{K})$ then there exist $\epsilon > 0$ and a positive constant $\mathcal{K}_1$ such that $\omega \in G_{q+\epsilon}(\mathcal{K}_1)$ and

$$G_q(\mathscr{K}) \subset G_{q+\epsilon}(\mathscr{K}_1). \tag{2.22}$$

The self-improving property of the Gehring class has applications in different fields, especially in studying the optimal regularity of solutions to some elliptic PDEs (see, for example, Kenig [28]) where the solution of the Dirichlet problem $div\,A(\sigma)\nabla\theta = 0$ on the unit disc D, with $\theta|_D = \varphi$, can be expressed in terms of G^q conditions on the boundary ∂D for the harmonic measures associated to $A(\sigma)$, with $1/p+1/q = 1$. We refer the reader to the book [28] for more applications of these classes on extrapolation theory, vector-valued inequalities, and estimates for certain classes of nonlinear partial differential equations. The relation between Gehring and Muckenhoupt classes (inclusion properties) was given by Coifman and Fefferman in [15]. They proved that any Gehring class is contained in some Muckenhoupt class, and vice versa. In other words, they proved the following inclusions:

$$G_q(\mathscr{K}) \subset A_p(\mathscr{K}_1) \tag{2.23}$$

and

$$A_{p_1}(\mathscr{K}_1) \subset G_{q_1}(\mathscr{K}). \tag{2.24}$$

The sharp results for the reverse Hölder inequalities can be found in Martio–Sbordone [33]. The proof of Gehring's inequality is based on the use of the Calderón–Zygmund decomposition and the scale structure of L^p-spaces. For further studies of the Muckenhoupt and Gehring classes, we refer the reader to [5, 11, 18–20, 25–27, 33–37, 40, 41, 47, 56, 57, 65–68].

In recent years the study of regularity and boundedness of discrete operator on $l^p_\vartheta(\mathbb{Z}_+)$ analogues for $L^p_w(\mathbb{R}_+)$-regularity and boundedness has been considered by some authors; see, for example, [7, 29, 31] and the references therein. One of the reasons for this surge of interest in discrete cases is due to the fact that the discrete operators may even behave differently from their continuous counterparts as is exhibited by the discrete spherical maximal operator [32]. In some special cases it is possible to translate or adapt almost straightforwardly the objects and results from the continuous setting to the discrete setting, or vice versa, however, in some other cases that is far from trivial [13, 14, 16].

For example, in the simplest cases of l^p-bounds for discrete analogues of classical operators such as Calderón–Zygmund singular integral operators, fractional integral operators, and the maximal Hardy–Littlewood operator follow from known L^p-bounds for the original operators in the Euclidean setting, via elementary comparison arguments (see [42–44]). But l^p-bounds for discrete analogues of more complicated operators are not implied by results in the continuous setting, and, moreover, the discrete analogues are resistant to conventional methods. The main challenge is that there are no general methods to study these questions. These methods have to be developed starting from the basic definitions. In [6] the authors mentioned that the study of discrete inequalities is not an easy task and more difficult to analyze than its integral counterparts and discovered that the conditions do not correspond, in any natural way, with those that are obtained by discretizing the results for functions, but the converse is true. This means that what goes for sums goes, with the obvious modifications, for integrals which in fact proved the first part of the basic principle of Hardy, Littlewood, and Polya [24, p. 11]. Indeed, the proofs for series translate immediately and become much simpler when applied to integrals. This fact motivated a lot of authors to study the characterizations of discrete weights and use the new characterizations to formulate some conditions for the boundedness of the discrete operators and prove some embedding theorems for Lorentz spaces. For further studies of the discrete classes, we refer the reader to [12, 45, 46, 48, 50, 51, 53–55, 57, 58, 61–63].

In the following, for the sake of completeness, we present the background and the basic definitions for discrete weights. Throughout this chapter, $\mathbb{Z}_+$ stands for the set of nonnegative integers, i. e., $\mathbb{Z}_+ = \{1, 2, \ldots\}$. By an interval $\mathbb{J}$ we mean a finite subset of $\mathbb{Z}_+$ consisting of consecutive integers and, for $J \subset \mathbb{J}$, the number $|J|$ stands for its cardinality. A discrete weight on $\mathbb{Z}_+$ is a sequence $\vartheta = \{\vartheta(n)\}_{n=1}^{\infty}$ of nonnegative real numbers. The space $l_\vartheta^p(\mathbb{Z}_+)$, for $1 \leq p < \infty$, is the space containing all real-valued sequences u defined on $\mathbb{Z}^+$ and satisfying the condition

$$\|u\|_{l_\vartheta^p(\mathbb{Z}_+)} := \left(\sum_{n=1}^{\infty} |u(n)|^p \vartheta(n)\right)^{1/p} < \infty,$$

where ϑ is a discrete weight. We shall denote by $A = 2^{\mathbb{Z}_+}$ the power set of $\mathbb{Z}_+$. A discrete weight ϑ belongs to the discrete Muckenhoupt class $\mathscr{A}_1(\mathscr{C})$ on $\mathbb{Z} \subset \mathbb{Z}_+$ for $p > 1$ and $\mathscr{C} > 1$ if the inequality

$$\frac{1}{|J|} \sum_{k \in J} \vartheta(k) \leq \inf_{k \in J} \vartheta(k), \quad \text{for all } k \in J, \tag{2.25}$$

holds for every subinterval $J \subset \mathbb{J}$, with $|J|$ being the cardinality of the set J. Sometimes it is convenient to consider a symmetric form of which it is equivalent to (2.25). A discrete weight ϑ belongs to the discrete Muckenhoupt class $\mathscr{A}_2(\mathscr{C})$ on the interval $\mathbb{J} \subseteq \mathbb{Z}_+$ for $p > 1$ and $\mathscr{C} > 1$ if the inequality

$$\sum_{k \in J} \vartheta(k) \sum_{k \in J} \vartheta^{-1}(k) \leq A|J|^2, \tag{2.26}$$

holds for every subinterval $J \subset \mathbb{J}$. This class has been used in harmonic analysis by some authors. For example, in [4], Ariño and Muckenhoupt proved that if ϑ is nonincreasing and satisfies (2.25), then the space $\Lambda(\vartheta^{-q^*/q}, q^*)$ is the dual space of the discrete classical Lorentz space

$$\Lambda(\vartheta, q) = \left\{ x : \|x\|_{\vartheta,q} = \left(\sum_{n=1}^{\infty} |x^*(n)|^q \vartheta(n) \right)^{1/q} < \infty \right\},$$

where $x^*(n)$ is the nonincreasing rearrangement of $|x(n)|$ and q^* is the conjugate of q. In [39] Pavlov gave a full description of all complete interpolating sequences on the real line by using the integral from of (2.26). In particular, he proved that a sequence λ_n of real numbers is a complete interpolating sequence if and only if the weight $v = |F(x + iy)|^2$, $x, y \in \mathbb{R}$, satisfies the Muckenhoupt condition

$$\int_J v(x)dx \int_J v^{-1}(x)dx \leq \mathscr{C}|J|^2, \tag{2.27}$$

for some constant $\mathscr{C} > 0$ and some $y \neq 0$ for all intervals $J \subset \mathbb{R}$ of finite length $|J|$, where

$$F(z) = \lim_{R \to \infty} \prod_{|\lambda_n| < R} \left(1 - \frac{z}{\lambda_n} \right).$$

In [30], Lyubarskii and Seip claimed that the condition (2.27) can be replaced by a discrete version (2.26) and proved that a sequence λ_n of real numbers is a complete interpolating sequence if and only if there is a relatively dense subsequence λ_{n_k} such that the numbers $d(k) = |F'(\lambda_{n_k})|^2$ satisfy the discrete Muckenhoupt condition

$$\sum_{k \in J} d(k) \sum_{k \in J} d^{-1}(k) \leq \mathscr{C}|J|^2, \tag{2.28}$$

for some constant $\mathscr{C} > 0$ and all finite sets J of consecutive integers containing $|J|$ elements. Checking the Muckenhoupt condition (2.27) for a weight F given by an infinite product (covering in the Cauchy principle value sense) is particularly quite hard. However, condition (2.28) is relatively easier to verify since it involves only countably many sets J instead of all finite intervals. A discrete nonnegative sequence ϑ belongs to the discrete Muckenhoupt class $\mathscr{A}_p(\mathscr{C})$ on the interval $\mathbb{J} \subseteq \mathbb{Z}_+$ for $p > 1$ and $\mathscr{C} > 1$ if the inequality

$$\left(\frac{1}{|J|} \sum_{k \in J} \vartheta(k) \right) \left(\frac{1}{|J|} \sum_{k \in J} \vartheta^{\frac{-1}{p-1}}(k) \right)^{p-1} \leq \mathscr{C} \tag{2.29}$$

holds for every subinterval $J \subset \mathbb{J}$. For a given exponent $p > 1$, we define the $\mathscr{A}^p$-norm of the discrete weight ϑ by the following quantity:

$$[\mathscr{A}^p(\vartheta)] := \sup_{J \subset \mathbb{J}} \left(\frac{1}{|J|} \sum_{k \in J} \vartheta(k) \right) \left(\frac{1}{|J|} \sum_{k \in J} \vartheta^{\frac{-1}{p-1}}(k) \right)^{p-1} < \infty, \tag{2.30}$$

where the supremum is taken over all intervals $J \subset \mathbb{J}$. When we fix a constant $\mathscr{C} > 1$, the pair of real numbers $(p, \mathscr{C})$ defines the $\mathscr{A}_p$ discrete Muchenhoupt class $\mathscr{A}_p(\mathscr{C})$:

$$\vartheta \in \mathscr{A}_p(\mathscr{C}) \iff [\mathscr{A}_p(\vartheta)] \leq \mathscr{C},$$

and we will refer to $\mathscr{C}$ as the $\mathscr{A}_p$-constant of the class. The Hardy–Littlewood maximal sequence $\mathscr{M}f$ of the sequence f is defined by

$$(\mathscr{M}f)(n) := \sup_{n\in J} \frac{1}{|J|} \sum_J f(k). \tag{2.31}$$

The operator $\mathscr{M}: f \to \mathscr{M}f$ is called discrete Hardy–Littlewood maximal operator. Observe that $\mathscr{M}$ is merely sublinear, rather than linear, and it is a contraction on ℓ^∞. The boundedness of discrete Hardy–Littlewood maximal operator has been characterized in [54], and it has been proved that $\mathscr{M}f(n)$ is bounded on $\ell^p(\vartheta)$ if and only if $\vartheta \in \mathscr{A}_p$. In [61] the authors proved that if $q > 1$, $\mathscr{C} > 1$, and ϑ is a nondecreasing weight belonging to $\mathscr{A}_q(\mathscr{C})$, then $\vartheta \in \mathscr{A}_p(\mathscr{C}_1)$ for $p \in (p_0, q]$ where p_0 is a unique solution of an algebraic inequality. This result proves that if $\vartheta \in \mathscr{A}_q(\mathscr{C})$ then there exist an $\epsilon > 0$ and a constant $\mathscr{C}_1 = A_1(p, \mathscr{C})$ such that $\vartheta \in \mathscr{A}_{q-\epsilon}(\mathscr{C}_1)$, (self-improving property) and thus

$$\mathscr{A}_q(\mathscr{C}) \subset \mathscr{A}_{q-\epsilon}(\mathscr{C}_1). \tag{2.32}$$

In the following, we present some basic properties of discrete Muckenhoupt weights.

Theorem 2.12 ([61]). *Let ϑ be a nonnegative weight and p and q be positive real numbers. The following properties hold:*
(1) *$\vartheta \in \mathscr{A}_p$ if and only if $\vartheta^{1-p'} \in \mathscr{A}_{p'}$, with $\mathscr{A}_{p'}(\vartheta^{1-p'}) = [\mathscr{A}_p(\vartheta)]^{p'-1}$ where p' is the conjugate of p;*
(2) *$\mathscr{A}_p \subset \mathscr{A}_q$ for all $1 < p \leq q$;*
(3) *if $\vartheta \in \mathscr{A}_p$, then $\vartheta^\alpha \in \mathscr{A}_p$, for $0 \leq \alpha \leq 1$, with $\mathscr{A}_p(\vartheta^\alpha) = [\mathscr{A}_p(\vartheta)]^\alpha$;*
(4) *$\mathscr{A}_1 \subset \mathscr{A}_p \subset \mathscr{A}_\infty$, for all $1 < p < \infty$;*
(5) *$\mathscr{A}_\infty = \bigcup_{1<p} \mathscr{A}_p$ with $\mathscr{A}_\infty(\vartheta) = \lim_{p\to\infty} \mathscr{A}_p$ and $\mathscr{A}_1 \subset \bigcap_{p>1} \mathscr{A}_p$;*
(6) *if $\vartheta_1, \vartheta_2 \in \mathscr{A}_p$, then $\vartheta_1^\alpha \vartheta_2^{1-\alpha} \in \mathscr{A}_p$, $0 \leq \alpha \leq 1$, with a constant $\mathscr{A}_p(\vartheta_1^\alpha \vartheta_2^{1-\alpha}) = [\mathscr{A}_p(\vartheta_1)]^\alpha [\mathscr{A}_p(\vartheta_2)]^{1-\alpha}$.*

Theorem 2.13 ([61]). *Let ϑ be a nonnegative weight and p and q be positive real numbers. The following properties hold:*
(1) *$\vartheta \in \mathscr{A}_p$ if and only if there exists $\vartheta_1, \vartheta_2 \in \mathscr{A}_1$ such that $\vartheta = \vartheta_1 \vartheta_2^{1-p}$, $1 < p < \infty$;*
(2) *if $\vartheta \in \mathscr{A}_p$, then $\vartheta^\tau \in \mathscr{A}_p$ for some $\tau > 1$;*
(3) *if $\vartheta \in \mathscr{A}_p$, $p > 1$, then $\vartheta \in \mathscr{A}_{p-\epsilon}$, for some $\epsilon > 0$;*
(4) *$\vartheta \in \mathscr{A}_p$ if and only if ϑ and $\vartheta^{\frac{1}{1-p}}$ are in $\mathscr{A}_\infty$.*

A discrete nonnegative weight ϑ belongs to the discrete Gehring class $\mathscr{G}_q(\mathscr{K})$ for a given exponent $q > 1$ and a constant $\mathscr{K} > 1$ (or satisfies the reverse Hölder inequality) on the interval $\mathbb{J} \subset \mathbb{Z}_+$ if, for every subinterval $J \subseteq \mathbb{J}$, we have

$$\left(\frac{1}{|J|}\sum_{k\in J}\vartheta^{q}(k)\right)^{\frac{1}{q}} \leq \mathscr{K}\frac{1}{|J|}\sum_{k\in J}\vartheta(k).$$

For a given exponent $q > 1$, we define the $\mathscr{G}_q$-norm of ϑ as

$$[\mathscr{G}^{q}(\vartheta)] := \sup_{J\subset I}\left[\left(\frac{1}{|J|}\sum_{k\in J}\vartheta(k)\right)^{-1}\left(\frac{1}{|J|}\sum_{k\in J}\vartheta^{q}(k)\right)^{\frac{1}{q}}\right]^{\frac{q}{q-1}},$$

where the supremum is taken over all intervals $J \subset \mathbb{J}$ and represents the best constant for which the $\mathscr{G}^{q}$-condition holds true independently of the interval $J \subseteq \mathbb{J}$. We say that ϑ is a discrete Gehring weight if its $\mathscr{G}_q$-norm is finite, i. e.,

$$\vartheta \in \mathscr{G}_q \iff [\mathscr{G}_q(\vartheta)] < \infty.$$

When we fix a constant $\mathscr{K} > 1$, the pair of real numbers $(q, \mathscr{K})$ defines the $\mathscr{G}^{p}$-discrete Gehring class $\mathscr{G}^{p}(\mathscr{K})$:

$$\vartheta \in \mathscr{G}_q(\mathscr{K}) \iff [\mathscr{G}_q(\vartheta)] \leq \mathscr{K},$$

and we will refer to $\mathscr{K}$ as the $\mathscr{G}_q$- constant of the class. In [50] the authors proved that if $q > 1$ and $\mathscr{K}_q > 1$, and ϑ is a nonincreasing sequence belonging to $\mathscr{G}_q(\mathscr{K}_q)$, then $\vartheta \in \mathscr{G}_p(\mathscr{K}_p)$ for $p \in [q, q^*)$ where q^* is a unique solution of an algebraic inequality. In [53] it has been proved that if ϑ is a nonincreasing sequence and satisfies (2.25) for $\mathscr{C} > 1$, then for $p \in [1, \mathscr{C}/(\mathscr{C}-1))$ the inequality

$$\frac{1}{|J|}\sum_{k\in J}\vartheta^{p}(k) \leq \mathscr{C}_1\left(\frac{1}{|J|}\sum_{k\in J}\vartheta(k)\right)^{p}, \quad \text{for } J \subset \mathbb{J}, \tag{2.33}$$

holds for every subinterval $J \subset \mathbb{J}$. This result proves that the Muckenhoupt $\mathscr{A}_1$ weight belongs to some Gehring class of weights satisfying reverse Hölder inequality (*a transition property*). A discrete nonnegative weight ϑ belongs to the discrete Gehring class $\mathscr{G}^{1}(\mathscr{K})$ for a constant $\mathscr{K} > 1$ on the interval $\mathbb{J} \subset \mathbb{Z}_+$ if, for every subinterval $J \subseteq \mathbb{J}$, we have

$$[\mathscr{G}_1(v)] = \sup_{J\subset\mathbb{J}}\exp\left(\frac{1}{|J|}\sum_{k\in J}\frac{\vartheta(k)}{\vartheta_J}\log\frac{\vartheta(k)}{\vartheta_J}\right) \leq \mathscr{K},$$

where $\vartheta_J = (1/|J|)\sum_{k\in J}\vartheta(k)$ and the supremum is taken over all $J \subset \mathbb{J}$. The class $\mathscr{G}_\infty$ consists of all weights v defined on $\mathbb{J} \subset \mathbb{Z}_+$ such that $\mathscr{G}_\infty(v)$-norm is finite, where

$$[\mathscr{G}_\infty(v)] = \sup_{J\subset\mathbb{J}}\frac{\operatorname{ess\,sup}_{k\in J}\vartheta(k)}{|J|\sum_{k\in J}\vartheta(k)}.$$

In the following, we present some basic properties of discrete Gehring weights.

Theorem 2.14 ([61]). *Let ϑ be a nonnegative weight and p and q be real positive numbers such that p, $q > 1$. The following properties hold:*
(1) $\mathcal{G}_q \subset \mathcal{G}_p$ *for all* $1 < p \le q$;
(2) $\mathcal{G}_\infty \subset \mathcal{G}_q \subset \mathcal{G}_1$ *for all* $1 < q \le \infty$;
(3) $\mathcal{G}_1 = \bigcup_{1<q\le\infty} \mathcal{G}_q$ *with* $\mathcal{G}_1(\vartheta) = \lim_{q\to 1} \mathcal{G}_q$;
(4) *if* $\vartheta \in \mathcal{G}_q$, *then* $\vartheta \in \mathcal{G}_{q+\epsilon}$, *for some* $\epsilon > 0$;
(5) $\mathcal{G}_1 = \mathcal{A}_\infty = \bigcup_{1<p\le\infty} \mathcal{A}_p = \bigcup_{1<q\le\infty} \mathcal{G}_q$.

For $\mathscr{C} \ge 1$ and $q > p > 1$, we denote by $\mathscr{B}^{p,q}(\mathscr{C})$ the class of all nonnegative weights v that satisfy a generalized reverse Hölder inequality

$$\left(\frac{1}{|J|}\sum_{n\in J} v^q(n)\right)^{1/q} \le \mathscr{C}\left(\frac{1}{|J|}\sum_{n\in J} v^p(n)\right)^{1/p}, \quad \text{for all } J \subset \mathbb{J}. \tag{2.34}$$

By recalling the classical Hölder inequality, it is clear that the definition of $\mathscr{B}^{p,q}$ is well posed only for $\mathscr{C} \ge 1$, where the equality prevails in case of constant sequences. The smallest constant, independent of the interval J, satisfying the inequality (2.34) is called the $\mathscr{B}^{p,q}$-norm of the weight v and will be denoted by $\mathscr{B}^{p,q}(v)$; it is given by

$$\mathscr{B}^{p,q}(v) := \sup_{J\subset\mathbb{J}}\left(\frac{1}{|J|}\sum_{n\in J} v^p(n)\right)^{-\frac{1}{p}}\left(\frac{1}{|J|}\sum_{n\in J} v^q(n)\right)^{\frac{1}{q}}. \tag{2.35}$$

We say that v is a $\mathscr{B}^{p,q}$-weight if its $\mathscr{B}^{p,q}$-norm is finite, i. e.,

$$v \in \mathscr{B}^{p,q} \iff \mathscr{B}^{p,q}(v) < +\infty.$$

When we fix a constant $\mathscr{C} > 1$, the triple of real numbers $(p, q, \mathscr{C})$ defines the $\mathscr{B}^{p,q}$ discrete class:

$$v \in \mathscr{B}^{p,q}(\mathscr{C}) \iff \mathscr{B}^{p,q}(v) \le \mathscr{C},$$

and we will refer to $\mathscr{C}$ as the $\mathscr{B}^{p,q}$-constant of the class. Moreover,

$$v \in \mathscr{B}^{p,q}(\mathscr{C}) \iff v^p \in \mathscr{B}^{1,q/p}(\mathscr{C}^p) \iff v^q \in \mathscr{B}^{p/q,1}(\mathscr{C}^q),$$

and the following properties hold:

$$\begin{aligned} \mathscr{B}^{p,q}(\mathscr{C}) &\subset \mathscr{B}^{p,r}(\mathscr{C}), \quad \text{for } p < r \le q, \\ \mathscr{B}^{p,q}(\mathscr{C}) &\subset \mathscr{B}^{r,q}(\mathscr{C}), \quad \text{for } p \le r < q. \end{aligned}$$

It is immediate to observe that the classes $\mathscr{A}^p$ and $\mathscr{G}^q$ are special cases of the $\mathscr{B}^{p,q}(\mathscr{C})$ of discrete weights as follows:

(1) $\mathscr{A}_p(\mathscr{C}) = \mathscr{B}^{\frac{1}{1-p},1}(\mathscr{C}) \Longleftrightarrow \mathscr{A}_{\frac{p-1}{p}}(\mathscr{C}) = \mathscr{B}^{1,p}(\mathscr{C})$,

(2) $\mathscr{G}_q(\mathscr{C}) = \mathscr{B}^{1,q}(\mathscr{C})$.

The natural question that arises here is: *Is it possible to prove the properties of general class of weights on time scales which are special cases contain the properties of the continuous and discrete Muckenhoupt and Gehring weights?*

The objective of this chapter is to provide an affirmative answer to this question and prove some properties and some relations between Muckenhoupt and Gehring classes in the context of time scales and use these properties to prove that self-improving properties hold and then extend the results to prove some higher integrability on time scales.

Now, we present the definitions of the Muckenhoupt and Gehring weights on time scales. We assume that ω is a nonnegative locally Δ-integrable weight defined on $\mathbb{I} = [0,\infty)_{\mathbb{T}} = [0,\infty) \cap \mathbb{T}$ and p is a positive real number. The nonnegative weight ω is said to belong to the Muckenhoupt class $\mathbb{A}_p(\mathscr{C})$ on time scales on the interval $\mathbb{I}_0$ for $p > 1$ and $\mathscr{C} > 1$ (independent of p) if the inequality

$$\frac{1}{|I|}\int_I \omega(s)\Delta s \le \mathscr{C}\left(\frac{1}{|I|}\int_I \omega^{\frac{1}{1-p}}(s)\Delta s\right)^{1-p} \tag{2.36}$$

holds for every subinterval $I \subset \mathbb{I}_0$. For $p > 1$, we define the $\mathbb{A}_p$-norm on time scales by

$$[\mathbb{A}_p(\omega)] := \sup_{I \subset \mathbb{I}}\left(\frac{1}{|I|}\int_I \omega(s)\Delta s\right)\left(\frac{1}{|I|}\int_I \omega^{\frac{-1}{p-1}}(s)\Delta s\right)^{p-1} < \infty.$$

The weight ω is said to belong to the Muckenhoupt class $\mathbb{A}_1(\mathscr{C})$ on time scales on the interval $\mathbb{I}_0$ if the inequality

$$\frac{1}{|I|}\int_I \omega(s)\Delta s \le \mathscr{C}\inf_{x\in I}\omega(x), \quad \text{for } \mathscr{C} > 1,$$

holds for every subinterval $I \subset \mathbb{I}$. We define the $\mathbb{A}_1$-norm on time scales by

$$[\mathbb{A}_1(\omega)] := \sup_{I \subset \mathbb{I}_0}\left(\frac{1}{|I|}\int_I \omega(x)\Delta x\right)\left(\inf_{x\in I}\omega(x)\right)^{-1}.$$

The weight ω is said to belong to the Muckenhoupt class $\mathbb{A}_\infty(\mathscr{C})$ on time scales on the interval $\mathbb{I}$ if the inequality

$$\left(\frac{1}{|I|}\int_I \omega(s)\Delta s\right)\left(\exp\left(\frac{1}{|I|}\int_I \log\frac{1}{\omega(s)}\Delta s\right)\right) \le \mathscr{C}, \quad \mathscr{C} > 1,$$

holds for every subinterval $I \subset \mathbb{I}$, and we define the $\mathbb{A}_\infty$-norm on time scales by

$$[\mathbb{A}_\infty(\omega)] := \sup_{I\subset I_0}\left(\frac{1}{|I|}\int_I \omega(s)\Delta s\right)\left(\exp\left(\frac{1}{|I|}\int_I \log\frac{1}{\omega(s)}\Delta s\right)\right).$$

The weight ω is said to belong to the Gehring class $\mathbb{G}_q(K)$ on time scales on the interval $\mathbb{I}$ for $q > 1$ and $K > 1$ (independent of q) if the inequality

$$\left(\frac{1}{|I|}\int_I \omega^q(s)\Delta s\right)^{\frac{1}{q}} \leq K\frac{1}{|I|}\int_I \omega(s)\Delta s \tag{2.37}$$

holds for every subinterval $I \subset \mathbb{I}$, and the $\mathbb{G}_q$-norm is defined by

$$[\mathbb{G}_q(\omega)] := \sup_{I\subset I_0}\left[\left(\frac{1}{|I|}\int_I \omega^q(s)\Delta s\right)^{\frac{1}{q}}\left(\frac{1}{|I|}\int_I \omega(s)\Delta s\right)^{-1}\right]^{\frac{q}{q-1}} < \infty.$$

The weight ω is said to belong to the Gehring class $\mathbb{G}_\infty(K)$ on time scales on the interval $\mathbb{I}$ if the inequality

$$[\mathbb{G}_\infty(\omega)] = \sup_{I\subset\mathbb{I}_0}\left(\sup_{s\in I}\omega(s)\left(\frac{1}{|I|}\int_I \omega(s)\Delta s\right)^{-1}\right) \leq K, \quad \text{for } K > 0,$$

holds for every subinterval $I \subset \mathbb{I}$. The weight ω is said to belong to the Gehring class $\mathbb{G}_1(K)$ on the interval $\mathbb{I}$ if the inequality

$$[\mathbb{G}_1(\omega)] = \exp\left(\frac{1}{|I|}\int_I \frac{\omega(s)}{\frac{1}{|I|}\int_I \omega(s)\Delta s}\log\left(\frac{\omega(s)}{\frac{1}{|I|}\int_I \omega(s)\Delta s}\right)\Delta s\right) \leq K,$$

holds for every interval $I \subset \mathbb{I}$. We say that a nonnegative weight ω belongs to the class $\mathscr{U}_p^q(B)$ if it satisfies the reverse Hölder inequality

$$\left[\frac{1}{|I|}\int_I \omega^q(s)\Delta s\right]^{\frac{1}{q}} \leq B\left[\frac{1}{|I|}\int_I \omega^p(s)\Delta s\right]^{\frac{1}{p}},$$

with a positive constant $B > 1$ for $1 \leq p < q$ for every interval $I \subset \mathbb{I}$. When we fix a constant $\mathscr{C} > 1$, the triple of real numbers $(p, q, \mathscr{C})$ defines the $\mathscr{U}_p^q$ class:

$$\omega \in \mathscr{U}_p^q(\mathscr{C}) \iff [\mathscr{U}_p^q(\omega)] \leq \mathscr{C},$$

and we will refer to $\mathscr{C}$ as the $\mathscr{U}_p^q$-constant of the class. It is immediate to observe that the classes $\mathbb{A}^p$ and $\mathbb{G}^q$ are special cases of the class $\mathscr{U}_p^q$ of weights as follows:

$$\mathbb{A}^p := \mathscr{U}_{\frac{1}{1-p}}^1 \quad \text{and} \quad \mathbb{G}^q := \mathscr{U}_1^q.$$

2.3 Properties of Muckenhoupt and Gehring weights

In this section, we state some basic properties of the Muckenhoupt $\mathbb{A}_p$-weights and Gehring $\mathbb{G}_q$-weights on time scales. The results are adapted from [49]. The results are particular cases when $\mathbb{T} = \mathbb{R}$ and cover the results due to Cruz-Uribe [17], Johnson and Neugebauer [26], and Popoli [41]. Throughout this chapter, we assume that the functions in the statements of the theorems are nonnegative and rd-continuous, while the integrals considered are assumed to exist and be finite. Therefore, these conditions will be omitted, for brevity.

Definition 2.1. We define the operator $\mathcal{M}_q\omega : \mathbb{I} \to \mathbb{R}^+$ by

$$\mathcal{M}_q\omega := \left(\frac{1}{|I|}\int_I \omega^q(s)\Delta s\right)^{\frac{1}{q}}, \tag{2.38}$$

for any real number $q \neq 0$ and every $I \subset \mathbb{I}$.

In the following lemma, we state some basic properties of the operator $\mathcal{M}_q$ which will be needed in the proof of the main results. The results are adapted from [49].

Lemma 2.4. *Assume that $p, q \neq 0$ are real numbers and $\mathcal{M}_q$ is defined as in* (2.38). *Then the following properties hold:*
(1) $\mathcal{M}_{-1}\omega(s) \leq \mathcal{M}_1\omega(s)$;
(2) $\mathcal{M}_q\omega(s) \geq \mathcal{M}_1\omega(s)$ *for all* $q \geq 1$;
(3) $\mathcal{M}_q\omega(s) \leq \mathcal{M}_1\omega(s)$ *for all* $q < 1$;
(4) $\mathcal{M}_p\omega(s) \leq \mathcal{M}_q\omega(s)$ *for all* $p \leq q$.

Lemma 2.5. *Let q be a positive real numbers. If $\omega \in \mathbb{G}_q(K)$ for $q > 1$ and $K > 1$, then $\mathcal{M}_q\omega \leq K\mathcal{M}_1\omega$ and, consequently, $\mathcal{M}_q\omega \leq K\mathcal{M}_p\omega$ for all $p \geq 1$.*

Lemma 2.6. *If $\omega \in \mathbb{A}_p(\mathcal{C})$ and $p > 1$, then*

$$\mathcal{M}_1\omega \leq \mathcal{C}\exp(\mathcal{M}_1 \log \omega) \tag{2.39}$$

holds.

The next lemma gives the inclusion of the Gehring classes of weights $\mathbb{G}_q$ into the $\mathbb{G}_1$-class.

Lemma 2.7. *Assume that $q > 1$ is a nonnegative number. If $\omega \in \mathbb{G}_q$, then*

$$\exp\left(\frac{1}{|I|}\int_I \frac{\omega(s)}{\frac{1}{|I|}\int_I \omega(s)\Delta s}\log\left(\frac{\omega(s)}{\frac{1}{|I|}\int_I \omega(s)\Delta s}\right)\Delta s\right) < \infty \tag{2.40}$$

holds for all $I \subset \mathbb{I}$.

In the following theorems, we present some basic inclusion properties of Muckenhoupt and Gehring classes on time scales.

Theorem 2.15. *Let p and q be positive real numbers. Then the following inclusion properties of Muckenhoupt classes hold:*
(1) $\mathbb{A}_p \subset \mathbb{A}_q$ *for all* $1 < p \le q$;
(2) *let* $1 < p < \infty$, *then* $\mathbb{A}_1 \subset \mathbb{A}_p \subset \mathbb{A}_\infty$;
(3) $\mathbb{A}_\infty = \bigcup_{1<p}\mathbb{A}_p$ *with* $\mathbb{A}_\infty = \lim_{p\to\infty} \mathbb{A}_p$ *and* $\mathbb{A}_1 \subset \bigcap_{p>1}\mathbb{A}_p$.

Theorem 2.16. *Let p and q be nonnegative real numbers. Then the following inclusion properties of Gehring classes hold:*
(1) $\mathbb{G}_q \subset \mathbb{G}_p$ *for all* $1 \le p \le q$;
(2) $\mathbb{G}_\infty \subset \mathbb{G}_q \subset \mathbb{G}_1$ *for all* $1 \le p \le \infty$;
(3) $\mathbb{G}_1 = \bigcup_{1<q\le\infty} \mathbb{G}_q, 1 < q < \infty$ *with* $\mathbb{G}_1(\omega) = \lim_{q\to 1} \mathbb{G}_q(\omega)$, *that is,* $\omega \in \mathbb{G}_p$, *which is the desired result.*

Here, we prove some additional properties of the Muckenhoupt classes of weights on time scales.

Theorem 2.17. *Assume that p and q be positive real numbers. Then the following properties hold:*
(1) $\omega \in \mathbb{A}_p$ *if and only if* $\omega^{1-p'} \in \mathbb{A}_{p'}$ *with* $[\mathbb{A}_{p'}(\omega^{1-p'})] = [\mathbb{A}_p(\omega)]^{p'-1}$, *where* p' *is the conjugate of* p;
(2) *if* $\omega \in \mathbb{A}_p$ *for* $1 \le p < \infty$, *then for each* $0 < \epsilon < 1$ *we have* $\omega^\epsilon \in \mathbb{A}_{\epsilon p+1-\epsilon}$.

Proof. (1) From the definition of the class $\mathbb{A}_p$ and since $1 - p' = 1/(1-p) < 0$, we have, for $\mathscr{C} > 1$, that

$$\begin{aligned}
\omega \in \mathbb{A}_p &\iff \left(\frac{1}{|I|}\int_I \omega(s)\Delta s\right) \le \mathscr{C}\left(\frac{1}{|I|}\int_I \omega^{\frac{1}{1-p}}(s)\Delta s\right)^{1-p} \\
&\iff \left(\frac{1}{|I|}\int_I \omega(s)\Delta s\right)^{\frac{1}{1-p}} \ge \mathscr{C}^{\frac{1}{1-p}}\frac{1}{|I|}\int_I \omega^{\frac{1}{1-p}}(s)\Delta s \\
&\iff \frac{1}{|I|}\int_I \omega^{1-p'}(s)\Delta s \le \mathscr{C}^{p'-1}\left(\frac{1}{|I|}\int_I \left(\omega^{1-p'}(s)\right)^{\frac{1}{1-p'}}\Delta s\right)^{1-p'} \\
&\iff \omega^{1-p'} \in \mathbb{A}_{p'},
\end{aligned}$$

with $[\mathbb{A}_{p'}(\omega^{1-p'})] = [\mathbb{A}_p(\omega)]^{p'-1}$. This is the desired result.

(2) Let $1 \le p < \infty$, $0 < \epsilon < 1$, and $r = \epsilon p + 1 - \epsilon$. Then $r - 1 = \epsilon(p-1)$ and, by applying Lemma 2.4 for $\epsilon < 1$, we have

$$\left(\frac{1}{|I|}\int_I \omega^{\epsilon}(s)\Delta s\right)\left(\frac{1}{|I|}\int_I (\omega^{\epsilon}(s))^{\frac{-1}{r-1}}\Delta s\right)^{r-1}$$
$$=\left(\frac{1}{|I|}\int_I \omega^{\epsilon}(s)\Delta s\right)\left(\frac{1}{|I|}\int_I \omega^{\frac{-\epsilon}{r-1}}(s)\Delta s\right)^{r-1}$$
$$\leq\left(\frac{1}{|I|}\int_I \omega(s)\Delta s\right)^{\epsilon}\left(\frac{1}{|I|}\int_I \omega^{\frac{-1}{p-1}}(s)\Delta s\right)^{\epsilon(p-1)}\leq \mathscr{C}^{\epsilon},$$

whereupon $\omega^{\epsilon} \in \mathbb{A}_{\epsilon p+1-\epsilon}$. This is the desired result, completing our proof. □

In the next theorem, we discuss the power rule for weights in the Muckenhoupt classes on time scales.

Theorem 2.18. *Assume that $1 < p < \infty$ is a positive real number. Then the following properties hold:*
(1) *If $\omega \in \mathbb{A}_p$, then $\omega^{\alpha} \in \mathbb{A}_p$ for $0 \leq \alpha \leq 1$, with $[\mathbb{A}_p(\omega^{\alpha})] \leq [\mathbb{A}_p(\omega)]^{\alpha}$;*
(2) *If $\omega_1, \omega_2 \in \mathbb{A}_p$, then $\omega_1^{\alpha}\omega_2^{1-\alpha} \in \mathbb{A}_p$ for $0 \leq \alpha \leq 1$, with*

$$[\mathbb{A}_p(\omega_1^{\alpha}\omega_2^{1-\alpha})] \leq [\mathbb{A}_p(\omega_1)]^{\alpha}[\mathbb{A}_p(\omega_2)]^{1-\alpha}.$$

Proof. (1) For $0 \leq \alpha \leq 1$ and $\omega \in \mathbb{A}_p$, we have $1/(p-1) \geq \alpha/(p-1) > 0$ and, by Lemma 2.4, for $\alpha < 1$ and for all $I \subset I_0$, we have

$$\left(\frac{1}{|I|}\int_I \omega^{\alpha}(s)\Delta s\right)\left(\frac{1}{|I|}\int_I (\omega^{\alpha})^{\frac{-1}{p-1}}(s)\Delta s\right)^{p-1}$$
$$=\left(\frac{1}{|I|}\int_I \omega^{\alpha}(s)\Delta s\right)\left(\frac{1}{|I|}\int_I \omega^{\frac{-\alpha}{p-1}}(s)\Delta s\right)^{p-1}$$
$$\leq\left(\frac{1}{|I|}\int_I \omega(s)\Delta s\right)^{\alpha}\left(\frac{1}{|I|}\int_I \omega^{\frac{-1}{p-1}}(s)\Delta s\right)^{\alpha(p-1)}$$
$$=\left[\left(\frac{1}{|I|}\int_I \omega(s)\Delta s\right)\left(\frac{1}{|I|}\int_I \omega^{\frac{-1}{p-1}}(s)\Delta s\right)^{(p-1)}\right]^{\alpha}\leq \mathscr{C}^{\alpha},$$

that is, $\omega^{\alpha} \in \mathbb{A}_p$, with $[\mathbb{A}_p(\omega^{\alpha})] \leq [\mathbb{A}_p(\omega)]^{\alpha}$. This is the desired result.

(2) Since $\omega_1, \omega_2 \in \mathbb{A}_p$, we get that

$$\frac{1}{|I|}\int_I \omega_1(s)\Delta s\left(\frac{1}{|I|}\int_I \omega_1^{\frac{-1}{p-1}}(s)\Delta s\right)^{p-1}\leq \mathscr{C}_1 \tag{2.41}$$

and

$$\frac{1}{|I|}\int_I \omega_2(s)\Delta s\left(\frac{1}{|I|}\int_I \omega_2^{\frac{-1}{p-1}}(s)\Delta s\right)^{p-1} \le \mathscr{C}_2, \tag{2.42}$$

where $\mathscr{C}_1, \mathscr{C}_2 > 1$. By applying the Hölder inequality (note that $0 \le \alpha \le 1$) with $1/\alpha > 1$ and $1/(1-\alpha)$, and using (2.41) and (2.42), we have

$$\begin{aligned}
\frac{1}{|I|}\int_I \omega_1{}^{\alpha}(s)\omega_2{}^{1-\alpha}(s)\Delta s &\le \left(\frac{1}{|I|}\int_I \omega_1(s)\Delta s\right)^{\alpha}\left(\frac{1}{|I|}\int_I \omega_2(s)\Delta s\right)^{1-\alpha} \\
&\le \left(\left(\frac{\mathscr{C}_1}{|I|}\int_I \omega_1^{\frac{1}{1-p}}(s)\Delta s\right)^{1-p}\right)^{\alpha}\left(\left(\frac{\mathscr{C}_2}{|I|}\int_I \omega_2^{\frac{1}{1-p}}(s)\Delta s\right)^{1-p}\right)^{1-\alpha} \\
&= \mathscr{C}_1{}^{\alpha}\mathscr{C}_2{}^{1-\alpha}\left(\left(\frac{1}{|I|}\int_I \omega_1^{\frac{1}{1-p}}(s)\Delta s\right)^{\alpha}\left(\frac{1}{|I|}\int_I \omega_2^{\frac{1}{1-p}}(s)\Delta s\right)^{1-\alpha}\right)^{1-p}.
\end{aligned} \tag{2.43}$$

By applying the Hölder inequality with $1/\alpha$ and $1/(1-\alpha)$ on the term

$$\frac{1}{|I|}\int_I \omega_1{}^{\frac{\alpha}{1-p}}(s)\omega_2{}^{\frac{1-\alpha}{1-p}}(s)\Delta s,$$

we have

$$\frac{1}{|I|}\int_I \omega_1{}^{\frac{\alpha}{1-p}}(s)\omega_2{}^{\frac{1-\alpha}{1-p}}(s)\Delta s \le \left(\frac{1}{|I|}\int_I \omega_1^{\frac{1}{1-p}}(s)\Delta s\right)^{\alpha}\left(\frac{1}{|I|}\int_I \omega_2^{\frac{1}{1-p}}(s)\Delta s\right)^{1-\alpha}. \tag{2.44}$$

By substituting (2.44) into (2.43) and since $1-p < 0$, we have

$$\begin{aligned}
\frac{1}{|I|}\int_I \omega_1{}^{\alpha}(s)\omega_2{}^{1-\alpha}(s)\Delta s &\le \mathscr{C}_1{}^{\alpha}\mathscr{C}_2{}^{1-\alpha}\left[\frac{1}{|I|}\int_I \omega_1{}^{\frac{\alpha}{1-p}}(s)\omega_2{}^{\frac{1-\alpha}{1-p}}(s)\Delta s\right]^{1-p} \\
&= \mathscr{C}_1{}^{\alpha}\mathscr{C}_2{}^{1-\alpha}\left[\frac{1}{|I|}\int_I (\omega_1{}^{\alpha}(s)\omega_2{}^{1-\alpha}(s))^{\frac{1}{1-p}}\Delta s\right]^{1-p}.
\end{aligned}$$

This proves that $\omega_1, \omega_2 \in \mathbb{A}_p$ implies $\omega_1^{\alpha}\omega_2^{1-\alpha} \in \mathbb{A}_p$ for $0 \le \alpha \le 1$, with

$$[\mathbb{A}_p(\omega_1^{\alpha}\omega_2^{1-\alpha})] \le [\mathbb{A}_p(\omega_1)]^{\alpha}[\mathbb{A}_p(\omega_2)]^{1-\alpha}.$$

The proof is complete. □

Theorem 2.19. *Assume that p is a nonnegative real number. If $\omega \in \mathbb{A}_p$, then $\frac{1}{\omega} \in \mathbb{G}_{p'-1}$.*

Proof. Let $\omega \in \mathbb{A}_p$. Then there exists a constant $\mathscr{C} > 1$ such that the inequality

$$\frac{1}{|I|}\int_I \omega(s)\Delta s \le \mathscr{C}\left(\frac{1}{|I|}\int_I \omega^{1/(1-p)}(s)\Delta s\right)^{1-p} \tag{2.45}$$

holds for all $I \subset I_0$. From the property (1) in Lemma 2.4, we have

$$\left(\frac{1}{|I|}\int_I \frac{1}{\omega(s)}\Delta s\right)^{-1} \leq \frac{1}{|I|}\int_I \omega(s)\Delta s,$$

and (2.45) becomes

$$\left(\frac{1}{|I|}\int_I \left(\frac{1}{\omega}\right)^{1/(p-1)}(s)\Delta s\right)^{p-1} \leq \mathscr{C}\frac{1}{|I|}\int_I \frac{1}{\omega(s)}\Delta s.$$

That is, $\frac{1}{\omega} \in \mathbb{G}_{p'-1}$. The proof is completed. □

Theorem 2.20. *Suppose that* $1 < p_1 < p_2 < \infty, 0 < \delta < 1$, *and* $\omega_1, \omega_2 \in \mathbb{A}_p$. *Then the following properties hold:*

(1) *If* $p = \delta p_1 + (1-\delta)p_2$, *then*

$$[\mathbb{A}_p(\omega_1^{\delta}\omega_2^{1-\delta})] \leq [\mathbb{A}_{p_1}(\omega_1)]^{\delta}[\mathbb{A}_{p_2}(\omega_2)]^{1-\delta};$$

(2) *If* $p = (\frac{\delta}{p_1} + \frac{1-\delta}{p_2})^{-1}$, *then*

$$[\mathbb{A}_p(\omega_1^{\delta p/p_1}\omega_2^{(1-\delta)p/p_2})] \leq [\mathbb{A}_{p_1}(\omega_1)]^{\delta p/p_1}[\mathbb{A}_{p_2}(\omega_2)]^{(1-\delta)p/p_2}.$$

Proof. (1) Since $\omega_1, \omega_2 \in \mathbb{A}_p$, we have

$$\begin{aligned}\mathbb{A}_p(\omega_1^{\delta}\omega_2^{1-\delta}) &= \left(\frac{1}{|I|}\int_I \omega_1^{\delta}(s)\omega_2^{1-\delta}(s)\Delta s\right)\left(\frac{1}{|I|}\int_I [\omega_1^{\delta}(s)\omega_2^{1-\delta}(s)]^{\frac{-1}{p-1}}\Delta s\right)^{p-1} \\ &= \left(\frac{1}{|I|}\int_I \omega_1^{\delta}(s)\omega_2^{1-\delta}(s)\Delta s\right)\left(\frac{1}{|I|}\int_I \omega_1^{\frac{-\delta}{p-1}}(s)\omega_2^{\frac{-(1-\delta)}{p-1}}(s)\Delta s\right)^{p-1}. \qquad (2.46)\end{aligned}$$

By applying the Hölder inequality with $1/\delta > 1$ and $1/(1-\delta)$, we obtain

$$\frac{1}{|I|}\int_I \omega_1^{\delta}(s)\omega_2^{1-\delta}(s)\Delta s \leq \left(\frac{1}{|I|}\int_I \omega_1(s)\Delta s\right)^{\delta}\left(\frac{1}{|I|}\int_I \omega_2(s)\Delta s\right)^{1-\delta}. \qquad (2.47)$$

Since $1 < p_1 < p_2 < \infty$, $0 < \delta < 1$ $(0 < 1-\delta < 1)$, we can easily see that

$$p = \delta p_1 + (1-\delta)p_2 > \delta p_1 + (1-\delta)p_1 = p_1 > 1,$$

and, by using the fact that $(1-\delta)p_2 > (1-\delta)$, we have

$$p = \delta p_1 + (1-\delta)p_2 > \delta p_1 + 1 - \delta = \delta(p_1 - 1) + 1,$$

and then

$$(p-1)/[\delta(p_1-1)] > 1. \tag{2.48}$$

From (2.48) and by applying the Hölder inequality with $(p-1)/[\delta(p_1-1)] > 1$ and $(p-1)/[(1-\delta)(p_2-1)]$, and taking into account that $p = \delta p_1 + (1-\delta)p_2$, we obtain

$$\begin{aligned}
&\frac{1}{|I|}\int_I \omega_1^{\frac{-\delta}{p-1}}(s)\omega_2^{\frac{-(1-\delta)}{p-1}}(s)\Delta s \\
&\le \left(\frac{1}{|I|}\int_I \omega_1^{\frac{-1}{p_1-1}}(s)\Delta s\right)^{\frac{\delta(p_1-1)}{p-1}}\left(\frac{1}{|I|}\int_I \omega_2^{\frac{-1}{p_2-1}}(s)\Delta s\right)^{\frac{(1-\delta)(p_2-1)}{p-1}}.
\end{aligned} \tag{2.49}$$

By using (2.47) and (2.49), then (2.46) becomes

$$\begin{aligned}
&\mathbb{A}_p(\omega_1^{\delta}\omega_2^{1-\delta}) \\
&\le \left(\frac{1}{|I|}\int_I \omega_1(s)\Delta s\right)^{\delta}\left(\frac{1}{|I|}\int_I \omega_2(s)\Delta s\right)^{1-\delta} \\
&\quad\times\left(\left(\frac{1}{|I|}\int_I \omega_1^{\frac{-1}{p_1-1}}(s)\Delta s\right)^{\frac{\delta(p_1-1)}{p-1}}\left(\frac{1}{|I|}\int_I \omega_2^{\frac{-1}{p_2-1}}(s)\Delta s\right)^{\frac{(1-\delta)(p_2-1)}{p-1}}\right)^{p-1} \\
&= \left[\left(\frac{1}{|I|}\int_I \omega_1(s)\Delta s\right)\left(\frac{1}{|I|}\int_I \omega_1^{\frac{-1}{p_1-1}}(s)\Delta s\right)^{p_1-1}\right]^{\delta} \\
&\quad\times\left[\left(\frac{1}{|I|}\int_I \omega_2(s)\Delta s\right)\left(\frac{1}{|I|}\int_I \omega_2^{\frac{-1}{p_2-1}}(s)\Delta s\right)^{p_2-1}\right]^{1-\delta} \\
&\le [\mathbb{A}_{p_1}(\omega_1)]^{\delta}[\mathbb{A}_{p_2}(\omega_2)]^{1-\delta}.
\end{aligned}$$

This is the desired result. The proof of (2) is similar to the proof of (1) and hence is omitted. The proof is completed. □

Theorem 2.21. *Assume that $1 < p < \infty$ is a positive real number. Then the following properties hold:*

(1) *if $\omega_1, \omega_2 \in \mathbb{A}_p$, then $\omega_1^{\delta/r}\omega_2^{(1-\delta)/p} \in \mathbb{A}_p$ for $p > 1, 0 < r < 1$ with $\delta = (1-\frac{1}{p})/(\frac{1}{r}-\frac{1}{p})$, and*

$$[\mathbb{A}_p(\omega_1^{\delta/r}\omega_2^{(1-\delta)/p})] \le \mathbb{A}_p[\omega_1]^{\delta/r}\mathbb{A}_p[\omega_2]^{(1-\delta)/p};$$

(2) *$\omega \in \mathbb{A}_p$ if and only if ω and $\omega^{\frac{1}{1-p}}$ are in $\mathbb{A}_\infty$.*

Proof. (1) Assume that $\omega_1, \omega_2 \in \mathbb{A}_p$, then

$$\begin{aligned}
&\mathbb{A}_p(\omega_1^{\delta/r}\omega_2^{(1-\delta)/p}) \\
&= \left(\frac{1}{|I|}\int_I \omega_1^{\delta/r}(s)\omega_2^{(1-\delta)/p}(s)\Delta s\right)\left(\frac{1}{|I|}\int_I [\omega_1^{\delta/r}(s)\omega_2^{(1-\delta)/p}(s)]^{\frac{-1}{p-1}}\Delta s\right)^{p-1} \\
&= \left(\frac{1}{|I|}\int_I \omega_1^{\delta/r}(s)\omega_2^{(1-\delta)/p}(s)\Delta s\right)\left(\frac{1}{|I|}\int_I \omega_1^{\frac{-\delta}{r(p-1)}}(s)\omega_2^{\frac{-(1-\delta)}{p(p-1)}}(s)\Delta s\right)^{p-1}.
\end{aligned} \tag{2.50}$$

Note that $0 < \delta < 1$ and $\delta/r + (1-\delta)/p = 1$, so that, by letting $\gamma = \delta/r$, we have $1-\gamma = (1-\delta)/p$, and (2.50) can be written as

$$\mathbb{A}_p(\omega_1^{\gamma}\omega_2^{1-\gamma}) = \left(\frac{1}{|I|}\int_I \omega_1^{\gamma}(s)\omega_2^{1-\gamma}(s)\Delta s\right)\left(\frac{1}{|I|}\int_I \omega_1^{\frac{-\gamma}{p-1}}(s)\omega_2^{\frac{-(1-\gamma)}{p-1}}(s)\Delta s\right)^{p-1}. \tag{2.51}$$

By applying the Hölder inequality with exponents $1/\gamma > 1$ and $1/(1-\gamma)$, we obtain

$$\left(\frac{1}{|I|}\int_I \omega_1^{\gamma}(s)\omega_2^{1-\gamma}(s)\Delta s\right) \le \left(\frac{1}{|I|}\int_I \omega_1(s)\Delta s\right)^{\gamma}\left(\frac{1}{|I|}\int_I \omega_2(s)\Delta s\right)^{1-\gamma} \tag{2.52}$$

and

$$\frac{1}{|I|}\int_I \omega_1^{\frac{-\gamma}{p-1}}(s)\omega_2^{\frac{-(1-\gamma)}{p-1}}(s)\Delta s \le \left(\frac{1}{|I|}\int_I \omega_1^{\frac{-1}{p-1}}(s)\Delta s\right)^{\gamma}\left(\frac{1}{|I|}\int_I \omega_2^{\frac{-1}{p-1}}(s)\Delta s\right)^{1-\gamma}. \tag{2.53}$$

By substituting (2.52) and (2.53) into (2.51), we have

$$\begin{aligned}
\mathbb{A}_p(\omega_1^{\gamma}\omega_2^{1-\gamma}) &\le \left(\frac{1}{|I|}\int_I \omega_1(s)\Delta s\right)^{\gamma}\left(\frac{1}{|I|}\int_I \omega_2(s)\Delta s\right)^{1-\gamma} \\
&\quad \times \left(\left(\frac{1}{|I|}\int_I \omega_1^{\frac{-1}{p-1}}(s)\Delta s\right)^{\gamma}\left(\frac{1}{|I|}\int_I \omega_2^{\frac{-1}{p-1}}(s)\Delta s\right)^{1-\gamma}\right)^{p-1} \\
&= \left(\left(\frac{1}{|I|}\int_I \omega_1(s)\Delta s\right)\left(\frac{1}{|I|}\int_I \omega_1^{\frac{-1}{p-1}}(s)\Delta s\right)^{p-1}\right)^{\gamma} \\
&\quad \times \left(\left(\frac{1}{|I|}\int_I \omega_2(s)\Delta s\right)\left(\frac{1}{|I|}\int_I \omega_2^{\frac{-1}{p-1}}(s)\Delta s\right)^{p-1}\right)^{1-\gamma} \\
&\le [\mathbb{A}_p(\omega_1)]^{\gamma}[\mathbb{A}_p(\omega_2)]^{1-\gamma} = [\mathbb{A}_p(\omega_1)]^{\delta/r}[\mathbb{A}_p(\omega_2)]^{(1-\delta)/p}.
\end{aligned}$$

(2) Using property (3) in Theorem 2.15, since $\mathbb{A}_\infty = \bigcup_{1\le p<\infty}\mathbb{A}_p$, it is clear that $\omega \in \mathbb{A}_p$, for some $p > 1$, if and only if $\omega \in \mathbb{A}_\infty$. Now, we have by property (1) in Theorem 2.17 that

$\omega \in \mathbb{A}_p$ if and only if $\omega^{1-p'} = \omega^{1/(1-p)} \in \mathbb{A}_{p'}$. That is, since $\mathbb{A}_{p'} \subset \mathbb{A}_\infty$, $\omega \in \mathbb{A}_p$ if and only if $\omega^{1/(1-p)} \in \mathbb{A}_\infty$. The proof is complete. □

2.4 Some fundamental relations

In this section, we prove some fundamental relations connecting different Muckenhoupt and Gehring classes. The results are adapted from [49].

Theorem 2.22. *Assume that ω is a positive weight and p is a positive real number. Then*

$$\max\{[\mathbb{A}_\infty(\omega)], [A_\infty(\omega^{1-p'})]^{p-1}\} \leq [\mathbb{A}_p(\omega)] \leq [\mathbb{A}_\infty(\omega)][\mathbb{A}_\infty(\omega^{1-p'})]^{p-1}. \tag{2.54}$$

Proof. For $p \leq q$, we have $[\mathbb{A}_p(\omega)] \geq [\mathbb{A}_q(\omega)]$, and thus,

$$[\mathbb{A}_\infty(\omega)] \leq [\mathbb{A}_p(\omega)]. \tag{2.55}$$

Furthermore, for $q < \infty$, we have

$$\begin{aligned}
&[\mathbb{A}_q(\omega^{1-p'})]^{p-1} \\
&= \sup_{I \subset I_0}\left\{\left(\frac{1}{|I|}\int_I \omega^{1-p'}(s)\Delta s\right)\left(\frac{1}{|I|}\int_I \omega^{(1-p')(1-q')}(s)\Delta s\right)^{q-1}\right\}^{p-1} \\
&= \sup_{I \subset I_0}\left\{\left(\frac{1}{|I|}\int_I \omega^{1-p'}(s)\Delta s\right)^{p-1}\left(\frac{1}{|I|}\int_I \omega^{(1-p')(1-q')}(s)\Delta s\right)^{(q-1)(p-1)}\right\} \\
&= \sup_{I \subset I_0} \frac{(\frac{1}{|I|}\int_I \omega(s)\Delta s)(\frac{1}{|I|}\int_I \omega^{\frac{-1}{p-1}}(s)\Delta s)^{p-1}}{(\frac{1}{|I|}\int_I \omega(s)\Delta s)(\frac{1}{|I|}\int_I \omega^{(1-p')(1-q')}(s)\Delta s)^{-(q-1)(p-1)}} \leq \mathbb{A}_p(\omega).
\end{aligned} \tag{2.56}$$

Taking the limit in (2.56) as q tends to ∞, we have

$$[\mathbb{A}_\infty(\omega^{1-p'})]^{p-1} \leq [\mathbb{A}_p(\omega)]. \tag{2.57}$$

From (2.55) and (2.57), then

$$\max\{[\mathbb{A}_\infty(\omega)], [\mathbb{A}_\infty(\omega^{1-p'})]^{p-1}\} \leq [\mathbb{A}_p(\omega)].$$

Now, for the second inequality, we have

$$\frac{1}{|I|}\int_I \omega(s)\Delta s\left(\frac{1}{|I|}\int_I \omega^{1-p'}(s)\Delta s\right)^{p-1}$$

$$= \frac{1}{|I|}\int_I \omega(s)\Delta s\left(\frac{1}{|I|}\int_I \omega^{1-q'}(s)\Delta s\right)^{q-1}$$

$$\times\left(\frac{1}{|I|}\int_I \omega^{1-p'}(s)\Delta s\left(\frac{1}{|I|}\int_I \omega^{1-q'}(s)\Delta s\right)^{\frac{1-q}{p-1}}\right)^{p-1}. \tag{2.58}$$

Since $1-q$ and $1-q' < 0$, by Lemma 2.4 for $1-q' < q'-1$, we have

$$\frac{1}{|I|}\int_I \omega^{1-p'}(s)\Delta s\left(\frac{1}{|I|}\int_I \omega^{1-q'}(s)\Delta s\right)^{\frac{1-q}{p-1}}$$

$$\le \frac{1}{|I|}\int_I \omega^{1-p'}(s)\Delta s\left(\frac{1}{|I|}\int_I \omega^{q'-1}(s)\Delta s\right)^{\frac{q-1}{p-1}}$$

$$= \frac{1}{|I|}\int_I \omega^{1-p'}(s)\Delta s\left(\frac{1}{|I|}\int_I \omega^{(1-p')(q'-1)/(1-p')}(s)\Delta s\right)^{\frac{q-1}{p-1}}. \tag{2.59}$$

By setting $r-1 = (q-1)/(p-1)$, we have

$$r'-1 = \frac{1}{r-1} = \frac{p-1}{q-1} = \frac{q'-1}{p'-1}.$$

Hence, from (2.58) and (2.59), we have

$$\frac{1}{|I|}\int_I \omega(s)\Delta s\left(\frac{1}{|I|}\int_I \omega^{1-p'}(s)\Delta s\right)^{p-1}$$

$$\le \frac{1}{|I|}\int_I \omega(s)\Delta s\left(\frac{1}{|I|}\int_I \omega^{1-q'}(s)\Delta s\right)^{q-1}$$

$$\times\left[\frac{1}{|I|}\int_I \omega^{1-p'}(s)\Delta s\left(\frac{1}{|I|}\int_I \omega^{(1-p')(1-r')}(s)\Delta s\right)^{r-1}\right]^{p-1}.$$

Taking supremum over all $I \subset I_0$, we have

$$[\mathbb{A}_p(\omega)] \le [\mathbb{A}_q(\omega)][\mathbb{A}_r(\omega^{1-p'})]^{p-1}. \tag{2.60}$$

Now by taking the limit on the both sides of (2.60) as q tends to ∞, we get that

$$[\mathbb{A}_p(\omega)] \le [\mathbb{A}_\infty(\omega)][\mathbb{A}_\infty(\omega^{1-p'})]^{p-1}.$$

The proof is complete. □

Theorem 2.23. *Assume that $p, r > 1$. Then, $\omega \in \mathbb{A}_p \cap \mathbb{G}_r$ if and only if $\omega^r \in \mathbb{A}_q$ for $q = r(p-1)+1$.*

Proof. First, assume that $\omega \in \mathbb{A}_p \cap \mathbb{G}_r$. Then $\omega \in \mathbb{A}_p$ and $\omega \in \mathbb{G}_r$. In other words, there exists a constant $\mathscr{C} > 1$ such that

$$\left(\frac{1}{|I|}\int_I \omega(s)\Delta s\right)\left(\frac{1}{|I|}\int_I \omega^{\frac{-1}{p-1}}(s)\Delta s\right)^{p-1} \le \mathscr{C}, \tag{2.61}$$

and there exists a constant $K > 1$ such that

$$\left(\frac{1}{|I|}\int_I \omega^r(s)\Delta s\right)^{1/r} \le K\left(\frac{1}{|I|}\int_I \omega(s)\Delta s\right). \tag{2.62}$$

From (2.62), we see that

$$\frac{1}{|I|}\int_I \omega^r(s)\Delta s \le K^r\left(\frac{1}{|I|}\int_I \omega(s)\Delta s\right)^r. \tag{2.63}$$

If $q = r(p-1)+1$, then $1/(p-1) = r/(q-1)$, and from (2.61), we have

$$\left(\frac{1}{|I|}\int_I \omega(s)\Delta s\right)\left(\frac{1}{|I|}\int_I \omega^{\frac{-r}{q-1}}(s)\Delta s\right)^{\frac{q-1}{r}} \le \mathscr{C}.$$

Thus

$$\left(\frac{1}{|I|}\int_I (\omega^r)^{\frac{-1}{q-1}}(s)\Delta s\right)^{q-1} \le \mathscr{C}^r\left(\frac{1}{|I|}\int_I \omega(s)\Delta s\right)^{-r}. \tag{2.64}$$

From (2.63) and (2.64), we see that

$$\begin{aligned}
&\left(\frac{1}{|I|}\int_I \omega^r(s)\Delta s\right)\left(\frac{1}{|I|}\int_I (\omega^r)^{\frac{-1}{q-1}}(s)\Delta s\right)^{q-1} \\
&\quad \le \mathscr{C}^r K^r\left(\frac{1}{|I|}\int_I \omega(s)\Delta s\right)^r\left(\frac{1}{|I|}\int_I \omega(s)\Delta s\right)^{-r} \\
&\quad = \mathscr{C}^r K^r,
\end{aligned}$$

which implies that $\omega^r \in \mathbb{A}_q$. Conversely, if $\omega^r \in \mathbb{A}_q$ for $q = r(p-1)+1$, then there exist $\mathscr{C}_1 > 1$ such that

$$\left(\frac{1}{|I|}\int_I \omega^r(s)\Delta s\right)\left(\frac{1}{|I|}\int_I (\omega^r)^{\frac{-1}{q-1}}(s)\Delta s\right)^{q-1} \le \mathscr{C}_1.$$

Since $q-1=r(p-1)$, we obtain

$$\left(\frac{1}{|I|}\int_I \omega^r(s)\Delta s\right)\left(\frac{1}{|I|}\int_I \omega^{\frac{-1}{p-1}}(s)\Delta s\right)^{r(p-1)} \leq \mathscr{C}_1$$

and

$$\left(\frac{1}{|I|}\int_I \omega^r(s)\Delta s\right)^{\frac{1}{r}}\left(\frac{1}{|I|}\int_I \omega^{\frac{-1}{p-1}}(s)\Delta s\right)^{p-1} \leq \mathscr{C}_1^{\frac{1}{r}}. \tag{2.65}$$

From (2.65), by using Lemma 2.4, we get

$$\left(\frac{1}{|I|}\int_I \omega^r(s)\Delta s\right)^{\frac{1}{r}} \geq \frac{1}{|I|}\int_I \omega(s)\Delta s. \tag{2.66}$$

From (2.65) and (2.66), we get

$$\left(\frac{1}{|I|}\int_I \omega(s)\Delta s\right)\left(\frac{1}{|I|}\int_I \omega^{\frac{-1}{p-1}}(s)\Delta s\right)^{p-1} \leq \mathscr{C}_1^{\frac{1}{r}}, \tag{2.67}$$

which implies that $\omega \in \mathbb{A}_p$, and, by using Lemma 2.4 with $-1/(p-1) < 0$, we find

$$\left(\frac{1}{|I|}\int_I \omega(s)\Delta s\right)^{\frac{-1}{p-1}} \leq \frac{1}{|I|}\int_I \omega^{\frac{-1}{p-1}}(s)\Delta s$$

and

$$\left(\frac{1}{|I|}\int_I \omega(s)\Delta s\right)^{-1} \leq \left(\frac{1}{|I|}\int_I \omega^{\frac{-1}{p-1}}(s)\Delta s\right)^{p-1}.$$

From this and (2.65), we get that

$$\left(\frac{1}{|I|}\int_I \omega(s)\Delta s\right)^{-1}\left(\frac{1}{|I|}\int_I \omega^r(s)\Delta s\right)^{\frac{1}{r}} \leq \mathscr{C}^{\frac{1}{r}},$$

or equivalently,

$$\left(\frac{1}{|I|}\int_I \omega^r(s)\Delta s\right)^{\frac{1}{r}} \leq \mathscr{C}^{\frac{1}{r}}\left(\frac{1}{|I|}\int_I \omega(s)\Delta s\right), \tag{2.68}$$

which implies that $\omega \in \mathbb{G}_r$. From (2.67) and (2.68), we get $\omega \in \mathbb{A}_p \cap \mathbb{G}_r$. The proof is complete. □

Theorem 2.24. *Assume that p is a positive real number. Then the following properties hold:*

(1) *If* $1 < r < \infty$, *then*

$$\frac{[\mathbb{A}_{\infty}(\omega^r)]^{1/r}}{[\mathbb{A}_{\infty}(\omega)]} \le [\mathbb{G}_r(\omega)] \le [\mathbb{A}_{\infty}(\omega^r)]^{1/r};$$

(2) *If* $\omega \in \bigcap_{p>1} \mathbb{A}_p$, *then* $1/\omega \in \bigcap_{r<\infty} \mathbb{G}_r$.

Proof. (1) From the definition of $\mathbb{G}_r(\omega)$, we have for all $I \subset I_0$ that

$$\left(\frac{1}{|I|}\int_I \omega^r(s)\Delta s\right) \le [\mathbb{G}_r(\omega)]^r \left(\frac{1}{|I|}\int_I \omega(s)\Delta s\right)^r.$$

By multiplying both sides by $(1/|I| \int_I \omega^{-r/(p-1)}(s)\Delta s)^{p-1}$, for $p < \infty$, we get that

$$\begin{aligned}&\left(\frac{1}{|I|}\int_I \omega^r(s)\Delta s\right)\left(\frac{1}{|I|}\int_I \omega^{\frac{-r}{p-1}}(s)\Delta s\right)^{p-1} \\ &\quad \le [\mathbb{G}_r(\omega)]^r \left(\frac{1}{|I|}\int_I \omega(s)\Delta s\left(\frac{1}{|I|}\int_I \omega^{\frac{-r}{p-1}}(s)\Delta s\right)^{\frac{p-1}{r}}\right)^r. \qquad (2.69)\end{aligned}$$

Taking supremum over all $I \subset I_0$ in (2.69), we have

$$\begin{aligned}&\sup_I \left(\frac{1}{|I|}\int_I \omega^r(s)\Delta s\right)\left(\frac{1}{|I|}\int_I \omega^{\frac{-r}{p-1}}(s)\Delta s\right)^{p-1} \\ &\quad \le [\mathbb{G}_r(\omega)]^r \sup_I \left(\frac{1}{|I|}\int_I \omega(s)\Delta s\left(\frac{1}{|I|}\int_I \omega^{\frac{-r}{p-1}}(s)\Delta s\right)^{\frac{p-1}{r}}\right)^r,\end{aligned}$$

or equivalently,

$$[\mathbb{A}_p(\omega^r)] \le [\mathbb{G}_r(\omega)]^r (\mathbb{A}_{\frac{r+p-1}{r}}(\omega))^r.$$

As p tends to ∞, we have that

$$\frac{[\mathbb{A}_{\infty}(\omega^r)]^{1/r}}{[\mathbb{A}_{\infty}(\omega)]} \le [\mathbb{G}_r(\omega)],$$

which is the left-side inequality. For the second inequality, from the definition of $[\mathbb{A}_p(\omega^r)]^{1/r}$, we have for all $I \subset I_0$ that

$$\frac{(\frac{1}{|I|}\int_I \omega^r(s)\Delta s)^{1/r}}{\frac{1}{|I|}\int_I \omega(s)\Delta s} = \frac{(\frac{1}{|I|}\int_I \omega^r(s)\Delta s)^{1/r}(\frac{1}{|I|}\int_I \omega^{\frac{-r}{p-1}}(s)\Delta s)^{\frac{p-1}{r}}}{\frac{1}{|I|}\int_I \omega(s)\Delta s(\frac{1}{|I|}\int_I \omega^{\frac{-r}{p-1}}(s)\Delta s)^{\frac{p-1}{r}}}. \tag{2.70}$$

By Lemma 2.4 and since $-r/(p-1) < 0 < 1$, we see that

$$\frac{1}{|I|}\int_I \omega^{\frac{-r}{p-1}}(s)\Delta s \geq \left(\frac{1}{|I|}\int_I \omega(s)\Delta s\right)^{\frac{-r}{p-1}},$$

which implies that

$$\frac{1}{\frac{1}{|I|}\int_I \omega(s)\Delta s(\frac{1}{|I|}\int_I \omega^{\frac{-r}{p-1}}(s)\Delta s)^{-\frac{p-1}{r}}} \leq 1.$$

Using this in (2.70), we get

$$\frac{(\frac{1}{|I|}\int_I \omega^r(s)\Delta s)^{1/r}}{\frac{1}{|I|}\int_I \omega(s)\Delta s} \leq \left(\frac{1}{|I|}\int_I \omega^r(s)\Delta s\left(\frac{1}{|I|}\int_I \omega^{\frac{-r}{p-1}}(s)\Delta s\right)^{\frac{p-1}{r}}\right)^{1/r}.$$

Taking the supremum in (2.70) over all $I \subset I_0$, we obtain $\mathbb{G}_r(\omega) \leq [\mathbb{A}_p(\omega^r)]^{1/r}$. As p tends to ∞, we have

$$\mathbb{G}_r(\omega) \leq [\mathbb{A}_\infty(\omega^r)]^{1/r}.$$

(2) If $\omega \in \bigcap_{p>1}\mathbb{A}_p$, then

$$\left(\frac{1}{|I|}\int_I \omega^{\frac{-1}{p-1}}(s)\Delta s\right)^{p-1} \leq \mathscr{C}\left(\frac{1}{|I|}\int_I \omega(s)\Delta s\right)^{-1} \tag{2.71}$$

holds for all $p > 1$. From (2.71), by using Lemma 2.4, we get that

$$\left(\frac{1}{|I|}\int_I \left(\frac{1}{\omega(s)}\right)^{\frac{1}{p-1}}\Delta s\right)^{p-1} \leq \mathscr{C}\left(\frac{1}{|I|}\int_I \frac{1}{\omega(s)}\Delta s\right).$$

Hence, $(1/\omega) \in \mathbb{G}_r$ for all $0 < r = 1/(p-1) < \infty$, we have $1/\omega \in \bigcap_{r<\infty}\mathbb{G}_r$. The proof is complete. □

2.5 Self-improving properties

In this section, we will employ some dynamic inequalities on time scales to prove the self-improving properties of the Gehring weights on a time scale $\mathbb{T}$. The results in this section are adapted from [10] and [60].

Definition 2.2. For any nonnegative weight ω which is nonnegative, and nonincreasing, we define $\mathscr{A}\omega$ by

$$\mathscr{A}\omega(t) = \frac{1}{t}\int_0^t \omega(s)\Delta s, \quad \text{for all } t \in [0,\infty)_{\mathbb{T}}, \tag{2.72}$$

and

$$\mathscr{A}^{\sigma}\omega(t) = \frac{1}{\sigma(t)}\int_0^{\sigma(t)} \omega(s)\Delta s, \quad \text{for all } t \in [0,\infty)_{\mathbb{T}}.$$

Some simple facts about $\mathscr{A}\omega$ are given next. Since ω is nonincreasing, we see that

$$\mathscr{A}\omega(t) = \frac{1}{t}\int_0^t \omega(s)\Delta s \geq \frac{1}{t}\int_0^t \omega(t)\Delta s = \omega(t), \quad \text{for all } t \in [0,\infty)_{\mathbb{T}}.$$

This gives us the following result.

Lemma 2.8. *If ω is nonincreasing, then $\mathscr{A}\omega \geq \omega$.*

The following lemma shows that $\mathscr{A}\omega$ inherits the nonincreasing nature of ω.

Lemma 2.9. *If ω is nonincreasing, then so is $\mathscr{A}\omega$.*

Proof. Taking into account Lemma 2.8 and the quotient rule in (2.12), we have that

$$\left(\mathscr{A}\omega(t)\right)^{\Delta} = \frac{-1}{\sigma(t)t}\int_0^t \omega(s)\Delta s + \frac{1}{\sigma(t)}\omega(t) = \frac{1}{\sigma(t)}\left[\omega(t) - \mathscr{A}\omega(t)\right] \leq 0,$$

which completes the proof. □

Remark 2.2. As a consequence of Lemma 2.8, we notice that if ω is nonincreasing, then $\mathscr{A}\omega^q \geq \omega^q$ for $q > 0$. In addition, by Lemma 2.9, we see that if ω is nonincreasing, then so is $\mathscr{A}\omega^q$ for $q > 0$.

Now, we state the Hardy inequality in a finite interval (see [2, Corollary 1.5.1]).

Theorem 2.25. *If $q > 1$ and ω is nonnegative and nonincreasing, then*

$$\mathscr{A}\left[(\mathscr{A}\omega)^{\sigma}\right]^q \leq \left(\frac{q}{q-1}\right)^q \mathscr{A}\omega^q. \tag{2.73}$$

We assume that there exists a constant $\lambda \geq 1$ such that

$$\sigma(t) \leq \lambda t, \quad \text{for all } t \in \mathbb{T}. \tag{2.74}$$

We now apply the time scales chain rule to obtain some estimates that will be used later.

Lemma 2.10. *Let $x(t) = t$. If $0 < \gamma < 1$, then*

$$(x^{1-\gamma})^{\Delta} \geq \frac{1-\gamma}{\sigma^{\gamma}}, \tag{2.75}$$

and if $\gamma > 1$ and (2.74) holds, then

$$(x^{1-\gamma})^{\Delta} \geq \frac{(1-\gamma)\lambda^{\gamma}}{\sigma^{\gamma}}. \tag{2.76}$$

Proof. By the chain rule, we obtain

$$\begin{aligned}(x^{1-\gamma})^{\Delta}(t) &= (1-\gamma)x^{\Delta}(t)\int_0^1 \frac{dh}{(hx(\sigma(t)) + (1-h)x(t))^{\gamma}} \\ &= (1-\gamma)\int_0^1 \frac{dh}{(h\sigma(t) + (1-h)t)^{\gamma}}.\end{aligned}$$

Thus, if $0 < \gamma < 1$, then

$$(x^{1-\gamma})^{\Delta}(t) \geq (1-\gamma)\int_0^1 \frac{dh}{(h\sigma(t) + (1-h)\sigma(t))^{\gamma}} = \frac{1-\gamma}{(\sigma(t))^{\gamma}},$$

which is (2.75), and if $\gamma > 1$ and (2.74) holds, then

$$(x^{1-\gamma})^{\Delta}(t) \geq (1-\gamma)\int_0^1 \frac{dh}{(ht + (1-h)t)^{\gamma}} = \frac{1-\gamma}{t^{\gamma}} \geq \frac{(1-\gamma)\lambda^{\gamma}}{(\sigma(t))^{\gamma}},$$

which is (2.76). □

Lemma 2.11. *If ω is nonnegative and nondecreasing and $\gamma > 1$, then*

$$(W^{\gamma})^{\Delta} \geq \gamma W^{\Delta} W^{\gamma-1}. \tag{2.77}$$

Proof. Again we apply the chain rule to see that

$$\begin{aligned}(W^{\gamma})^{\Delta} &= \gamma W^{\Delta}\int_0^1 (hW^{\sigma} + (1-h)W)^{\gamma-1} dh \\ &\geq \gamma W^{\Delta}\int_0^1 (hW + (1-h)W)^{\gamma-1} dh = \gamma W^{\Delta} W^{\gamma-1},\end{aligned}$$

which shows (2.77). □

Theorem 2.26. *Assume that ω is a nonnegative and nonincreasing and $p > 1$. Then, for any $q \in (0, p)$, we have*

$$\mathscr{A}\omega^p \le \frac{q}{p}[\mathscr{A}\omega^q]^{p/q} + \frac{(p-q)\lambda^{p/q}}{p}\mathscr{A}[(\mathscr{A}\omega^q)^\sigma]^{p/q}. \tag{2.78}$$

Proof. From the Hardy inequality (see (2.73)), we see that the second integral on the right-hand side of (2.78) is finite. Now, we consider this integral. Then, for $0 < q < p$, we put

$$\gamma = \frac{p}{q} > 1 \quad \text{and} \quad W(t) = \int_0^t \omega^q(s)\Delta s.$$

Using the notation from Lemma 2.10, we have

$$\begin{aligned}
&\frac{(p-q)\lambda^{p/q}}{pt}\int_0^t\left[\frac{1}{\sigma(s)}\int_0^{\sigma(s)}\omega^q(\tau)\Delta\tau\right]^{p/q}\Delta s \\
&= \frac{(\gamma-1)\lambda^\gamma}{\gamma t}\int_0^t\left[\frac{1}{\sigma(s)}\int_0^{\sigma(s)}\omega^q(\tau)\Delta\tau\right]^{\gamma}\Delta s \\
&\ge -\frac{1}{\gamma t}\int_0^t W^\gamma(\sigma(s))(x^{1-\gamma})^\Delta(s)\Delta s \\
&= \lim_{s\to 0^+}\frac{W^\gamma(s)x^{1-\gamma}(s)}{\gamma t} - \frac{W^\gamma(t)x^{1-\gamma}(t)}{\gamma t} + \frac{1}{\gamma t}\int_0^t (W^\gamma)^\Delta(s)x^{1-\gamma}(s)\Delta s \\
&= \frac{1}{\gamma t}\int_0^t s^{1-\gamma}(W^\gamma)^\Delta(s)\Delta s + \frac{1}{\gamma t}\lim_{s\to 0^+}\left[s\left(\frac{W(s)}{s}\right)^\gamma\right] - \frac{1}{\gamma}\left(\frac{W(t)}{t}\right)^\gamma \\
&\ge \frac{1}{\gamma t}\int_0^t \frac{\gamma W^\Delta(s)W^{\gamma-1}(s)}{s^{\gamma-1}}\Delta s - \frac{1}{\gamma}\left(\frac{W(t)}{t}\right)^\gamma \\
&= \frac{1}{t}\int_0^t \omega^q(s)[\mathscr{A}\omega^q(s)]^{\gamma-1}\Delta s - \frac{1}{\gamma}[\mathscr{A}\omega^q(t)]^\gamma \\
&\ge \frac{1}{t}\int_0^t \omega^q(s)[\omega^q(s)]^{\gamma-1}\Delta s - \frac{1}{\gamma}[\mathscr{A}\omega^q(t)]^\gamma \\
&= \frac{1}{t}\int_0^t [\omega^q(s)]^\gamma\Delta s - \frac{1}{\gamma}[\mathscr{A}\omega^q(t)]^\gamma = \mathscr{A}\omega^p(t) - \frac{q}{p}[\mathscr{A}\omega^q(t)]^{p/q},
\end{aligned}$$

from which (2.78) follows. □

Now, we are ready to state and prove our first time scales version of the Gehring self-improving property for monotone functions.

Theorem 2.27 (Gehring inequality I). *Assume that ω is a nonnegative and nonincreasing weight and $q > 1$ such that*

$$\mathscr{A}\omega^q \leq \kappa[\mathscr{A}\omega]^q \quad \textit{for some } \kappa > 0, \tag{2.79}$$

then

$$\mathscr{A}\omega^p \leq \tilde{\kappa}[\mathscr{A}\omega]^p, \tag{2.80}$$

where $p > q$ and

$$\tilde{\kappa} := \frac{q\kappa^{p/q}}{p - (p-q)(\lambda\kappa)^{p/q}(\frac{p}{p-1})^p} > 0. \tag{2.81}$$

Proof. Assuming (2.79), we find

$$\begin{aligned}
&\frac{1}{t}\int_0^t \omega^p(s)\Delta s \\
&\quad \leq \frac{q}{p}\left[\frac{1}{t}\int_0^t \omega^q(s)\Delta s\right]^{p/q} + \frac{(p-q)\lambda^{p/q}}{pt}\int_0^t \left[\frac{1}{\sigma(s)}\int_0^{\sigma(s)} \omega^q(\tau)\Delta\tau\right]^{p/q}\Delta s \\
&\quad \leq \frac{q}{p}\kappa^{p/q}\left[\frac{1}{t}\int_0^t \omega(s)\Delta s\right]^p + \frac{(p-q)\lambda^{p/q}}{pt}\int_0^t \kappa^{p/q}\left[\frac{1}{\sigma(s)}\int_0^{\sigma(s)} \omega(\tau)\Delta\tau\right]^p\Delta s \\
&\quad \leq \frac{q}{p}\kappa^{p/q}\left[\frac{1}{t}\int_0^t \omega(s)\Delta s\right]^p + \frac{(p-q)(\lambda\kappa)^{p/q}}{pt}\left(\frac{p}{p-1}\right)^p\int_0^t \omega^p(s)\Delta s
\end{aligned}$$

so that, due to (2.81),

$$\frac{1}{t}\int_0^t \omega^p(s)\Delta s \leq \tilde{\kappa}\left[\frac{1}{t}\int_0^t \omega(s)\Delta s\right]^p,$$

from which (2.80) follows. □

It is natural to ask what happens if in (2.21) we fix $p > 1$ and consider the improvement to this inequality that would result from lowering the exponent on the right-hand side. The following result gives an answer.

Theorem 2.28. *Suppose that the assumptions of Theorem 2.27 hold and define $\tilde{\kappa}$ as in (2.81). Then, for all $0 < r < 1$, we have*

$$\mathscr{A}\omega^p \leq \overline{\kappa}[\mathscr{A}\omega^r]^{p/r}, \quad \text{where } \overline{\kappa} := \tilde{\kappa}^{1/\theta} \text{ with } \theta := \frac{1-\frac{1}{p}}{\frac{1}{r}-\frac{1}{p}}. \tag{2.82}$$

Proof. Note first that $\theta \in (0,1)$ and

$$\frac{1-\theta}{p} + \frac{\theta}{r} = 1.$$

Then, by the Hölder inequality with exponents $p/(1-\theta)$ and r/θ, we have

$$\begin{aligned}
\left[\frac{1}{t}\int_0^t \omega^p(s)\Delta s\right]^{1/p} &\leq \frac{\tilde{\kappa}^{1/p}}{t}\int_0^t \omega(s)\Delta s \\
&= \frac{\tilde{\kappa}^{1/p}}{t}\int_0^t \omega^{1-\theta}(s)\omega^{\theta}(s)\Delta s \\
&\leq \frac{\tilde{\kappa}^{1/p}}{t}\left[\int_0^t \omega^p(s)\Delta s\right]^{(1-\theta)/p} [\int_0^t \omega^r(s)\Delta s)^{\theta/r} \\
&= \tilde{\kappa}^{1/p}\left[\frac{1}{t}\int_0^t \omega^p(s)\Delta s\right]^{(1-\theta)/p}\left[\frac{1}{t}\int_0^t \omega^r(s)\Delta s\right]^{\theta/r}
\end{aligned}$$

so that, by dividing, we find

$$\left[\frac{1}{t}\int_0^t \omega^p(s)\Delta s\right]^{\theta/p} \leq \tilde{\kappa}^{1/p}\left[\frac{1}{t}\int_0^t \omega^r(s)\Delta s\right]^{\theta/r},$$

i. e., (2.82) is true. □

We say that $\omega : [0,\infty)_{\mathbb{T}} \to \mathbb{R}$ belongs to $\mathbb{L}^p_\Delta([0,\infty)_{\mathbb{T}})$ provided

$$\left(\int_0^\infty |\omega(t)|^p \Delta t\right)^{\frac{1}{p}} < \infty.$$

By Theorem 2.28, under the assumptions of Theorem 2.27, if $\omega \in \mathbb{L}^r_\Delta[0,\infty)_{\mathbb{T}}$ for $0 < r < 1$, then $\omega \in \mathbb{L}^p_\Delta[0,\infty)_{\mathbb{T}}$ for $p > 1$. But in the general case when $p \neq r$, $\mathbb{L}^p_\Delta[0,\infty)_{\mathbb{T}}$ neither includes nor is included in $\mathbb{L}^r_\Delta[0,\infty)_{\mathbb{T}}$. The following theorem gives some results for $\mathbb{L}^p_\Delta[0,\infty)_{\mathbb{T}}$-interpolation.

Theorem 2.29. *Suppose that* $0 < p_0 < p_1 < \infty$ *and that* $0 < \theta < 1$.

(i) *If* $p = (1-\theta)p_0 + \theta p_1$ *and* $\omega \in \mathbb{L}_\Delta^{p_0}[0,\infty)_\mathbb{T} \cap \mathbb{L}_\Delta^{p_1}[0,\infty)_\mathbb{T}$, *then* $\omega \in \mathbb{L}_\Delta^{p}[0,\infty)_\mathbb{T}$ *and*

$$\mathscr{A}\omega^p \le [\mathscr{A}\omega^{p_0}]^{1-\theta}[\mathscr{A}\omega^{p_1}]^{\theta}.$$

(ii) *If* $p = \frac{1}{\frac{1-\theta}{p_0}+\frac{\theta}{p_1}}$ *and* $\omega \in \mathbb{L}_\Delta^{p_0}[0,\infty)_\mathbb{T} \cap \mathbb{L}_\Delta^{p_1}[0,\infty)_\mathbb{T}$, *then* $\omega \in \mathbb{L}_\Delta^{p}[0,\infty)_\mathbb{T}$ *and*

$$\mathscr{A}\omega^p \le [\mathscr{A}\omega^{p_0}]^{(1-\theta)p/p_0}[\mathscr{A}\omega^{p_1}]^{\theta p/p_1}.$$

Proof. For (i), we apply the Hölder inequality with exponents $1/(1-\theta)$ and $1/\theta$ to see that

$$\begin{aligned}\frac{1}{t}\int_0^t \omega^p(s)\Delta s &= \frac{1}{t}\int_0^t \omega^{(1-\theta)p_0}(s)\omega^{\theta p_1}(s)\Delta s\\ &\le \left[\frac{1}{t}\int_0^t \omega^{p_0}(s)\Delta s\right]^{1-\theta}\left[\frac{1}{t}\int_0^t \omega^{p_1}(s)\Delta s\right]^{\theta},\end{aligned}$$

which shows (i). For (ii), we apply the Hölder inequality with exponents $1/(1-\gamma)$ and $1/\gamma$, where

$$\gamma := \frac{\theta p}{p_1} \quad \text{so that } 1-\gamma = \frac{(1-\theta)p}{p_0},$$

to see that

$$\begin{aligned}\frac{1}{t}\int_0^t \omega^p(s)\Delta s &= \frac{1}{t}\int_0^t \omega^{(1-\theta)p}(s)\omega^{\theta p}(s)\Delta s\\ &\le \left[\frac{1}{t}\int_0^t \omega^{(1-\theta)p/(1-\gamma)}(s)\Delta s\right]^{1-\gamma}\left[\frac{1}{t}\int_0^t \omega^{\theta p/\gamma}(s)\Delta s\right]^{\gamma}\\ &= \left[\frac{1}{t}\int_0^t \omega^{p_0}(s)\Delta s\right]^{(1-\theta)p/p_0}\left[\frac{1}{t}\int_0^t \omega^{p_1}(s)\Delta s\right]^{\theta p/p_1},\end{aligned}$$

which shows (ii). □

In the following, we give a new proof of the Gehring mean inequality on time scales proving that if the weight belongs to the $\mathbb{A}_1$-Muckenhoupt class, then it will be a member of a Gehring class. The inequality will be proved by using a condition similar to that for the class $\mathscr{A}_1$ of Muckenhoupt. In fact, we do not assume that the reverse Hölder inequality holds.

Theorem 2.30 (Gehring inequality II). *Assume* (2.74). *If* ω *is nonnegative and nonincreasing such that*

$$\mathscr{A}\omega^\sigma \le \nu\omega \quad \text{for some } \nu > 1, \tag{2.83}$$

then

$$\mathscr{A}(\omega^p)^\sigma \le \tilde{\nu}[\mathscr{A}\omega^\sigma]^p \quad \text{and} \quad \tilde{\nu} := \frac{\alpha}{\alpha - p(\alpha - 1)} > 0, \tag{2.84}$$

for $p \in [1, \alpha/(\alpha - 1))$, *where* $\alpha = \lambda\nu$.

Proof. For this proof, we put

$$W(t) = \int_0^t \omega^\sigma(s)\Delta s, \quad l(t) = \log(t), \quad L(t) = \log(\omega(t)).$$

By the chain rule, we get

$$\begin{aligned}
\frac{1}{\alpha} l^\Delta(t) &= \frac{1}{\lambda\nu}\int_0^1 \frac{dh}{h\sigma(t) + (1-h)t} \\
&\le \frac{1}{\lambda\nu}\int_0^1 \frac{dh}{h\sigma(t) + (1-h)\frac{\sigma(t)}{\lambda}} \le \frac{1}{\lambda\nu}\cdot\frac{\lambda}{\sigma(t)} = \frac{1}{\nu\sigma(t)} \\
&\le \frac{\omega(\sigma(t))}{W(\sigma(t))} = \frac{W^\Delta(t)}{W(\sigma(t))} = W^\Delta(t)\int_0^1 \frac{dh}{hW(\sigma(t)) + (1-h)W(\sigma(t))} \\
&\le W^\Delta(t)\int_0^1 \frac{dh}{hW(\sigma(t)) + (1-h)W(t)} = L^\Delta(t),
\end{aligned}$$

and hence, by integrating,

$$\log\left(\frac{t}{\sigma(s)}\right)^{1/\alpha} = \frac{1}{\alpha}l(t) - \frac{1}{\alpha}l(\sigma(s)) \le L(t) - L(\sigma(s)) = \log\left(\frac{W(t)}{W(\sigma(s))}\right)$$

so that

$$\omega(\sigma(s)) \le \frac{1}{\sigma(s)}\int_0^{\sigma(s)} \omega(\sigma(\tau))\Delta\tau = \frac{W(\sigma(s))}{\sigma(s)} \le \left(\frac{\sigma(s)}{t}\right)^{1/\alpha}\frac{W(t)}{\sigma(s)},$$

and, by integrating again, putting $\gamma = p(1 - 1/\alpha) \in (0, 1)$, and using the notation from Lemma 2.10, we obtain

$$\frac{1}{t}\int_0^t \omega^p(\sigma(s))\Delta s \le \frac{W^p(t)}{t^{1+p/\alpha}}\int_0^t \frac{\Delta s}{(\sigma(s))^{p(1-1/\alpha)}}$$

$$\le \frac{W^p(t)}{(1-\gamma)t^{1+p/\alpha}}\int_0^t (x^{1-\gamma})^{\Delta}(s)\Delta s$$

$$= \frac{t^{1-\gamma}W^p(t)}{(1-\gamma)t^{1+p/\alpha}} = \frac{1}{1-\gamma}\left(\frac{W(t)}{t}\right)^p,$$

proving (2.84). □

In the following, we consider the class $\mathscr{U}_p^q(B)$ of all nonnegative weights ω that satisfy the reverse Hölder inequality

$$\left[\frac{1}{\sigma(x)-a}\int_a^{\sigma(x)} \omega^q(t)\Delta t\right]^{\frac{1}{q}} \le B\left[\frac{1}{\sigma(x)-a}\int_a^{\sigma(x)} \omega^p(t)\Delta t\right]^{\frac{1}{p}}, \tag{2.85}$$

for $1 \le p < q$ where the constant $B > 1$. We say that ω is a $\mathscr{U}_p^q$-weight if its $\mathscr{U}_p^q$-norm is finite, i. e.,

$$\omega \in \mathscr{U}_p^q \iff [\mathscr{U}_p^q(\omega)] < +\infty.$$

When we fix a constant $\mathscr{C} > 1$, the triple of real numbers $(p, q, \mathscr{C})$ defines the $\mathscr{U}_p^q$ class:

$$\omega \in \mathscr{U}_p^q(\mathscr{C}) \iff [\mathscr{U}_p^q(\omega)] \le \mathscr{C},$$

and we will refer to $\mathscr{C}$ as the $\mathscr{U}_p^q$-constant of the class. It is immediate to observe that the classes $\mathbb{A}^p$ and $\mathbb{G}^q$ are special cases of the class $\mathscr{U}_p^q$ of weights as follows:

$$\mathbb{A}^p := \mathscr{U}_{\frac{1}{1-p}}^{1} \quad \text{and} \quad \mathbb{G}^q := \mathscr{U}_1^q.$$

In order to establish the main results, we need a new version of the refinement of Hardy inequality on a time scale. To prove this inequality, we will use the following elementary inequality:

$$(\vartheta + \upsilon)^p \ge \vartheta^p + p\vartheta^{p-1}\upsilon, \quad \text{where } p > 1 \text{ or } p < 0. \tag{2.86}$$

Recall that this relation is a variant of the well-known Bernoulli inequality and it is valid for all $\vartheta \ge 0$ and $\vartheta + \nu \ge 0$, if $p > 1$, or for $\vartheta > 0$ and $\vartheta + \nu > 0$, if $p < 0$. The equality in (2.86) holds if and only if $\nu = 0$. We will assume that the forward jump operator is uniformly bounded from above by a linear function. More precisely, we suppose that there exists a real number $m \ge 1$ such that

$$\sigma(t) - a \le m(t-a), \quad \text{for } t > a. \tag{2.87}$$

It should be noticed here that the condition $\sigma(t) \leq \lambda t$ may be removed if the graininess weight $\mu(t)$ on the time scale $\mathbb{T}$ satisfies the relation $\mu(t) = O(t)$. Indeed, if $\mu(t) = O(t)$, then there exists $\lambda > 1$ such that $0 < \mu(t)/t \leq \lambda - 1$, for all $t \in \mathbb{T}$. Hence, $1 \leq (t + \mu(t))/t \leq \lambda$ and therefore, $1 \leq \sigma(t)/t \leq \lambda$, for all $t \in \mathbb{T}$. Note also that if $\mathbb{T} = \mathbb{R}$, then $\sigma(t) = t$, while for $\mathbb{T} = \mathbb{N}$, we have $\sigma(t) = t + 1$.

In the following, we prove a time scale version of an integral inequality due to Hardy, Littlewood, and Pólya [23] and the new refinement of Hardy-type inequality on time scales that will play important roles in the proof of the main results of higher integrability.

Theorem 2.31. *Assume that $\phi : [0, \infty) \to \mathbb{R}$ is a differentiable convex function. If ω is a nonnegative decreasing function, then the inequality*

$$\phi(0) + \int_a^b \phi'((x-a)\omega(x))\omega(x)\Delta x \leq \phi\left(\int_a^b \omega(x)\Delta x\right) \tag{2.88}$$

holds.

Proof. Let $x \in [a, b]_{\mathbb{T}}$. Since ω is a decreasing function, it follows that $(x - a)\omega(x) \leq \int_a^x \omega(t)\Delta t$. On the other hand, by defining

$$W(x) = \int_a^x \omega(t)\Delta t,$$

we see that $\omega^{\Delta}(x) = \omega(x) \geq 0$, which implies that ω is an increasing function. Now, due to convexity of the weight ϕ, we have

$$\phi'((x-a)\omega(x))\omega(x) \leq \phi'\left(\int_a^x \omega(t)\Delta t\right)\omega(x) = \phi'(W(x))\omega(x). \tag{2.89}$$

Further, taking into account the chain rule, it follows that

$$\phi^{\Delta}(W(x)) = \phi'(W(\zeta))W^{\Delta}(x), \text{ where } \zeta \in [x, \sigma(x)],$$

and consequently,

$$\phi^{\Delta}(W(x)) \geq \phi'(W(x))W^{\Delta}(x) = \phi'(W(x))\omega(x), \tag{2.90}$$

since ω is an increasing function. Now, considering the relations (2.89) and (2.90), we obtain the inequality

$$\phi'((x-a)\omega(x))\omega(x) \leq \phi^{\Delta}(W(x)).$$

Finally, integrating the latter inequality from a to b yields the relation

$$\int_a^b \phi'((x-a)\omega(x))\omega(x)\Delta x \leq \int_a^b (\phi(W(x)))^{\Delta}\Delta x = \phi(W(b)) - \phi(0),$$

which proves our assertion. The proof is complete. □

In the sequel, we consider a special case of Theorem 2.31, when the weight $\phi : [0,\infty) \to \mathbb{R}^+$ is defined by $\phi(\vartheta) = \vartheta^p$, for $p \geq 1$. Clearly, this weight is differentiable and convex, so we have the following consequence.

Corollary 2.2. *Assume that ω is a nonnegative decreasing function. If $p \geq 1$, then the inequality*

$$\int_a^b (x-a)^{p-1}\omega^p(x)\Delta x \leq \frac{1}{p}\left(\int_a^b \omega(x)\Delta x\right)^p \tag{2.91}$$

holds.

Remark 2.3. Applying the Hölder inequality with exponents $1/p$ and $(p-1)/p$ to the right-hand side of (2.91), we obtain the following inequality:

$$\int_a^b (x-a)^{p-1}\omega^p(x)\Delta x \leq \frac{(b-a)^{p-1}}{p}\int_a^b \omega^p(x)\Delta x.$$

Our next intention is to rewrite the inequality (2.91) in a form which will be more suitable in our further discussion. First of all, it should be noticed here that if a nonnegative weight ω is a decreasing function, then the weight $(x-a)^{\gamma-1}\omega$, where $\gamma \leq 1$, is also decreasing. Therefore, considering the relation (2.91) with $(x-a)^{\gamma-1}\omega$ instead of ω, we obtain the following result.

Corollary 2.3. *Assume that ω is a decreasing function. If $p \geq 1$ and $\gamma \leq 1$, then the inequality*

$$\int_a^b (x-a)^{p\gamma-1}\omega^p(x)\Delta x \leq \frac{1}{p}\left(\int_a^b (x-a)^{\gamma-1}\omega(x)\Delta x\right)^p \tag{2.92}$$

holds.

Remark 2.4. Let r and s be positive real numbers such that $r \leq s$. Considering the relation (2.92) with the weight ω^r instead of ω, and with parameters $p = s/r \geq 1$, $\gamma = 1/p \leq 1$, we see that the inequality

$$\left(\int_a^b \omega^s(x)\Delta x\right)^{\frac{r}{s}} \leq \frac{r}{s}\int_a^b (x-a)^{\frac{r}{s}-1}\omega^r(x)\Delta x \tag{2.93}$$

holds. The above inequality will be an important relation that will be used later in the proof of our main higher integrability theorem. On the other hand, it is important in its own right, since it is the time scale version of an inequality due to Hardy, Littlewood, and Pólya (for more details, see [23]).

Our next intention is to give a time scale extension and refinement of the famous Hardy inequality (2.73). In order to summarize our further discussion, we first define an operator $\mathscr{H}$ by

$$\mathscr{H}(x) := \frac{1}{x-a}\int_a^x \omega(t)\Delta t, \quad \text{for all } x \in [a,\infty)_{\mathbb{T}}, \tag{2.94}$$

where $\omega : [a,\infty)_{\mathbb{T}} \to \mathbb{R}$ is a nonnegative function. Next, we give several simple facts about the operator $\mathscr{H}$ following from its definition.

Lemma 2.12. *If ω is a nonnegative and decreasing function, then $\mathscr{H}(x) \geq \omega(x)$ for $x \in [a,\infty)_{\mathbb{T}}$.*

Proof. Since ω is decreasing, it follows that

$$\mathscr{H}(x) = \frac{1}{x-a}\int_a^x \omega(t)\Delta t \geq \frac{1}{x-a}\int_a^x \omega(x)\Delta t = \omega(x), \quad x \in [a,\infty)_{\mathbb{T}},$$

which completes the proof. □

It should be also noticed here that $\mathscr{H}$ inherits the decreasing nature of the weight ω.

Lemma 2.13. *If ω is a nonnegative decreasing function, then so is $\mathscr{H}$.*

Proof. By utilizing the quotient rule, it follows that

$$\mathscr{H}^{\Delta}(x) = \frac{\omega(x)(x-a) - \int_a^x \omega(t)\Delta t}{(x-a)(\sigma(x)-a)}, \quad x \in [a,\infty)_{\mathbb{T}}.$$

Hence, by virtue of Lemma 2.12, we have $(\sigma(x)-a)\mathscr{H}^{\Delta}(x) = \omega(x) - \mathscr{H}(x) \leq 0$, for $x \in [a,\infty)_{\mathbb{T}}$, which proves our assertion. □

Now, we are ready to state and prove a refinement of the Hardy inequality on time scales.

Theorem 2.32. *Let $\mathbb{T}$ be a time scale with $a, b \in \mathbb{T}$, and let ω be a nonnegative decreasing function. Further, assume that* (2.87) *holds. If $\alpha \leq 1$ and $\beta > 1$, then the inequality*

$$\int_a^b (x-a)^{\alpha-1}\left(\mathscr{H}^\sigma(x)\right)^\beta \Delta x + \left(\frac{\beta}{\beta-\alpha}\right)(b-a)^\alpha \mathscr{H}^\beta(b)$$
$$\leq \left(\frac{\beta}{\beta-\alpha}\right)^\beta \int_a^b (x-a)^{\alpha-1}\omega^\beta(x)\Delta x \tag{2.95}$$

holds, where $\mathscr{H}$ is defined in (2.94).

Proof. Taking into account the chain rule and utilizing the fact that $\mathscr{H}$ is decreasing on $[a,b]_{\mathbb{T}}$ by Lemma 2.13, we obtain the inequality

$$\begin{aligned}(\mathscr{H}^\beta(x))^\Delta &= \beta\int_0^1 [h\mathscr{H}^\sigma(x)+(1-h)\mathscr{H}(x)]^{\beta-1} dh \mathscr{H}^\Delta(x)\\ &\leq \beta\int_0^1 [h\mathscr{H}^\sigma(x)+(1-h)\mathscr{H}^\sigma(x)]^{\beta-1} dh \mathscr{H}^\Delta(x)\\ &= \beta\mathscr{H}^\Delta(x)(\mathscr{H}^\sigma(x))^{\beta-1}.\end{aligned}$$

On the other hand, since $(x-a)\mathscr{H}(x) = \int_a^x \omega(t)\Delta t$, the product rule implies the equality

$$(x-a)\mathscr{H}^\Delta(x) + \mathscr{H}^\sigma(x) = \omega(x),$$

and then we have

$$(x-a)^\alpha \mathscr{H}^\Delta(x) = (x-a)^{\alpha-1}[\omega(x) - \mathscr{H}^\sigma(x)]. \tag{2.96}$$

Further, applying integration by parts formula with

$$\vartheta(x) = \int_a^x (t-a)^{\alpha-1}\Delta t \quad \text{and} \quad \upsilon(x) = \mathscr{H}^\beta(x),$$

it follows that

$$\begin{aligned}&\int_a^b (x-a)^{\alpha-1}\left(\mathscr{H}^\sigma(x)\right)^\beta \Delta x\\ &\quad= \vartheta(b)\mathscr{H}^\beta(b) - \lim_{x\to a^+} \vartheta(x)\mathscr{H}^\beta(x) - \int_a^b \vartheta(x)(\mathscr{H}^\beta(x))^\Delta \Delta x\\ &\quad\geq \frac{(b-a)^\alpha \mathscr{H}^\beta(b)}{\alpha} - \frac{\beta}{\alpha}\int_a^b (x-a)^\alpha \mathscr{H}^\Delta(x)(\mathscr{H}^\sigma(x))^{\beta-1}\Delta x\\ &\quad\quad - \lim_{x\to a^+} \vartheta(x)\mathscr{H}^\beta(x)\end{aligned}$$

$$= \frac{(b-a)^{\alpha}\mathcal{H}^{\beta}(b)}{\alpha} - \frac{\beta}{\alpha}\int_a^b (x-a)^{\alpha-1}[\omega(x) - \mathcal{H}^{\sigma}(x)](\mathcal{H}^{\sigma}(x))^{\beta-1}\Delta x$$
$$= \frac{(b-a)^{\alpha}\mathcal{H}^{\beta}(b)}{\alpha} - \frac{\beta}{\alpha}\int_a^b (x-a)^{\alpha-1}\omega(x)(\mathcal{H}^{\sigma}(x))^{\beta-1}\Delta x$$
$$+ \frac{\beta}{\alpha}\int_a^b (x-a)^{\alpha-1}(\mathcal{H}^{\sigma}(x))^{\beta}\Delta x - \lim_{x\to a^+}\vartheta(x)\mathcal{H}^{\beta}(x).$$

Now, our intention is to show that $\lim_{x\to a^+}\vartheta(x)\mathcal{H}^{\beta}(x) = 0$. Taking into account the definitions of the functions $\vartheta(x)$ and $\mathcal{H}^{\beta}(x)$, we have

$$\begin{aligned}
\vartheta(x)\mathcal{H}^{\beta}(x) &= \int_a^x (t-a)^{\alpha-1}\Delta t\left(\frac{1}{x-a}\int_a^x \omega(t)\Delta t\right)^{\beta} \\
&= \left(\frac{1}{x-a}\right)^{\beta}\int_a^x (t-a)^{\alpha-1}\Delta t\left(\int_a^x \omega(t)\Delta t\right)^{\beta} \\
&\le \omega^{\beta}(a)\left(\frac{1}{x-a}\right)^{\beta}\int_a^x (t-a)^{\alpha-1}\Delta t\left(\int_a^x \Delta t\right)^{\beta} \\
&= \omega^{\beta}(a)\int_a^x (t-a)^{\alpha-1}\Delta t \\
&\le \omega^{\beta}(a)\int_a^x \frac{1}{(\sigma(t)-a)^{1-\alpha}}\left(\frac{\sigma(t)-a}{t-a}\right)^{1-\alpha}\Delta t \\
&\le m^{1-\alpha}\omega^{\beta}(a)\int_a^x \frac{1}{(\sigma(t)-a)^{1-\alpha}}\Delta t. \qquad (2.97)
\end{aligned}$$

Since $\alpha \le 1$, another application of the chain rule yields the estimate

$$\begin{aligned}
((t-a)^{\alpha})^{\Delta} &= \alpha\int_0^1 [h(\sigma(t)-a) + (1-h)(t-a)]^{\alpha-1}dh \\
&\ge \alpha\int_0^1 [h(\sigma(t)-a) + (1-h)(\sigma(t)-a)]^{\alpha-1}dh \\
&= \alpha(\sigma(t)-a)^{\alpha-1},
\end{aligned}$$

which implies the inequality

$$\int_a^x (\sigma(t) - a)^{\alpha-1}\Delta t \le \int_a^x \frac{((t-a)^\alpha)^\Delta}{\alpha}\Delta t = \frac{(x-a)^\alpha}{\alpha}.$$

Thus, utilizing the above inequality and (2.97), we obtain the estimate

$$\vartheta(x)\mathscr{H}^\beta(x) \le m^{1-\alpha}\omega^\beta(a)\frac{(x-a)^\alpha}{\alpha},$$

and consequently,

$$\lim_{x\to a^+} \vartheta(x)\mathscr{H}^\beta(x) = 0.$$

Clearly, from the above discussion, we obtain the inequality

$$\left(\frac{\beta-\alpha}{\alpha}\right)\int_a^b (x-a)^{\alpha-1}(\mathscr{H}^\sigma(x))^\beta\Delta x + \frac{(b-a)^\alpha\mathscr{H}^\beta(b)}{\alpha}$$
$$\le \frac{\beta}{\alpha}\int_a^b (x-a)^{\alpha-1}\omega(x)(\mathscr{H}^\sigma(x))^{\beta-1}\Delta x.$$

Now, applying the Hölder inequality with exponents $1/\beta$ and $(\beta-1)/\beta$ to the right-hand side of the latter inequality yields the relation

$$\left(\frac{\beta-\alpha}{\alpha}\right)\int_a^b (x-a)^{\alpha-1}(\mathscr{H}^\sigma(x))^\beta\Delta x + \frac{(b-a)^\alpha\mathscr{H}^\beta(b)}{\alpha}$$
$$\le \frac{\beta}{\alpha}\left\{\int_a^b (x-a)^{\alpha-1}\omega^\beta(x)\Delta x\right\}^{\frac{1}{\beta}}\left\{\int_a^b (x-a)^{\alpha-1}(\mathscr{H}^\sigma(x))^\beta\Delta x\right\}^{\frac{\beta-1}{\beta}},$$

which can be rewritten in the following form:

$$\left(\frac{\beta}{\beta-\alpha}\right)^\beta\int_a^b (x-a)^{\alpha-1}\omega^\beta(x)\Delta x$$
$$\ge \left[\left\{\int_a^b (x-a)^{\alpha-1}(\mathscr{H}^\sigma(x))^\beta\Delta x\right\}^{\frac{1}{\beta}} + \frac{\frac{(b-a)^\alpha}{(\beta-\alpha)}\mathscr{H}^\beta(b)}{\{\int_a^b (x-a)^{\alpha-1}(\mathscr{H}^\sigma(x))^\beta\Delta x\}^{\frac{\beta-1}{\beta}}}\right]^\beta.$$

Finally, applying the Bernoulli inequality (2.86), with

$$\vartheta = \left\{\int_a^b (x-a)^{\alpha-1}(\mathscr{H}^\sigma(x))^\beta\Delta x\right\}^{\frac{1}{\beta}} \quad \text{and}$$

$$\upsilon = \frac{\frac{(b-a)^{\alpha}}{(\beta-\alpha)}\mathcal{H}^{\beta}(b)}{\{\int_a^b (x-a)^{\alpha-1}(\mathcal{H}^{\sigma}(x))^{\beta}\Delta x\}^{\frac{\beta-1}{\beta}}},$$

to the right-hand side of the latter inequality, we obtain

$$\begin{aligned}&\left(\frac{\beta}{\beta-\alpha}\right)^{\beta}\int_a^b (x-a)^{\alpha-1}\omega^{\beta}(x)\Delta x\\ &\quad\geq \int_a^b (x-a)^{\alpha-1}\big(\mathcal{H}^{\sigma}(x)\big)^{\beta}\Delta x + \frac{\beta}{\beta-\alpha}(b-a)^{\alpha}\mathcal{H}^{\beta}(b),\end{aligned}$$

which represents the desired inequality (2.95). The proof is now complete. □

The established inequality (2.95) is both a refinement and a time scale extension of the Hardy inequality (2.73) for a class of decreasing functions. To see this, let $\alpha = q/p$ and $\beta = q$, where $p \geq q > 1$. In this setting, Theorem 2.32 reduces to the following corollary.

Corollary 2.4. *Let $\mathbb{T}$ be a time scale with $a, b \in \mathbb{T}$, let ω be a nonnegative decreasing function. Suppose that* (2.87) *holds. If $p \geq q > 1$, then the inequality*

$$\begin{aligned}&\int_a^b (x-a)^{\frac{q}{p}-1}\big(\mathcal{H}^{\sigma}(x)\big)^{q}\Delta x + \frac{p}{p-1}(b-a)^{\frac{q}{p}}\mathcal{H}^{q}(b)\\ &\quad\leq \left(\frac{p}{p-1}\right)^{q}\int_a^b (x-a)^{\frac{q}{p}-1}\omega^{q}(x)\Delta x \qquad (2.98)\end{aligned}$$

holds, where $\mathcal{H}$ is defined in (2.94).

Clearly, if $\mathbb{T} = \mathbb{R}$, $a = 0$ and $p = q$, then $\sigma(x) = x$ and $\mathcal{H}(x) = (1/x)\int_0^x \omega(t)dt$. So the inequality (2.98) provides a refinement of the classical Hardy inequality (2.73) for a class of decreasing functions. In the next few remarks, we will compare our Theorem 2.32 with some results known from the literature.

Remark 2.5. If $\alpha = 1$, Theorem 2.32 provides the following refinement of the Hardy-type inequality:

$$\int_a^b \big(\mathcal{H}^{\sigma}(x)\big)^{\beta}\Delta x + \frac{\beta(b-a)}{\beta-1}\mathcal{H}^{\beta}(b) \leq \left(\frac{\beta}{\beta-1}\right)^{\beta}\int_a^b \omega^{\beta}(x)\Delta x.$$

Remark 2.6. If $\mathbb{T} = \mathbb{R}$, our inequality (2.95) can be regarded as a refinement of the inequality

$$\int_a^b (x-a)^{\alpha-1}\mathscr{H}^{\beta}(x)dx \le \left(\frac{\beta}{\beta-\alpha}\right)^{\beta}\int_a^b (x-a)^{\alpha-1}\omega^{\beta}(x)dx,$$

established by Popoli [40, Theorem 2.1], for $\alpha \le 1$ and $\beta > 1$. Clearly, $\mathscr{H}$ is defined here by $\mathscr{H}(x) := (1/(x-a))\int_a^x \omega(t)dt$.

In the sequel, we first state and prove several lemmas, interesting in their own right, which will be utilized in establishing the higher integrability theorem.

Lemma 2.14. *Assume that φ, ψ are nonnegative functions. Then*

$$\int_a^b \varphi(t)\left(\int_t^b \psi(x)\Delta x\right)\Delta t = \int_a^b \psi(t)\left(\int_a^{\sigma(t)} \varphi(x)\Delta x\right)\Delta t.$$

Proof. Define $\Psi(t) = \int_t^b \psi(x)\Delta x$. Applying the integration by parts formula to the term $\int_a^b \varphi(t)\Psi(t)\Delta t$ with $\vartheta(t) = \Psi(t)$ and $\upsilon^{\Delta}(t) = \varphi(t)$, we get

$$\int_a^b \varphi(t)\left(\int_t^b \psi(x)\Delta x\right)\Delta t = \int_a^b \varphi(t)\Psi(t)\Delta t = \Psi(t)\upsilon(t)|_a^b - \int_a^b \Psi^{\Delta}(t)\upsilon^{\sigma}(t)\Delta t,$$

where $\upsilon(t) = \int_a^t \varphi(x)\Delta x$. Now, since $\upsilon(a) = 0$ and $\Psi(b) = 0$, it follows that

$$\begin{aligned}\int_a^b \varphi(t)\left(\int_t^b \psi(x)\Delta x\right)\Delta t &= \int_a^b [-\Psi^{\Delta}(t)]\upsilon^{\sigma}(t)\Delta t = \int_a^b \psi(t)\upsilon^{\sigma}(t)\Delta t \\ &= \int_a^b \psi(t)\left(\int_a^{\sigma(t)} \varphi(x)\Delta x\right)\Delta t,\end{aligned}$$

which completes the proof. □

Lemma 2.15. *Assume that w is a nonnegative increasing function. Further, suppose that there exists $m \ge 1$ such that*

$$w^{\sigma}(x) \le m w(x) \quad \textit{for all } x \in [a,b]_{\mathbb{T}}. \tag{2.99}$$

If $\lambda < 0$, then the inequality

$$\int_a^b w^{\lambda}(x)w^{\Delta}(x)\mathscr{H}^{\sigma}(x)\Delta x \ge \frac{1}{\lambda m}\left[w^{\lambda}(b)\int_a^b \varphi(x)\Delta x - \int_a^b w^{\lambda}(x)\varphi(x)\Delta x\right], \tag{2.100}$$

holds, where $\mathscr{H}(x) := (1/w(x))\int_a^x \varphi(t)\Delta t$.

Proof. Taking into account the definition of $\mathscr{H}$, assumption (2.99), and Lemma 2.14, we obtain

$$\begin{aligned}\int_a^b w^{\lambda}(x)w^{\Delta}(x)\mathscr{H}^{\sigma}(x)\Delta x &= \int_a^b w^{\Delta}(x)w^{\lambda-1}(x)\left(\frac{w}{w^{\sigma}}\int_a^{\sigma(x)}\varphi(t)\Delta t\right)\Delta x \\ &\geq \frac{1}{m}\int_a^b w^{\Delta}(x)w^{\lambda-1}(x)\left(\int_a^{\sigma(x)}\varphi(t)\Delta t\right)\Delta x \\ &= \frac{1}{m}\int_a^b \varphi(x)\left(\int_x^b w^{\Delta}(t)w^{\lambda-1}(t)\Delta t\right)\Delta x.\end{aligned}$$

Further, since $\lambda < 0$ and $w^{\Delta}(x) > 0$, by applying the chain role, it follows that

$$\begin{aligned}\left(w^{\lambda}(t)\right)^{\Delta} &= \lambda\int_0^1 \left[hw^{\sigma}(t) + (1-h)w(t)\right]^{\lambda-1} dhw^{\Delta}(t) \\ &\geq \lambda\int_0^1 \left[hw(t) + (1-h)w(t)\right]^{\lambda-1} dhw^{\Delta}(t) = \lambda w^{\Delta}(t)w^{\lambda-1}(t),\end{aligned}$$

which implies the estimate $w^{\Delta}(t)w^{\lambda-1}(t) \geq \frac{1}{\lambda}(w^{\lambda}(t))^{\Delta}$. Hence, we obtain

$$\begin{aligned}\int_a^b w^{\lambda}(x)w^{\Delta}(x)\mathscr{H}^{\sigma}(x)\Delta x &\geq \frac{1}{\lambda m}\int_a^b \varphi(x)\left(\int_x^b \left(w^{\lambda}(t)\right)^{\Delta}\Delta t\right)\Delta x \\ &= \frac{1}{\lambda m}\left[w^{\lambda}(b)\int_a^b \varphi(x)\Delta x - \int_a^b w^{\lambda}(x)\varphi(x)\Delta t\right],\end{aligned}$$

which proves our assertion. The proof is complete. □

We will also employ the following lemma which has been proved in [40, Lemma 2.2].

Lemma 2.16. *Let $C > 1$, $q > p > 0$, and let L be defined by*

$$L(p,q,x,C) = 1 - C^{q}(1-x)\left(\frac{q}{q-px}\right)^{\frac{q}{p}}, \tag{2.101}$$

where $x \in [0,1]$. Then, there exists a unique solution x_q of the equation $L(p,q,x,C) = 0$. In addition, $L(p,q,x,C) > 0$ if and only if $x \in (x_q, 1]$.

Before we state and prove our main theorem, we first need the following auxiliary result.

Theorem 2.33. *Let* $0 < p < q$, $K > 1$. *Further, assume that* ω *is a nonnegative decreasing weight satisfying* (2.85). *If*

$$\mathscr{L}(p,q,\alpha,K) = 1 - K^q m(1-\alpha)\left(\frac{q}{q-p\alpha}\right)^{\frac{q}{p}} \tag{2.102}$$

and the condition (2.87) *holds, then the inequality*

$$\int_a^b (x-a)^{\alpha-1}\omega^q(x)\Delta x \le \frac{(b-a)^{\alpha-1}}{\mathscr{L}(p,q,\alpha,K)}\int_a^b \omega^q(x)\Delta x \tag{2.103}$$

holds for all $\alpha \in (\alpha_q, 1]$, *where* α_q *is the unique root of equation* (2.101) *with* $C = Km^{1/q}$.

Proof. Without loss of generality, we can suppose that $\alpha \in (\alpha_q, 1)$ since for $\alpha = 1$ the inequality (2.103) holds trivially.

Now, since $\omega : [a,b]_{\mathbb{T}} \to \mathbb{R}$ satisfies the reverse Hölder inequality, it follows that

$$W_q^\sigma(x) \le K^q [W_p^\sigma(x)]^{\frac{q}{p}},$$

where $W_p(x) = (1/(x-a))\int_a^x \omega^p(t)\Delta t$ and $W_q(x) = (1/(x-a))\int_a^x \omega^q(t)\Delta t$. By integrating, it follows that

$$\int_a^b (x-a)^{\alpha-1} W_q^\sigma(x)\Delta x \le K^q \int_a^b (x-a)^{\alpha-1}(W_p^\sigma(x))^{\frac{q}{p}}\Delta x. \tag{2.104}$$

Further, it should be noticed that if $w(x) = x - a$, then the condition (2.99) reduces to (2.87). Therefore, applying Lemma 2.15 with $\lambda = \alpha - 1 < 0$, $w(x) = x - a$ and $\varphi(x) = \omega^q(x)$ to the left-hand side of (2.104), it follows that

$$\begin{aligned}
&\int_a^b (x-a)^{\alpha-1} W_q^\sigma(x)\Delta x \\
&\quad \ge \frac{1}{(\alpha-1)m}\left[(b-a)^{\alpha-1}\int_a^b \omega^q(x)\Delta x - \int_a^b (x-a)^{\alpha-1}\omega^q(x)\Delta x\right].
\end{aligned}$$

On the other hand, applying Theorem 2.32 with $\beta = q/p > 1$ and $\omega = \omega^p$ to the right-hand side of (2.104), we have

$$\begin{aligned}
\int_a^b (x-a)^{\alpha-1}(W_p^\sigma(x))^{\frac{q}{p}}\Delta x \le{}& \left(\frac{q}{q-p\alpha}\right)^{\frac{q}{p}}\int_a^b (x-a)^{\alpha-1}\omega^q(x)\Delta x \\
&- \left(\frac{q}{q-p\alpha}\right)(b-a)^{\alpha-\frac{q}{p}}\left(\int_a^b \omega^p(x)\Delta x\right)^{\frac{q}{p}}
\end{aligned}$$

$$\leq \left(\frac{q}{q-p\alpha}\right)^{\frac{q}{p}} \int_a^b (x-a)^{\alpha-1}\omega^q(x)\Delta x.$$

Now, taking into account (2.104) and the previous two estimates, we obtain the inequality

$$\begin{aligned}&\frac{1}{(\alpha-1)m}\left[(b-a)^{\alpha-1}\int_a^b \omega^q(x)\Delta x - \int_a^b (x-a)^{\alpha-1}\omega^q(x)\Delta x\right]\\&\leq K^q\left(\frac{q}{q-p\alpha}\right)^{\frac{q}{p}} \int_a^b (x-a)^{\alpha-1}\omega^q(x)\Delta x,\end{aligned}$$

which can be rewritten as

$$\begin{aligned}&(b-a)^{\alpha-1}\int_a^b \omega^q(x)\Delta x\\&\geq \left[1-K^q(1-\alpha)m\left(\frac{q}{q-p\alpha}\right)^{\frac{q}{p}}\right]\int_a^b (x-a)^{\alpha-1}\omega^q(x)\Delta x\\&= \mathscr{L}(p,q,\alpha,K)\int_a^b (x-a)^{\alpha-1}\omega^q(x)\Delta x.\end{aligned}$$

Finally, by Lemma 2.16, there exists a unique $\alpha_q \in (0,1)$ such that $\mathscr{L}(p,q,\alpha,K) > 0$ for $\alpha \in (\alpha_q, 1]$. This provides the inequality (2.103), and the proof is complete. □

Remark 2.7. If $p = 1$ and $\alpha = \frac{q}{r}$, provided that $r \geq q$, the inequality (2.103) reduces to

$$\int_a^b (x-a)^{\frac{q}{r}-1}\omega^q(x)\Delta x \leq \frac{(b-a)^{\frac{q}{r}-1}}{\mathscr{L}(1,q,q/r,K)}\int_a^b \omega^q(x)\Delta x,$$

where $\mathscr{L}(1,q,q/r,K)$ is defined by (2.102). In particular, if $\mathbb{T} = \mathbb{R}$ this inequality provides a relation established by D'Apuzzo and Sbordone [18, Lemma 3.2].

Finally, we are able to state and prove the higher integrability theorem for decreasing functions on time scales.

Theorem 2.34. *Let $0 < p < q$, $K > 1$. Further, suppose that ω is a nonnegative decreasing weight satisfying* (2.85). *If the condition* (2.87) *holds, then*

$$\left(\frac{1}{b-a}\int_a^b \omega^s(x)\Delta x\right)^{\frac{q}{s}} \leq \frac{q}{s\mathscr{L}(p,q,\frac{q}{s},K)}\left(\frac{1}{b-a}\int_a^b \omega^q(x)\Delta x\right), \tag{2.105}$$

for $q \leq s < q_0$, where q_0 is the unique solution of the equation

$$\left(\frac{x}{x-q}\right)^{\frac{1}{q}} = Km^{\frac{1}{q}}\left(\frac{x}{x-p}\right)^{\frac{1}{p}}. \tag{2.106}$$

Proof. Utilizing Theorem 2.33 with $\alpha = q/s$ and $\alpha_q = q/q_0$, we obtain the inequality

$$\int_a^b (x-a)^{\frac{q}{s}-1}\omega^q(x)\Delta x \le \frac{(b-a)^{\frac{q}{s}-1}}{\mathscr{L}(p,q,\frac{q}{s},K)}\int_a^b \omega^q(x)\Delta x,$$

where $q \le s < q_0$. In addition, applying the inequality (2.93) with $r = q$ to the left-hand side of the previous inequality, we have

$$\left(\int_a^b \omega^s(x)\Delta x\right)^{\frac{q}{s}} \le \frac{q(b-a)^{\frac{q}{s}-1}}{s\mathscr{L}(p,q,\frac{q}{s},K)}\int_a^b \omega^q(x)\Delta x, \quad \text{for } q \le s < q_0,$$

and thus,

$$\left(\frac{1}{b-a}\int_a^b \omega^s(x)\Delta x\right)^{\frac{q}{s}} \le \frac{q}{s\mathscr{L}(p,q,\frac{q}{s},K)}\left(\frac{1}{b-a}\int_a^b \omega^q(x)\Delta x\right),$$

which provides the inequality (2.105). Clearly, q_0 is the unique solution of the equation

$$\mathscr{L}\left(p,q,\frac{q}{x},K\right) = 0,$$

which reduces to (2.106). The proof is now complete. □

Remark 2.8. Since ω is a decreasing function, we have $\omega^\sigma(x) \le \omega(x)$, so the relation (2.105) implies that the inequality

$$\left(\frac{1}{b-a}\int_a^b \omega^s(\sigma(x))\Delta x\right)^{\frac{q}{s}} \le \frac{q}{s\mathscr{L}(p,q,\frac{q}{s},K)}\left(\frac{1}{b-a}\int_a^b \omega^q(x)\Delta x\right) \tag{2.107}$$

holds for $q \le s < q_0$, where q_0 is the unique solution of the equation (2.106) and $\mathscr{L}(p,q,q/s,K)$ is defined in (2.102).

We conclude this section with a discrete version of the inequality (2.107) from the previous remark. Namely, if $\mathbb{T} = \mathbb{N}$, then $\sigma(n) = n+1$, so the condition (2.87) is satisfied for $m = 2$. In this case, the constant $\mathscr{L}$ defined by (2.102) reduces to

$$\overline{\mathscr{L}}\left(p,q,\frac{q}{s},K\right) \equiv 1 - 2K^q\left(1-\frac{q}{s}\right)\left(\frac{s}{s-p}\right)^{\frac{q}{p}},$$

so we have the following result.

Corollary 2.5. *Let $0 < p < q$, $K > 1$, and let $\omega(n)$ be a nonnegative decreasing sequence such that*

$$\left(\frac{1}{n+1-a}\sum_{k=a}^{n}\omega^{q}(k)\right)^{\frac{1}{q}} \leq K\left(\frac{1}{n+1-a}\sum_{k=a}^{n}\omega^{p}(k)\right)^{\frac{1}{p}}$$

holds for $n \in [a,b]_{\mathbb{N}}$. Then, $\omega \in l^{s}[a,b]_{\mathbb{N}}$ for $q \leq s < q_0$ and

$$\left(\frac{1}{b-a}\sum_{n=a}^{b-1}\omega^{s}(n+1)\right)^{\frac{q}{s}} \leq \frac{q}{s\overline{\mathscr{L}}(p,q,\frac{q}{s},K)}\left(\frac{1}{b-a}\sum_{n=a}^{b-1}\omega^{q}(n)\right),$$

where q_0 is the unique solution of the equation

$$\left(\frac{x}{x-q}\right)^{\frac{1}{q}} = 2^{\frac{1}{q}}K\left(\frac{x}{x-p}\right)^{\frac{1}{p}}.$$

Now, we derive the self-improving properties of the two classes

$$\mathscr{A}^{p} := \mathscr{U}^{1}_{\frac{1}{1-p}} \quad \text{and} \quad \mathscr{H}^{q} := \mathscr{U}^{q}_{1}.$$

Theorem 2.35. *Let $p > 1$ and ω be any nonnegative and nondecreasing weight belonging to $\mathbb{A}^{p}(B)$ for $B > 1$. Then $\omega \in \mathbb{A}^{\eta}(B_2'')$ for $\eta \in (\eta^{-}, p]$, where η^{-} is the root of the equation*

$$\frac{p-\eta}{p-1}(B\eta)^{\frac{1}{p-1}} = 1. \tag{2.108}$$

Proof. Since $\mathbb{A}^{p} := \mathscr{U}^{1}_{\frac{1}{1-p}}$, equation (2.106) becomes

$$\left(\frac{x}{x-1}\right)\left(\frac{(p-1)x}{(p-1)x+1}\right)^{p-1} = Bm^{\frac{1}{q}}.$$

By applying the transform $\eta \to 1/(1-x)$, we see that η^{-} is determined from the equation

$$\frac{p-\eta}{p-1}(Bm^{\frac{1}{q}}\eta)^{\frac{1}{p-1}} = 1, \tag{2.109}$$

which is the desired equation (2.108). The proof is complete. □

Theorem 2.36. *Let $q > 1$ and g be any nonnegative and nondecreasing weight belongs to $\mathbb{G}^{q}(B)$ for $B > 1$. Then $g \in \mathbb{G}^{\eta}(B_1'')$ for $\eta \in [q, \eta^{+})$, where η^{+} is the root of the equation*

$$\left(\frac{x-1}{x}\right)\left(\frac{x}{x-q}\right)^{\frac{1}{q}} = Bm^{\frac{1}{q}}, \tag{2.110}$$

Proof. Since $\mathbb{G}^q := \mathscr{U}_1^q$, equation (2.106) becomes

$$\left(\frac{x}{x-1}\right)^{-1}\left(\frac{x}{x-q}\right)^{\frac{1}{q}} = Bm^{\frac{1}{q}},$$

which is the desired equation (2.21). The proof is complete. □

2.6 Higher integrability theorems

In this section, we apply the self-improving properties of the Gehring weights to prove higher integrability theorems for monotone nonincreasing weights on time scales. In [38], Nania considered a new type of reverse inequalities of the form

$$\frac{1}{t}\int_0^t \omega^q(s)ds \le C\omega^{q-1}(t)\frac{1}{t}\int_0^t \omega(s)ds, \quad \text{for all } t \in I, \tag{2.111}$$

and proved a higher integrability theorem for nonincreasing functions with the constants $C > 1$ and $q > 1$. In particular, Nania proved that if (2.111) holds, then for every $p \in [q, q+\varepsilon]$, we have

$$\left(\frac{1}{|I|}\int_I \omega^p(t)dt\right)^{1/p} \le K\left(\frac{1}{|I|}\int_I \omega(s)ds\right), \tag{2.112}$$

where $\varepsilon = q/(\alpha - 1)$, $\alpha = Cq(q-1)$, as well as

$$K = \left[\frac{\alpha^{r+1}}{\alpha - r(\alpha-1)}\right]^{1/p} \quad \text{and} \quad r = p/q.$$

Nania proved (2.112) by employing the Hardy inequality

$$\frac{1}{|I|}\int_I\left(\frac{1}{t}\int_0^t \omega(s)ds\right)^p dt \le \left(\frac{p}{p-1}\right)^p \frac{1}{|I|}\int_I \omega^p(t)dt, \quad p > 1. \tag{2.113}$$

Alzer [3] improved the Nania result and proved that if ω is a nonnegative and nonincreasing weight on I which satisfies (2.111) for all $t \in I$, then

$$\left(\frac{1}{|I|}\int_I \omega^p(t)dt\right)^{1/p} \le K_1\left(\frac{1}{|I|}\int_I \omega^I(s)ds\right)^{1/I}, \tag{2.114}$$

holds with a new constant K_1 smaller than K and for all $p \in [I, I+\delta]$ and $\delta > \varepsilon$. Alzer proved his results by employing the Shum inequality (see [64])

$$\frac{1}{|I|}\int_I\left(\frac{1}{t}\int_0^t\omega(s)ds\right)^p dt+\frac{p}{p-1}\left(\frac{1}{|I|}\int_I\omega(t)dt\right)^p\leq\left(\frac{p}{p-1}\right)^p\frac{1}{|I|}\int_I\omega^p(t)dt. \tag{2.115}$$

Definition 2.3. For any weight $\omega : (0,T]_{\mathbb{T}} \to \mathbb{R}^+$ which is a nonnegative and nonincreasing function, we define $\mathscr{A}\omega : (0,T]_{\mathbb{T}} \to \mathbb{R}^+$ by

$$\mathscr{A}\omega=\frac{1}{t}\int_0^t\omega(s)\Delta s\quad\text{for all } t\in(0,T]_{\mathbb{T}}, \tag{2.116}$$

as well as

$$\mathscr{A}^\sigma\omega=\frac{1}{\sigma(t)}\int_0^{\sigma(t)}\omega(s)\Delta s,\quad\text{for all } t\in(0,T]_{\mathbb{T}},$$

and

$$\mathscr{A}^\sigma\omega^\sigma(t)=\frac{1}{\sigma(t)}\int_0^{\sigma(t)}\omega^\sigma(s)\Delta s,$$

and assume that there exists a constant $\lambda>1$ such that $\sigma(t)\leq\lambda t$.

We mentioned here that the condition $\sigma(t)\leq\lambda t$ may be removed if the graininess weight $\mu(t)$ on the time scale $\mathbb{T}$ satisfies $\mu(t)=O(t)$. Indeed, if $\mu(t)=O(t)$, then there exists $\lambda>1$ such that $0<\mu(t)/t\leq\lambda-1$ for all $t\in\mathbb{T}$. Hence $1\leq(t+\mu(t))/t\leq\lambda$, and thus $1\leq\sigma(t)/t\leq\lambda$ for all $t\in\mathbb{T}$. Note that when $\mathbb{T}=\mathbb{R}$ we have $\sigma(t)=t$, and when $\mathbb{T}=\mathbb{N}$, we have $\sigma(t)=t+1$.

Theorem 2.37. *Assume that ω be a nonnegative and nonincreasing. If there exists a constant $A>1$ such that*

$$\mathscr{A}^\sigma\omega^\sigma\leq A\omega^\sigma,\quad\textit{for all } t\in(0,\infty)_{\mathbb{T}}, \tag{2.117}$$

then

$$\mathscr{A}^\sigma(\omega^\sigma)^p\leq\frac{M}{M-p(M-1)}(\mathscr{A}^\sigma\omega^\sigma)^p\leq\frac{M}{M-p(M-1)}(\mathscr{A}^\sigma\omega)^p, \tag{2.118}$$

for $p\in[1,M/(M-1)]$, where $M:=A\lambda>1$.

Proof. Let $W(t):=\int_0^t\omega^\sigma(s)\Delta s$. From this and (2.117), we get that (note that ω is decreasing)

$$\frac{1}{A\sigma(t)}\leq\frac{\omega^\sigma(t)}{\int_0^{\sigma(t)}\omega^\sigma(s)\Delta s}=\frac{\omega^\Delta(t)}{\omega^\sigma(t)}.$$

Since $\sigma(t) \leq \lambda t$ for some constant $\lambda \geq 1$, we have that

$$\frac{1}{Mt} \leq \frac{W^{\Delta}(t)}{W^{\sigma}(t)}, \quad \text{where } M = A\lambda > 1.$$

From the chain rule, we see that

$$(\log W(t))^{\Delta} = \left\{ \int_0^1 \frac{1}{h\omega^{\sigma} + (1-h)\omega} dh \right\} W^{\Delta}(t). \tag{2.119}$$

Now, since $\omega^{\Delta}(t) = \omega^{\sigma}(t) > 0$, we obtain $\omega^{\sigma} \geq \omega(t)$ and this implies that

$$\frac{1}{(1-h)W(t) + hW^{\sigma}} \geq \frac{1}{(1-h)W^{\sigma} + hW^{\sigma}} = \frac{1}{W^{\sigma}}.$$

This and (2.119) yield

$$\frac{W^{\Delta}(t)}{W^{\sigma}} \leq \left\{ \int_0^1 \frac{1}{hW^{\sigma} + (1-h)W} dh \right\} W^{\Delta}(t) = (\log W(t))^{\Delta}. \tag{2.120}$$

Also, we have (since $\sigma(t) \geq t$) that

$$(\log t)^{\Delta} = \left\{ \int_0^1 \frac{1}{h\sigma + (1-h)t} dh \right\} \leq \frac{1}{t}. \tag{2.121}$$

Combining (2.120) and (2.121), we deduce

$$\frac{1}{M}(\log t)^{\Delta} \leq (\log W(t))^{\Delta}. \tag{2.122}$$

Integrating the latter inequality from $\sigma(y)$ to $\sigma(t)$ (where $0 < \sigma(y) < \sigma(t)$), we get that

$$\log\left(\frac{\sigma(t)}{\sigma(y)}\right)^{\frac{1}{M}} \leq \log \frac{W^{\sigma}(t)}{W^{\sigma}(y)}.$$

Hence,

$$\int_0^{\sigma(y)} \omega^{\sigma}(s)\Delta s \leq \left(\frac{\sigma(y)}{\sigma(t)}\right)^{\frac{1}{M}} \int_0^{\sigma(t)} \omega^{\sigma}(s)\Delta s.$$

Since ω is nonincreasing, we see that $\omega^{\sigma}(y) \leq \omega^{\sigma}(s)$, where $\sigma(y) \geq \sigma(s)$, and then

$$\omega^{\sigma}(y) \leq \frac{1}{\sigma(y)} \int_0^{\sigma(y)} \omega^{\sigma}(s)\Delta s \leq \frac{1}{\sigma(y)} \left(\frac{\sigma(y)}{\sigma(t)}\right)^{\frac{1}{M}} \int_0^{\sigma(t)} \omega(s)\Delta s.$$

Thus,

$$(\omega^{\sigma}(y))^{p} \leq \left(\frac{1}{\sigma(y)}\right)^{p}\left(\frac{\sigma(y)}{\sigma(t)}\right)^{\frac{p}{M}}\left(\int_{0}^{\sigma(t)} \omega(s)\Delta s\right)^{p}.$$

Integrating from 0 to $\sigma(t)$, we see that

$$\int_{0}^{\sigma(t)} (\omega^{\sigma}(y))^{p}\Delta y \leq \Theta\left(\int_{0}^{\sigma(t)} \omega^{\sigma}(s)\Delta s\right)^{p}, \tag{2.123}$$

where

$$\Theta = \int_{0}^{\sigma(t)} \left(\frac{1}{\sigma(y)}\right)^{p}\left(\frac{\sigma(y)}{\sigma(t)}\right)^{\frac{p}{M}}\Delta y = \left(\frac{1}{\sigma(t)}\right)^{\frac{p}{M}}\int_{0}^{\sigma(t)} \frac{1}{\sigma^{\gamma}(y)}\Delta y,$$

with $\gamma = p(M-1)/M < 1$. From the chain rule, we see that

$$\begin{aligned}
(s^{1-\gamma})^{\Delta} &= (1-\gamma)\int_{0}^{1} [h\sigma(s) + (1-h)s]^{-\gamma} dh \\
&= (1-\gamma)\int_{0}^{1} \frac{dh}{[h\sigma(s) + (1-h)s]^{\gamma}} \\
&\geq (1-\gamma)\int_{0}^{1} \frac{dh}{[h\sigma + (1-h)\sigma(s)]^{\gamma}} = \frac{(1-\gamma)}{\sigma^{\gamma}(s)}.
\end{aligned}$$

This implies that

$$\begin{aligned}
\Theta &= \left(\frac{1}{\sigma(t)}\right)^{\frac{p}{M}}\int_{0}^{\sigma(t)} \frac{1}{\sigma^{\gamma}(y)}\Delta y \\
&\leq \frac{1}{(1-\gamma)}\left(\frac{1}{\sigma(t)}\right)^{\frac{p}{M}}\int_{0}^{\sigma(t)} (s^{1-\gamma})^{\Delta}\Delta s \\
&= \frac{1}{(1-\gamma)}\left(\frac{1}{\sigma(t)}\right)^{\frac{p}{M}}(\sigma(t))^{1-\frac{p(M-1)}{M}} = \frac{M}{M-p(M-1)}\sigma^{1-p}(t).
\end{aligned}$$

Substituting into (2.123) yields

$$\int_{0}^{\sigma(t)} (\omega^{\sigma})^{p}\Delta y \leq \frac{M}{M-p(M-1)}\sigma^{1-p}(t)\left(\int_{0}^{\sigma(t)} \omega^{\sigma}(s)\Delta s\right)^{p},$$

and hence (since ω is nonincreasing), we get that

$$\frac{1}{\sigma(t)}\int_0^{\sigma(t)}(\omega^\sigma(y))^p\Delta y \le \frac{M}{M-p(M-1)}\left(\frac{1}{\sigma(t)}\int_0^{\sigma(t)}\omega^\sigma(s)\Delta s\right)^p$$
$$\le \frac{M}{M-p(M-1)}\left(\frac{1}{\sigma(t)}\int_0^{\sigma(t)}\omega(s)\Delta s\right)^p,$$

which is the desired inequality (2.118). The proof is complete. □

Notice that for all nonnegative and nonincreasing functions $\omega \in \mathbb{L}^q_\Delta[0,\infty)_\mathbb{T}$ with $q > 1$, we always have

$$\mathscr{A}\omega^q(t) = \frac{1}{t}\int_0^t \omega^q(s)\Delta s = \frac{1}{t}\int_0^t \omega^{q-1}(s)\omega(s)\Delta s \ge \frac{\omega^{q-1}(t)}{t}\int_0^t \omega(s)\Delta s. \tag{2.124}$$

Let us now consider the class of nonnegative and nonincreasing functions $\omega \in \mathbb{L}^q_\Delta[0,\infty)_\mathbb{T}$ that satisfy the reverse of (2.134), namely

$$\mathscr{A}\omega^q \le \eta\omega^{q-1}\mathscr{A}\omega, \quad \text{for some } \eta > 1. \tag{2.125}$$

Theorem 2.38. *Assume that ω is a nonnegative and nonincreasing weight such that* (2.125) *holds for $q > 1$. Then*

$$\mathscr{A}(\omega^p)^\sigma \le \tilde{\eta}[\mathscr{A}\omega^q]^{p/q}, \quad \tilde{\eta} := \frac{\lambda\eta_q^{1+p/q}}{\lambda\eta_q - \frac{p}{q}(\lambda\eta_q - 1)}, \quad \eta_q = \frac{\eta q}{q-1}, \tag{2.126}$$

for $p \in [q, q+c]$, $c \in (q, \eta)$.

Proof. In this proof, we write $W = \mathscr{A}\omega^q$ for brevity. By using the Hölder inequality with exponents $q/(q-1)$ and q, we obtain

$$\frac{1}{t}\int_0^t W(\sigma(s))\Delta s$$
$$\le \frac{\eta}{t}\int_0^t (\omega(\sigma(s)))^{q-1}\cdot\frac{1}{\sigma(s)}\int_0^{\sigma(s)}\omega(\tau)\Delta\tau\Delta s$$
$$\le \frac{\eta}{t}\left[\int_0^t (\omega(\sigma(s)))^q\Delta s\right]^{(q-1)/q}\left[\int_0^t\left(\frac{1}{\sigma(s)}\int_0^{\sigma(s)}\omega(\tau)\Delta\tau\right)^q\Delta s\right]^{1/q}$$
$$\le \frac{\eta q}{(q-1)t}\left[\int_0^t (\omega(s))^q\Delta s\right]^{(q-1)/q}\left[\int_0^t (\omega(s))^q\Delta s\right]^{1/q}$$

$$= \frac{\eta_q}{t}\int_0^t \omega^q(s)\Delta s = \eta_q W(t),$$

i. e.,

$$\mathscr{A} W^\sigma \le \eta_q W. \tag{2.127}$$

Since ω is also nonnegative and nonincreasing, it satisfies the assumptions of Theorem 2.30 and thus,

$$\frac{1}{t}\int_0^t [W(\sigma(s))]^r \Delta s \le \tilde{\eta}_q \left[\frac{1}{t}\int_0^t W(\sigma(s))\Delta s\right]^r, \tag{2.128}$$

with

$$\tilde{\eta}_q = \frac{\alpha_q}{\alpha_q - r(\alpha_q - 1)}, \quad \alpha_q = \lambda \eta_q, \quad r = \frac{p}{q} \in [1, \frac{\alpha_q}{\alpha_q - 1}).$$

Noting that

$$W(t) = \frac{1}{t}\int_0^t \omega^q(s)\Delta s \ge \omega^q(t), \tag{2.129}$$

we obtain

$$\begin{aligned}\frac{1}{t}\int_0^t (\omega(\sigma(s)))^p \Delta s &= \frac{1}{t}\int_0^t (\omega^q(\sigma(s)))^r \Delta s \le \frac{1}{t}\int_0^t (W(\sigma(s)))^r \Delta s \\ &\le \tilde{\eta}_q \left(\frac{1}{t}\int_0^t W^\sigma(s)\Delta s\right)^r \\ &\le \tilde{\eta}_q \eta_q^r [W(t)]^r = \tilde{\eta}[W(t)]^r = \tilde{\eta}\left[\frac{1}{t}\int_0^t \omega^q(s)\Delta s\right]^{p/q},\end{aligned}$$

proving (2.136). □

In Theorem 2.38, if $\mathbb{T} = \mathbb{R}$, then we have that $\sigma(t) = t$, $\alpha_q = \eta_q$, and we get the following result.

Corollary 2.6. *Let $\eta > 1$ and $q > 1$ and ω is a nonnegative nonincreasing weight satisfying*

$$\int_0^t \omega^q(x)dx \le \eta \omega^{q-1}(t)\int_0^t \omega(x)dx,$$

then

$$\frac{1}{t}\int_0^t \omega^p(x)\mathrm{d}x \le \tilde{\eta}\left(\frac{1}{t}\int_0^t \omega^q(x)\mathrm{d}x\right)^{p/q},$$

for $p \in [q, q+c]$, $c \in (q, \eta)$, *and*

$$\tilde{\eta} := \frac{(\frac{\eta q}{q-1})^{\frac{p}{q}+1}}{\frac{\eta q}{q-1} - \frac{p}{q}(\frac{\eta q}{q-1} - 1)}.$$

In the following, we prove a new extension of the Hardy inequality on time scales with reiteration term on a finite interval. The results are adapted from [59]. The proof depends on the applications of the Hölder inequality, the chain rules on time scales, and the algebraic inequality

$$(\vartheta + v)^p \ge \vartheta^p + p\vartheta^{p-1}v \quad \text{if } p < 0 \text{ or } p > 1. \tag{2.130}$$

This inequality is valid for all $\vartheta \ge 0$ and $\vartheta + v \ge 0$ (if $p > 0$), or $\vartheta > 0$ and $\vartheta + v > 0$ (if $p < 0$) and equality holds if only if $v = 0$.

Theorem 2.39. *Assume that* ω *is nonincreasing. Then for* $q > 1$, *we have*

$$\mathscr{A}(\mathscr{A}^\sigma\omega)^q + \left|\frac{q}{q-1}\right|(\mathscr{A}\omega)^q \le \left(\frac{q}{q-1}\right)^q \mathscr{A}(\omega^q). \tag{2.131}$$

Proof. In this proof, we write $\mathscr{W} = \mathscr{A}\omega$ for brevity. By the chain rule, we obtain

$$\begin{aligned}(\mathscr{W}^q)^\Delta &= q(\mathscr{W})^\Delta \int_0^1 (h\mathscr{W}^\sigma + (1-h)\mathscr{W})^{q-1} dh,\\ &\le q\mathscr{W}^\Delta \int_0^1 (h\mathscr{W}^\sigma + (1-h)\mathscr{W}^\sigma)^{q-1} dh = q\mathscr{W}^\Delta (\mathscr{W}^\sigma)^{q-1}.\end{aligned}$$

Moreover, since $t\mathscr{W}(t) = \int_0^t \omega(s)\Delta s$, the product rule in (2.12) yields that

$$t\mathscr{W}^\Delta(t) + \mathscr{W}^\sigma(t) = \omega(t), \quad \text{where } \mathscr{W}^\sigma = \mathscr{A}^\sigma\omega.$$

Now, putting $\vartheta = t$ and $v = \mathscr{W}^q$, and using integration by parts, we find that

$$\int_0^t (\mathscr{W}^\sigma)^q \Delta s = tv(t) - \lim_{s\to 0^+} sv(s) - \int_0^t sv^\Delta(s)\Delta s.$$

By applying the Hölder inequality with exponents q and $q/(q-1)$, we see that

$$tv(t) = \frac{1}{t^{q-1}}\left(\int_0^t \omega(s)\Delta s\right)^q \leq \frac{1}{t^{q-1}}\left(t^{\frac{q-1}{q}}\left(\int_0^t \omega^q(s)\Delta s\right)^{1/q}\right)^q = \int_0^t \omega^q(s)\Delta s,$$

implying (note that $tv(t) \geq 0$) $\lim_{t\to 0^+} tv(t) = 0$. This and the fact that $\mathscr{W}$ is decreasing imply that

$$\begin{aligned}\int_0^t (\mathscr{W}^\sigma)^q \Delta s &= t\mathscr{W}^q(t) - \int_0^t s v^\Delta(s)\Delta s \geq t\mathscr{W}^q(t) - q\int_0^t s\mathscr{W}^\Delta(s)(\mathscr{W}^\sigma(s))^{q-1}\Delta s \\ &= t\mathscr{W}^q(t) - q\int_0^t [\omega(s) - \mathscr{W}^\sigma(s)](\mathscr{W}^\sigma(s))^{q-1}\Delta s \\ &= t\mathscr{W}^q(t) - q\int_0^t (\mathscr{W}^\sigma(s))^{q-1}\omega(s)\Delta s + q\int_0^t (\mathscr{W}^\sigma(s))^q \Delta s.\end{aligned}$$

By applying the Hölder inequality with exponents q and $q/(q-1)$ on the term

$$\int_0^t (\mathscr{W}^\sigma(s))^{q-1}\omega(s)\Delta s,$$

we get

$$\begin{aligned}&(q-1)\int_0^t (\mathscr{W}^\sigma(s))^q \Delta s + t\mathscr{W}^q(t) \\ &\quad \leq q\int_0^t (\mathscr{W}^\sigma(s))^{q-1}\omega(s)\Delta s \leq q\left[\int_0^t \omega^q(s)\Delta s\right]^{1/q}\left[\int_0^t (\mathscr{W}^\sigma(s))^q \Delta s\right]^{(q-1)/q},\end{aligned}$$

i. e.,

$$\begin{aligned}&\left(\frac{q}{q-1}\right)^q \int_0^t \omega^q \Delta s \\ &\quad \geq \left[\left(\int_0^t (\mathscr{W}^\sigma(s))^q \Delta s\right)^{1/q} + \frac{1}{(q-1)}\frac{t\mathscr{W}^q(t)}{[\int_0^t (\mathscr{W}^\sigma(s))^q \Delta s]^{(q-1)/q}}\right]^q \\ &\quad \geq \int_0^t (\mathscr{W}^\sigma(s))^q \Delta s + \frac{q}{(q-1)}\frac{t\mathscr{W}^q(t)}{[\int_0^t (\mathscr{W}^\sigma(s))^q \Delta s]^{(q-1)/q}}\left[\int_0^t (\mathscr{W}^\sigma(s))^q \Delta s\right]^{(q-1)/q}.\end{aligned}$$

This implies

$$\frac{1}{t}\int_0^t \left(\mathscr{W}^\sigma(s)\right)^q \Delta s + \frac{q}{(q-1)}\mathscr{W}^q(t) \le \left(\frac{q}{q-1}\right)^q \frac{1}{t}\int_0^t \omega^q(s)\Delta s,$$

which is the desired inequality (2.131). The proof is complete. □

As a special case of Theorem 2.39, if $\mathbb{T} = \mathbb{R}$, we obtain the following Shum's inequality [64].

Corollary 2.7. *If $1 < q$, then for $0 < T < \infty$, we have*

$$\frac{1}{T}\int_0^T \left(\frac{1}{t}\int_0^t \omega(s)ds\right)^q \Delta t + \frac{q}{q-1}\left(\frac{1}{T}\int_0^T \omega(t)dt\right)^q$$
$$\le \left(\frac{q}{q-1}\right)^q \frac{1}{T}\int_0^T \omega^q(t)dt.$$

Remark 2.9. From the inequality (2.131), we get a Hardy-type inequality for decreasing functions on a finite time scales interval, namely

$$\frac{1}{T}\int_0^T \left(\frac{1}{\sigma(t)}\int_0^{\sigma(t)} \omega(s)\Delta s\right)^q \Delta t \le \left(\frac{q}{q-1}\right)^q \frac{1}{T}\int_0^T \omega^q(t)\Delta t. \tag{2.132}$$

In Theorem 2.39 if $\mathbb{T} = \mathbb{N}$, we have that $\sigma(n) = n+1$, and then we get the following result.

Corollary 2.8. *If $1 < q$ then for $0 < N < \infty$, we have*

$$\frac{1}{N}\sum_{n=0}^{N-1}\left(\frac{1}{n+1}\sum_{k=0}^{n}\omega(k)\right) + \frac{q}{(q-1)}\left(\frac{1}{N}\sum_{k=0}^{N-1}\omega(k)\right)^q \le \left(\frac{q}{q-1}\right)^q \frac{1}{N}\sum_{n=0}^{N-1}\omega^q(n).$$

From the proof of Theorem 2.39, we see that the inequality (2.131) is still satisfied if we replace $\mathscr{A}$ by $\mathscr{A}^\sigma$, and we thus have the following result.

Corollary 2.9. *Assume that ω is a nonnegative and nonincreasing weight. Then for $q > 1$, we have*

$$\mathscr{A}^\sigma(\mathscr{A}^\sigma\omega)^q(t) + \left|\frac{q}{q-1}\right|(\mathscr{A}^\sigma\omega(t))^q \le \left(\frac{q}{q-1}\right)^q \mathscr{A}^\sigma\omega^q(t). \tag{2.133}$$

In the following, we prove a higher integrability result for monotone decreasing functions by employing the inequalities (2.118) and (2.133). The results are adapted from [52]. For all nonnegative and nonincreasing functions ω, the following inequality holds:

$$\omega^{q-1}(\sigma(t))\mathscr{A}^\sigma\omega(t) \le \mathscr{A}^\sigma\omega^q(t). \tag{2.134}$$

For $C > 1$ and $q > 1$, let us consider the class $L^q(0,\infty)_{\mathbb{T}}$ of nonnegative nonincreasing functions which satisfy

$$\mathscr{A}^\sigma \omega^q(t) \le C(\omega^\sigma(t))^{q-1} \mathscr{A}^\sigma \omega(t). \tag{2.135}$$

Notice that the inequality (2.125) is the reverse of the inequality (2.134).

Theorem 2.40. *If ω is a positive decreasing weight satisfying* (2.135) *for $C > 1$, then*

$$\left(\mathscr{A}^\sigma (\omega^\sigma)^p\right)^{1/p} \le R\left(\mathscr{A}^\sigma \omega^q\right)^{1/q}, \tag{2.136}$$

for all $p \in [q, q+\delta]$, where

$$R := \left[\frac{\beta_\lambda^{r+1}}{\beta_\lambda - r(\beta_\lambda - 1)}\right]^{1/p}, \quad r = \frac{p}{q}, \quad \delta = \frac{q}{(\beta - 1)}, \quad \beta_\lambda = \lambda\beta,$$

and

$$\beta := \left[C^q\left(\frac{q}{q-1}\right)^q - \frac{qC^{q-1}}{q-1}\right]^{1/q} > 1.$$

Proof. Since ω is decreasing and $\sigma(t) \ge t$, we see that $(\omega^\sigma(t))^{q-1} \le \omega^{q-1}(t)$. From this and (2.125), we get that

$$\frac{1}{\sigma(t)} \int_0^{\sigma(t)} \omega^q(s)\Delta s \le C\omega^{q-1}(t) \frac{1}{\sigma(t)} \int_0^{\sigma(t)} \omega(s)\Delta s. \tag{2.137}$$

Let us integrate (2.137) between 0 and $\sigma(y) \in (0,\infty)_{\mathbb{T}}$, to get

$$\int_0^{\sigma(y)} \frac{1}{\sigma(t)} \int_0^{\sigma(t)} \omega^q(s)\Delta s\Delta t \le C \int_0^{\sigma(y)} \omega^{q-1}(t) \frac{1}{\sigma(t)} \int_0^{\sigma(t)} \omega(s)\Delta s\Delta t.$$

Applying the Hölder inequality with indices q and $q/(q-1)$, we obtain

$$\begin{aligned}\left(\int_0^{\sigma(y)} \frac{1}{\sigma(t)} \int_0^{\sigma(t)} \omega^q(s)\Delta s\Delta t\right) &\le C \int_0^{\sigma(y)} \omega^{q-1}(t) \frac{1}{\sigma(t)} \int_0^{\sigma(t)} \omega(s)\Delta s\Delta t \\ &\le C\left(\int_0^{\sigma(y)} (\omega^{q-1}(t))^{q/(q-1)} \Delta t\right)^{\frac{q-1}{q}} \\ &\quad \times \left(\int_0^{\sigma(y)} \left(\frac{1}{\sigma(t)} \int_0^{\sigma(t)} \omega(s)\Delta s\right)^q \Delta t\right)^{\frac{1}{q}}.\end{aligned} \tag{2.138}$$

This implies that

$$\left(\int_0^{\sigma(y)} \frac{1}{\sigma(t)} \int_0^{\sigma(t)} \omega^q(s)\Delta s\Delta t\right)^q$$
$$\leq C^q\left(\int_0^{\sigma(y)} \omega^q(t)\Delta t\right)^{q-1} \int_0^{\sigma(y)} \left(\frac{1}{\sigma(t)} \int_0^{\sigma(t)} \omega(s)\Delta s\right)^q \Delta t. \tag{2.139}$$

Applying (2.133) on the second term of the right-hand side, we see that

$$\int_0^{\sigma(y)} \left(\frac{1}{\sigma(t)} \int_0^{\sigma(t)} \omega(s)\Delta s\right)^q \Delta t$$
$$\leq \left(\frac{q}{q-1}\right)^q \int_0^{\sigma(y)} \omega^q(t)\Delta t - \frac{q}{q-1} \frac{(\int_0^{\sigma(y)} \omega(t)\Delta t)^q}{\sigma^{q-1}(y)},$$

This and (2.139) imply that

$$\left(\frac{1}{\sigma(y)} \int_0^{\sigma(y)} \frac{1}{\sigma(t)} \int_0^{\sigma(t)} \omega^q(s)\Delta s\Delta t\right)^q$$
$$\leq C^q\left(\int_0^{\sigma(y)} \omega^q(t)\Delta t\right)^{q-1} \left[\left(\frac{q}{q-1}\right)^q \int_0^{\sigma(y)} \omega^q(t)\Delta t - \frac{q}{q-1} \frac{(\int_0^{\sigma(y)} \omega(t)\Delta t)^q}{\sigma^{q-1}(y)}\right]$$
$$= \left(\frac{Cq}{q-1}\right)^q \left(\frac{1}{\sigma(y)} \int_0^{\sigma(y)} \omega^q(t)\Delta t\right)^q \left[1 - \left(\frac{q-1}{q}\right)^{q-1} \frac{(\frac{1}{\sigma(y)} \int_0^{\sigma(y)} \omega(t)\Delta t)^q}{\frac{1}{\sigma(y)}(\int_0^{\sigma(y)} \omega^q(t)\Delta t)}\right]. \tag{2.140}$$

Setting

$$\Lambda(t) := \frac{1}{t} \int_0^t \omega^q(s)\Delta s, \tag{2.141}$$

we get from (2.140) that

$$\left(\frac{1}{\sigma(y)} \int_0^{\sigma(y)} \Lambda^\sigma(t)\Delta t\right)^q$$
$$\leq \left(\frac{Cq}{q-1}\Lambda^\sigma(y)\right)^q \left[1 - \left(\frac{q-1}{q}\right)^{q-1} \frac{(\frac{1}{\sigma(y)} \int_0^{\sigma(y)} \omega(t)\Delta t)^q}{\frac{1}{\sigma(y)} \int_0^{\sigma(y)} \omega^q(t)\Delta t}\right]. \tag{2.142}$$

Since ω is nonincreasing, we have

$$\omega^{\sigma}(y) \le \frac{1}{\sigma(y)} \int_0^{\sigma(y)} \omega(t)\Delta t. \tag{2.143}$$

From (2.125) it follows that

$$\begin{aligned}\frac{(\frac{1}{\sigma(y)}\int_0^{\sigma(y)} \omega(t)\Delta t)^q}{\frac{1}{\sigma(y)}\int_0^{\sigma(y)} \omega^q(t)\Delta t} &\ge \frac{(\frac{1}{\sigma(y)}\int_0^{\sigma(y)} \omega(t)\Delta t)^q}{C\omega^{q-1}(\sigma(y))\frac{1}{\sigma(y)}\int_0^{\sigma(y)} \omega(s)\Delta s} \\ &= \frac{1}{C}\left(\frac{\frac{1}{\sigma(y)}\int_0^{\sigma(y)} \omega(t)\Delta t}{\omega^{\sigma}(y)}\right)^q.\end{aligned}$$

This and (2.143) imply

$$\left(\frac{1}{\sigma(y)} \int_0^{\sigma(y)} \omega(t)\Delta t\right)^q \ge \frac{1}{C}\frac{1}{\sigma(y)} \int_0^{\sigma(y)} \omega^q(t)\Delta t.$$

Substituting into (2.142), we have that

$$\left(\frac{1}{\sigma(y)} \int_0^{\sigma(y)} \Lambda^{\sigma}(t)\Delta t\right)^q \le \left[1 - \frac{1}{C}\left(\frac{q-1}{q}\right)^{q-1}\right]\left(\frac{Cq}{q-1}\Lambda^{\sigma}(y)\right)^q.$$

This leads to

$$\frac{1}{\sigma(y)} \int_0^{\sigma(y)} \Lambda^{\sigma}(t)\Delta t \le \beta\Lambda^{\sigma}(y), \tag{2.144}$$

with

$$\beta = \left[C^q\left(\frac{q}{q-1}\right)^q - C^{q-1}\left(\frac{q}{q-1}\right)\right]^{1/q}.$$

From, the inequality $x^q - y^q \ge qy^{q-1}(x-y)$ for $q \ge 1$, $x > y > 0$, we see that $x^q - 1 > q(x-1)$ for $q > 1$. This implies, by setting $x = (q/(q-1))$, that

$$\left(\frac{q}{q-1}\right)^q - \left(\frac{q}{q-1}\right) > 1 \quad \text{for } q > 1.$$

Since the weight

$$G(t) := t^q\left(\frac{q}{q-1}\right)^q - t^{q-1}\left(\frac{q}{q-1}\right)$$

is strictly increasing on $[1, \infty)$, we conclude for $C > 1$ that

$$\beta = \left[C^q\left(\frac{q}{q-1}\right)^q - C^{q-1}\left(\frac{q}{q-1}\right)\right]^{1/q} > 1.$$

Since ω^q is decreasing, it follows that $\Lambda(t)$ is also decreasing (see Lemma 2.15) and satisfies (2.144). So we can apply the inequality (2.118) with Λ instead of ω and β_λ instead of A to get for all $r \in [1, \beta_\lambda/(\beta_\lambda - 1))$ that

$$\mathscr{A}^\sigma(\Lambda^\sigma)^r \leq \frac{\beta_\lambda}{\beta_\lambda - r(\beta_\lambda - 1)}(\mathscr{A}^\sigma(\Lambda^\sigma))^r, \quad \text{where } \beta_\lambda = \lambda\beta.$$

By using the fact that

$$(\omega^\sigma(s))^{qr} \leq \left(\frac{1}{\sigma(s)}\int_0^{\sigma(s)} \omega^q(\vartheta)\Delta\vartheta\right)^r = (\Lambda^\sigma(s))^r,$$

we obtain

$$\begin{aligned}\frac{1}{\sigma(t)}\int_0^{\sigma(t)} \omega^{rq}(\sigma(s))\Delta s &\leq \frac{1}{\sigma(t)}\int_0^{\sigma(t)} (\Lambda^\sigma)^r \Delta s \\ &\leq \frac{\beta_\lambda^{r+1}}{\beta_\lambda - r(\beta_\lambda - 1)}\left(\frac{1}{\sigma(t)}\int_0^{\sigma(t)} \Lambda^\sigma(s)\Delta s\right)^r.\end{aligned}$$

From this and (2.144), we see that

$$\mathscr{A}^\sigma(\omega^\sigma)^{rq} \leq \frac{\beta_\lambda^{r+1}}{\beta_\lambda - r(\beta_\lambda - 1)}(\mathscr{A}^\sigma(\omega^q))^r.$$

Putting $r = p/q$, we get the desired inequality (2.136). The proof is complete. □

In Theorem 2.38 if $\mathbb{T} = \mathbb{R}$, we have that $\sigma(t) = t$ and $\lambda = 1$ and then we get the following result due to Alzer.

Corollary 2.10. *Let $C > 1$, $q > 1$, and let ω be a positive decreasing weight satisfying*

$$\frac{1}{t}\int_0^t \omega^q(s)ds \leq C\omega^{q-1}(t)\frac{1}{t}\int_0^t \omega(s)ds.$$

Then

$$\frac{1}{T}\int_0^T \omega^p(s)ds \leq M^*\left(\frac{1}{T}\int_0^T \omega^q(s)ds\right)^{p/q}$$

for all $p \in [q, q+\delta]$, *where* $\delta = q/(\beta - 1)$,

$$M^* = \frac{\beta^{r+1}}{\beta - r(\beta - 1)}, \quad \text{and} \quad \beta := \left(C^q\left(\frac{q}{q-1}\right)^q - C^{q-1}\frac{q}{q-1}\right)^{1/q}.$$

In Theorem 2.38 if $\mathbb{T} = \mathbb{N}$, we have that $\sigma(n) = n+1$ and, by considering $\lambda = 2$, we get the following.

Corollary 2.11. *Let* $C > 1$, $q > 1$, *and let* $a(n)$ *be a positive decreasing sequence which satisfies*

$$\frac{1}{n+1}\sum_{i=0}^{n} a^q(i) \le Ca^{q-1}(n+1)\frac{1}{n+1}\sum_{i=0}^{n} a(i),$$

then

$$\frac{1}{N+1}\sum_{i=1}^{N} a^p(i+1) \le M_*\left(\frac{1}{N+1}\sum_{i=0}^{N} a^q(i)\right)^{p/q},$$

for all $p \in [q, q+\delta]$, *where* $\delta = q/(\beta_* - 1)$,

$$M_* = \frac{\beta_*^{r+1}}{\beta_* - r(\beta_* - 1)}, \quad \text{and} \quad \beta_* := 2\left(C^q\left(\frac{q}{q-1}\right)^q - C^{q-1}\frac{q}{q-1}\right)^{1/q}.$$

Bibliography

[1] R. P. Agarwal, D. O'Regan and S. H. Saker, Dynamic Inequalities on Time Scales, Springer, Heidlelberg–New York–Dordrecht–London, 2014.

[2] R. P. Agarwal, D. O'Regan and S. H. Saker, Hardy Type Inequalities on Time Scales, Springer, Cham, 2016.

[3] H. Alzer, An improvement of the constants in a reverse integral inequality, *J. Math. Anal. Appl.*, 190 (1995), 774–779.

[4] M. A. Ariño and B. Muckenhoupt, A characterization of the dual of the classical Lorentz sequence space $d(w,q)$, *Proc. Am. Math. Soc.*, 112 (1991), 87–89.

[5] L. Basile, L. D'Apuzzo and M. Squillante, The limit class of Gehring type G^∞, *Boll. Unione Mat. Ital.*, 11.B (1997), 871–884.

[6] G. Bennett and K.-G. Grosse-Erdmann, Weighted Hardy inequalities for decreasing sequences and functions, *Math. Ann.*, 334 (2006), 489–531.

[7] J. Bober, E. Carneiro, K. Hughes and L. B. Pierce, On a discrete version of Tanaka's theorem for maximal functions, *Proc. Am. Math. Soc.*, 140 (2012), 1669–1680.

[8] M. Bohner and A. Peterson, Dynamic Equations on Time Scales: An Introduction with Applications, Birkhäuser, Boston, Mass, USA, 2001.

[9] M. Bohner and A. Peterson, Advances in Dynamic Equations on Time Scales, Birkhäuser, Boston, 2003.

[10] M. Bohner and S. H. Saker, Gehring inequalities on time scales, *J. Comput. Anal. Appl.*, 28 (2020), 11–23.

[11] B. Bojarski, C. Sbordone and I. Wik, The Muckenhoupt class $A_1(R)$, *Studia Mat. VII*, 10 (1992), 155–163.

[12] A. Böttcher and M. Seybold, Wackelsatz and Stechkin's inequality for discrete Muckenhoupt weights, Preprint no. 99-7, TU Chemnitz (1999).
[13] E. Carneiro and K. Hughes, On the endpoint regularity of discrete maximal operators, *Math. Res. Lett.*, 19 (2012), 1245–1262.
[14] E. Carneiro and J. Madrid, Derivative bounds for fractional maximal functions, *Trans. Am. Math. Soc.*, 369 (2017), 4063–4092.
[15] R. R. Coifman and C. Fefferman, Weighted norm inequalities for maximal functions and singular integrals, *Stud. Math.*, 51 (1974), 241–250.
[16] R. Corporente, A precise relation among A_∞ and G_1 constants in one dimension, *Rend. Accad. Sci. Fis. Mat.*, LXXII (2005), 65–70.
[17] D. Cruz-Uribe and C. J. Neugebauer, The structure of the reverse Hölder classes, *Trans. Am. Math. Soc.*, 347(8) (1995), 2941–2960.
[18] L. D'Apuzzo and C. Sbordone, Reverse Hölder inequalities. A sharp result, *Rend. Mat. Ser. VII*, 10 (1990), 357–366.
[19] M. Dindoš and T. Wall, The sharp A^p constant for weights in a reverse Hölder class, *Rev. Mat. Iberoam.*, 25 (2009), 559–594.
[20] M. Franciosi and G. Moscariello, Higher integrability results, *Manuscr. Math.*, 52 (1985), 151–170.
[21] F. W. Gehring, The L^p-integrability of the partial derivatives of a quasi-conformal mapping, *Acta Math.*, 130 (1973), 265–277.
[22] F. W. Gehring, The L^p-integrability of the partial derivatives of a quasi-conformal mapping, *Bull. Am. Math. Soc.*, 79 (1973), 465–466.
[23] G. H. Hardy, J. E. Littlewood and G. Pólya, Some simple inequalities satisfied by convex functions, *Messenger Math.*, 58 (1929), 145–152.
[24] G. H. Hardy, J. E. Littlewood and G. Polya, Inequalities, 2nd ed., Cambridge University Press, 1934.
[25] T. Iwaniec, On L^p-integrability in PDE and quasiconformal mapping for large exponents, *Ann. Acad. Sci. Fenn., Ser. A*, 7 (1982), 301–322.
[26] R. Johnson and C. J. Neugebauer, Homeomorphisms preserving A_p, *Rev. Mat. Iberoam.*, 3 (1987), 249–273.
[27] A. A. Korenovskii, The exact continuation of a reverse Hölder inequality and Muckenhoupt's conditions, *Math. Notes*, 52 (1992), 1192–1201.
[28] C. E. Kenig, Harmonic analysis techniques for second order elliptic boundary value problems, Conference Board of Mathematical Science, vol. 83, Am. Math. Soc., Providence, 1991.
[29] F. Liu, Endpoint regularity of discrete multisublinear fractional maximal operators associated with l^1-balls, *J. Inequal. Appl.*, 2018 (2018), 33.
[30] Y. I. Lyubarskii and K. Seip, Complete interpolating sequences for Paley–Wiener spaces and Muckenhoupt's *(Ap)* condition, *Rev. Mat. Iberoam.*, 13 (1997), 361–376.
[31] J. Madrid, Sharp inequalities for the variation of the discrete maximal function, *Bull. Aust. Math. Soc.*, 95 (2017), 94–107.
[32] A. Magyar, E. M. Stein and S. Wainger, Discrete analogues in harmonic analysis: Spherical averages, *Ann. Math.*, 155 (2002), 189–208.
[33] O. Martio and C. Sbordone, Quasiminimizers in one dimension: Integrability of the derivative, inverse weight and obstacle problems, *Ann. Mat. Pura Appl.*, 186 (2007), 579–590.
[34] N. A. Malaksiano, The exact inclusions of Gehring classes in Muckenhoupt classes, *Mat. Zametki*, 70(5) (2001), 742–750; (Russian), translation in *Math. Notes* 70(5–6) (2001), 673–681.
[35] N. A. Malaksiano, The precise embeddings of one-dimensional Muckenhoupt classes in Gehring classes, *Acta Sci. Math.*, 68(1–2) (2002), 237–248.
[36] M. Milman, A note on Gehring's lemma, *Ann. Acad. Sci. Fenn., Ser. A 1 Math.*, 21 (1996), 389–398.
[37] B. Muckenhoupt, Weighted norm inequalities for the Hardy maximal function, *Trans. Am. Math. Soc.*, 165 (1972), 207–226.
[38] L. Nania, On some reverse integral inequalities, *J. Aust. Math. Soc. A*, 49 (1990), 319–326.

[39] B. S. Pavlov, Basicity of an exponential system and Muckenhoupt's condition, *Dokl. Akad. Nauk SSSR*, 247(1) (1979), 37–40; English translation in *Sov. Math. Dokl.* 20(4) (1979), 655–659.

[40] A. Popoli, Optimal integrability in B_p^q classes, *Le Matematiche*, LII (1997), 157–168.

[41] A. Popoli, Sharp integrability exponents and constants for Muckenhoupt and Gehring weights as solution to a unique equation, *Ann. Acad. Sci. Fenn., Math.*, 43 (2018), 785–805.

[42] E. M. Stein and S. Wainger, Discrete analogues in harmonic analysis I: l^2-estimates for singular Radon transforms, *Am. J. Math.*, 121 (1999), 1291–1336.

[43] E. M. Stein and S. Wainger, Discrete analogues in harmonic analysis II: Fractional integration, *J. Anal. Math.*, 80 (2000), 335–355.

[44] E. M. Stein and S. Wainger, Two discrete fractional integral operators revisited, *J. Anal. Math.*, 87 (2002), 451–479.

[45] S. H. Saker and R. P. Agarwal, Discrete Hardy's type inequalities and structure of discrete class of weights satisfy reverse Hölder's inequality, *Math. Inequal. Appl.*, 24(2) (2021), 521–541.

[46] S. H. Saker and R. P. Agarwal, Theory of discrete Muckenhoupt weights and discrete Rubio de Francia extrapolation theorems, *Appl. Anal. Discrete Math.*, 15 (2021), 295–316.

[47] S. H. Saker, J. Alzabut, D. O'Regan and R. P. Agarwal, Self-improving properties of weighted Gehring classes with applications to partial differential equations, *Adv. Differ. Equ.*, 2021 (2021), 1–20.

[48] S. H. Saker, D. Baleanu and M. Krnić, Some further properties of discrete Muckenhoupt and Gehring weights, *J. Math. Inequal.*, 16 (2022), 1–18.

[49] S. H. Saker, M. A. Darwish, H. A. Elshamy and R. P. Agarwal, Fundamental properties of Muckenhoupt and Gehring weights on time scales, *J. Anal.* (submitted).

[50] S. H. Saker and M. Krnić, The weighted discrete Gehring classes, Muckenhoupt classes and their basic properties, *Proc. Am. Math. Soc.*, 149 (2021), 231–243.

[51] S. H. Saker, M. Krnić and J. Pečarić, Higher summability theorems from the weighted reverse discrete inequalities, *Appl. Anal. Discrete Math.*, 13 (2019), 423–439.

[52] S. H. Saker and I. Kubiaczyk, Reverse dynamic inequalities and higher integrability theorems, *J. Math. Anal. Appl.*, 471(1–2) (2019), 671–686.

[53] S. H. Saker and I. Kubiaczyk, Higher summability and discrete weighted Muckenhoupt and Gehring type inequalities, *Proc. Edinb. Math. Soc.*, 62 (2019), 949–973.

[54] S. H. Saker and R. R. Mahmoud, Boundedness of both discrete Hardy and Hardy–Littlewood maximal operators via Muckenhoupt weights, *Rocky Mt. J. Math.*, 51(2) (2021), 733–746.

[55] S. H. Saker, D. O'Regan and R. P. Agarwal, Self-improving properties of discrete Muckenhoupt weights, *Analysis*, 41 (2021), 169–178.

[56] S. H. Saker, D. O'Regan and R. P. Agarwal, A higher integrability theorem from a reverse weighted inequality, *Bull. Lond. Math. Soc.*, 51(6) (2019), 967–977.

[57] S. H. Saker, D. O'Regan and R. P. Agarwal, Self-improving properties of a generalized Muckenhoupt class, *Acta Math. Hungr.*, 164 (2021), 113–134.

[58] S. H. Saker, S. S. Rabie, G. AlNemer and M. Zakarya, On structure of discrete Muckenhoupt and discrete Gehring classes, *J. Inequal. Appl.*, 2020 (2020), 1–18.

[59] S. H. Saker, D. O'Regan, M. M. Osman and R. P. Agarwal, New dynamic inequalities for decreasing functions and theorems of higher integrability, *Ann. Appl. Math.*, 34(2) (2018), 165–177.

[60] S. H. Saker, M. M. Osman and M. Krnić, Higher integrability theorems on time scales from reverse Hölder's inequalities, *Appl. Anal. Discrete Math.*, 13 (2019), 819–838.

[61] S. H. Saker, S. S. Rabie, J. Alzabut, D. O'Regan and R. P. Agarwal, Some basic properties and fundamental relations for discrete Muckenhoupt and Gehring classes, *Adv. Differ. Equ.*, 2021(1) (2021), 1–22.

[62] S. H. Saker, S. S. Rabie and M. Krnić, Sharp self-improving property of discrete Muckenhoupt and Gehring weights, *Rev. R. Acad. Cienc. Exactas*, 115 (2021), 119.

[63] S. H. Saker, S. S. Rabie and R. P. Agarwal, Properties of a generalized class of weights satisfying reverse Hölder's inequality, *J. Funct. Spaces*, 2021 (2021), 1–16.

[64] D. T. Shum, On integral inequalities related to Hardy's, *Can. Math. Bull.*, 14 (1971), 225–230.
[65] C. Sbordone and I. Wik, Maximal functions and related weight classes, *Publ. Mat.*, 38 (1994), 127–155.
[66] C. Sbordone and G. Zecca, The L^p-Solvability of the Dirichlet problem for planar elliptic equations, sharp results, *J. Fourier Anal. Appl.*, 15 (2009), 871–903.
[67] V. Vasyunin, The exact constant in the inverse Hölder inequality for Muckenhoupt weights (Russian), *Algebra Anal.*, 15 (2003), 73–117; translation in *St. Petersburg Math. J.* 15(1) (2004), 49–79.
[68] L. Xing-Min, On the factorization of A_p-weights, *Proc. Am. Math. Soc.*, 121 (1994), 1075–1078.

Douglas R. Anderson and Masakazu Onitsuka

3 Ulam stability and instability of first-order linear 1- and 2-periodic dynamic equations on isolated time scales

Abstract: We apply a new definition of periodicity on isolated time scales introduced by Bohner, Mesquita, and Streipert to the study of Ulam stability. If the graininess (step size) of an isolated time scale is bounded by a finite constant, then the linear 1- and 2-periodic dynamic equations are Ulam stable if and only if the exponential function has modulus different from unity. If the graininess increases at least linearly to infinity, the 1- and 2-periodic dynamic equations are not Ulam stable. Applying these results, we give several interesting examples of first-order linear 1- or 2-periodic dynamic equations on specific isolated time scales such as h-difference equations, q-difference equations, triangular equations, Fibonacci equations, and harmonic equations. In some cases the minimum Ulam stability constant is found.

3.1 Introduction

Bohner, Mesquita, and Streipert [10] recently introduced the idea of periodicity of functions on isolated time scales by focusing on repeated area under the curve rather than repeated function values; see also the earlier paper by Bohner and Chieochan [9], which establishes this concept first for q-difference equations. The nature of this new notion of periodicity is distinct once the graininess (step size) of the isolated time scale is non-constant, making even 1-periodic functions new and interesting.

In this work we explore the Ulam stability and instability of first-order linear dynamic equations over isolated time scales, where the coefficient function is 1- or 2-periodic in the sense of [10]. Stability analysis that variously goes by the names Ulam stability [30], Hyers–Ulam stability [17], Hyers–Ulam–Rassias stability [27] is an area of interest that differs from Lyapunov stability analysis. Brillouët-Belluot, Brzdek, and Ciepliński [12] give an overview on some recent developments in Ulam-type stability, while Brzdek, Popa, Raşa, and Xu [13] have a book on the Ulam stability of operators.

Acknowledgement: The second author was supported by the Japan Society for the Promotion of Science (JSPS) KAKENHI (grant number JP20K03668).

Douglas R. Anderson, Department of Mathematics, Concordia College, Moorhead, MN 56562, USA, e-mail: andersod@cord.edu
Masakazu Onitsuka, Department of Applied Mathematics, Okayama University of Science, Okayama, 700-0005, Japan, e-mail: onitsuka@ous.ac.jp

https://doi.org/10.1515/9783111182971-003

Brzdek and Wójcik [14] write on approximate solutions of some difference equations. Popa [24, 25] studies Hyers–Ulam and Hyers–Ulam–Rassias stability of the linear recurrence, while Rasouli, Abbaszadeh, and Eshaghi [26] study approximately linear recurrences.

Jung and Nam [18] are interested in the Hyers–Ulam stability of the Pielou logistic difference equation. Nam [19–21] has explored the Hyers–Ulam stability of hyperbolic, elliptic, and loxodromic Möbius difference equations, respectively. Fukutaka and Onitsuka [15] and Onitsuka [22, 23] have studied Hyers–Ulam stability for first-order homogeneous linear differential equations with a periodic coefficient, first-order homogeneous linear difference equations, and second-order nonhomogeneous linear difference equations, respectively. Anderson and Onitsuka [5, 6] investigate Hyers–Ulam stability for Cayley quantum equations and its application to h-difference equations, while Anderson, Onitsuka, and Rassias [7] find the best constant for Ulam stability of first-order h-difference equations with periodic coefficient. András and Mészáros [8] study Ulam–Hyers stability of dynamic equations on time scales via Picard operators. Hua, Li, and Feng [16] investigate Hyers–Ulam stability of dynamic integral equation on time scales. Shen [28] has done work on the Ulam stability of first order linear dynamic equations on time scales, while Shen and Li [29] examine Hyers–Ulam stability of first order nonhomogeneous linear dynamic equations on time scales.

The outline of the paper is the following. In Section 3.2, a brief review of time scales and the corresponding notation is provided. In Section 3.3, we investigate the stability and instability of first-order dynamic equations on isolated time scales with a 1-periodic coefficient. In Section 3.4, we investigate the stability and instability of first-order dynamic equations on isolated time scales with a 2-periodic coefficient. In Section 3.5, we consider a certain dynamic equation with two variable coefficient functions, and show through a change of variables a connection to the earlier cases, establishing new Ulam stability results in the process for this equation. In Section 3.6, we provide examples of isolated time scales with 1- or 2-periodic coefficient functions and apply the results from Sections 3.3, 3.4 and 3.5 to these cases. In Section 3.7, we offer a conclusion, together with possible future directions.

3.2 Brief review of time scales

A time scale $\mathbb{T}$ is any closed subset of the real line. For $t \in \mathbb{T}$, define the forward and backward jump operators $\sigma, \rho : \mathbb{T} \to \mathbb{T}$ by

$$\sigma(t) := \inf\{s \in \mathbb{T} : s > t\} \quad \text{and} \quad \rho(t) := \sup\{s \in \mathbb{T} : s < t\},$$

respectively, and the graininess function $\mu : \mathbb{T} \to [0, \infty)$ by

$$\mu(t) := \sigma(t) - t.$$

A time scale $\mathbb{T}$ is isolated if $\sigma(t) > t$ and $\rho(t) < t$ both hold for all $t \in \mathbb{T}$. For $f : \mathbb{T} \to \mathbb{C}$, the derivative of f at $t \in \mathbb{T}$ is $f^\Delta(t) = \frac{f(\sigma(t))-f(t)}{\mu(t)}$. For an isolated time scale $\mathbb{T} = \{t_n\}_{n=0}^\infty$ that is unbounded above, the derivative can also be written as

$$f^\Delta(t_n) = \frac{f(t_{n+1}) - f(t_n)}{t_{n+1} - t_n}.$$

Throughout this work, we will assume $\mathbb{T}$ is an isolated time scale that is unbounded above.

3.3 Investigating stability of first-order dynamic equations with a 1-periodic coefficient

Consider an arbitrary isolated time scale $\mathbb{T} = \{t_n\}_{n=0}^\infty$. In this section we consider on $\mathbb{T}$ the Ulam stability of the first-order linear homogeneous periodic dynamic equation on isolated time scales, with a 1-periodic coefficient, represented by

$$z^\Delta(t) - p(t)z(t) = 0, \quad t \in \mathbb{T}, \tag{3.1}$$

where $p : \mathbb{T} \to \mathbb{C}$ is 1-periodic [9, Definition 3.1] if

$$p(t) = \sigma^\Delta(t)p(\sigma(t)), \quad \text{for all } t \in \mathbb{T}. \tag{3.2}$$

Let $p(t_0) = p_0$ for $p_0 \in \mathbb{C}\backslash\{\frac{-1}{\mu(t_0)}\}$. Through iteration of (3.2), one sees that p is given by

$$p(t) := \frac{\mu(t_0)p(t_0)}{\mu(t)} = \frac{\mu(t_0)p_0}{\mu(t)}, \quad t \in \mathbb{T}. \tag{3.3}$$

Remark 3.1 (Exponential function). Let $p(0) = p_0$ for $p_0 \in \mathbb{C}\backslash\{\frac{-1}{\mu(t_0)}\}$. For $t = t_n \in \mathbb{T}$ with $n \in \mathbb{N}_0$, set

$$e_p(t_n) := \prod_{j=0}^{n-1}[1 + \mu(t_j)p(t_j)], \quad \text{where } \prod_{j=0}^{-1}[1 + \mu(t_j)p(t_j)] \equiv 1. \tag{3.4}$$

By the 1-periodic nature of p given in (3.2) and (3.3), we have ($t = t_n$ and $n \in \mathbb{N}_0$)

$$e_p(t_n) = [1 + \mu(t_0)p_0]^n, \quad n \in \mathbb{N}_0. \tag{3.5}$$

Note that the condition $p(t_0) = p_0 \in \mathbb{C}\backslash\{\frac{-1}{\mu(t_0)}\}$ is the regressive condition, since $p(t_0) = p_0 \neq \frac{-1}{\mu(t_0)}$ is equivalent to $1 + \mu(t_0)p(t_0) \neq 0$. For regressive conditions on general time scales, see [11, Definition 2.25].

Throughout the remainder of this section, for notational convenience we define the base of the exponential function to be

$$P := 1 + \mu(t_0)p_0, \quad e_p(t_n) = P^n, \quad n \in \mathbb{N}_0. \tag{3.6}$$

Definition 3.1 (Ulam stability). Let $\eta : \mathbb{T} \to \mathbb{C}$ be a perturbation. Equation (3.1) has Ulam stability if and only if there exists a constant $L > 0$ such that, for arbitrary $\varepsilon > 0$, if a function $\psi : \mathbb{T} \to \mathbb{C}$ satisfies

$$\psi^{\Delta}(t) - p(t)\psi(t) = \eta(t), \quad |\eta(t)| \le \varepsilon, \quad t \in \mathbb{T}, \tag{3.7}$$

then there exists a solution $z : \mathbb{T} \to \mathbb{C}$ of (3.1) such that $|\psi(t) - z(t)| \le L\varepsilon$ for all $t \in \mathbb{T}$. If (3.1) is Ulam stable, the constant L is called an Ulam stability constant for (3.1) on $\mathbb{T}$.

Remark 3.2. According to [10, Theorem 5.1], the role of constant functions in the classical sense is assumed by 1-periodic functions on isolated time scales, where a function f is 1-periodic on $\mathbb{T}$ if and only if there exists a real constant c such that $f(t) = \frac{c}{\mu(t)}$ for all $t \in \mathbb{T} = \{t_n\}_{n=0}^{\infty}$, as in this paper.

Theorem 3.1. *Let $\mathbb{T} = \{t_n\}_{n=0}^{\infty}$ be an isolated time scale with graininess function $\mu > 0$ such that $\lim_{n\to\infty} t_n = \infty$. Let $p(t) = \frac{\mu_0 p_0}{\mu(t)}$ for all $t = t_n \in \mathbb{T}$, i. e., p is 1-periodic, where $p_0 \in \mathbb{C}\backslash\{\frac{-1}{\mu_0}\}$. If there exists a positive constant $\mu_{\max} \in (0, \infty)$ such that*

$$\mu(t_n) \le \mu_{\max}, \quad n \in \mathbb{N}_0, \tag{3.8}$$

then (3.1) is Ulam stable if and only if $\rho = |1 + \mu_0 p_0| \neq 1$, with Ulam constant $L = \frac{\mu_{\max}}{|1-\rho|}$.

Proof. The proof is a simpler case of the proof given later for the 2-periodic coefficient case in Theorem 3.3, and thus is omitted. □

Theorem 3.2. *Let $\mathbb{T} = \{t_n\}_{n=0}^{\infty}$ be an isolated time scale. Let $p(t) = \frac{\mu_0 p_0}{\mu(t)}$ for all $t = t_n \in \mathbb{T}$, i. e., p is 1-periodic. If there exist real constants $a > 0$ and $b \ge 0$ such that*

$$\mu(t_n) \ge an + b, \quad n \in \mathbb{N}_0, \tag{3.9}$$

then (3.1) is not Ulam stable.

Proof. Let $\varepsilon > 0$ be a fixed arbitrary constant throughout the proof. By the definition of Ulam stability given in Definition 3.1, we need to find a function $\psi : \mathbb{T} \to \mathbb{C}$ fulfilling (3.7), such that there is no solution z of (3.1) satisfying $|\psi(t) - z(t)| \le L\varepsilon$ for all $t \in \mathbb{T}$, for any constant $L \in \mathbb{R}$. Suppose a function ψ fulfills (3.7) for all $t \in \mathbb{T}$, and takes the form (for $t = t_n$ and $n \in \mathbb{N}_0$)

$$\psi(t_n) = \psi(t_0)e_p(t_n) + e_p(t_n)\sum_{j=0}^{n-1}\frac{\mu(t_j)\eta(t_j)}{e_p(t_{j+1})}, \quad \text{assuming} \sum_{j=0}^{-1}\frac{\mu(t_j)\eta(t_j)}{e_p(t_{j+1})} \equiv 0. \tag{3.10}$$

Take $\psi(t_0) = 0$, and let

$$\eta(t_j) = \frac{\varepsilon(aj+b)(1+\mu_0 p_0)^{j+1}}{\mu(t_j)|1+\mu_0 p_0|^{j+1}}, \quad |\eta(t_j)| \le \varepsilon, \quad j \in \mathbb{N}_0,$$

while recalling that $(aj+b) \le \mu(t_j)$ for $t_j \in \mathbb{T}$ and $j \in \mathbb{N}_0$, from (3.9). Moreover, for $\rho > 0$ and $\theta \in [0, 2\pi)$, express $(1+\mu_0 p_0) = \rho e^{i\theta}$, and thus

$$e_p(t_n) = (1+\mu_0 p_0)^n = \rho^n e^{in\theta}, \quad n \in \mathbb{N}_0.$$

Then we have from (3.10) that for $n \in \mathbb{N}_0$,

$$\begin{aligned}\psi(t_n) &= \varepsilon\rho^n e^{in\theta}\sum_{j=0}^{n-1}\frac{(aj+b)}{\rho^{j+1}} \\ &= \begin{cases} \frac{\varepsilon e^{in\theta}}{(1-\rho)^2}(a(n-n\rho-1+\rho^n)+b(1-\rho)(1-\rho^n)) & \text{if } \rho \neq 1, \\ \frac{\varepsilon n e^{in\theta}}{2}(a(n-1)+2b) & \text{if } \rho = 1. \end{cases}\end{aligned}$$

Then, ψ satisfies (3.7) for all $t = t_n \in \mathbb{T}$ and $n \in \mathbb{N}_0$, but

$$\begin{aligned}|\psi(t_n) - z(t_n)| &= \begin{cases} |\frac{\varepsilon}{(1-\rho)^2}(a(n-n\rho-1)+b(1-\rho)) \\ \quad + \rho^n(\frac{\varepsilon(a-b(1-\rho))}{(1-\rho)^2} - z_0)| & \text{if } \rho \neq 1, \\ |\frac{\varepsilon n}{2}(a(n-1)+2b) - z_0| & \text{if } \rho = 1 \end{cases} \\ &\to \infty\end{aligned}$$

as $n \to \infty$ for any choice of $z_0 \in \mathbb{C}$, since $z(t_n) = z_0(1+\mu_0 p_0)^n = z_0\rho^n e^{in\theta}$ for $n \in \mathbb{N}_0$ is the general solution of (3.1). Therefore, equation (3.1) is not Ulam stable on isolated time scales with graininess satisfying (3.9), and a 1-periodic coefficient function. □

Remark 3.3. The linear growth condition on the graininess μ in (3.9) can be significantly weakened to the condition

$$\mu(t_n) \ge an^\gamma + b, \tag{3.11}$$

for any real $\gamma > 0$.

3.4 Investigating stability of first-order dynamic equations with a 2-periodic coefficient

In this section we consider the Ulam stability of the first-order linear homogeneous dynamic equation (3.1) on $\mathbb{T} = \{t_n\}_{n=0}^{\infty}$ for all $n \in \mathbb{N}_0$ with a 2-periodic coefficient $p : \mathbb{T} \to \mathbb{C}$. Here, p is 2-periodic [9, Definition 3.1] if

$$p(t_n) = \frac{\mu(t_{n+2})p(t_{n+2})}{\mu(t_n)}, \quad \forall t = t_n \in \mathbb{T}. \tag{3.12}$$

Let $p(t_k) = p_k$, $\mu(t_k) = \mu_k$, and $p_k \in \mathbb{C}\backslash\{\frac{-1}{\mu_k}\}$ for $k \in \{0, 1\}$. Through iteration of (3.12), one sees that p is given by

$$p(t_n) := \frac{1}{\mu(t_n)} \begin{cases} \mu_0 p_0 & \text{if } n \equiv 0 \pmod 2, \\ \mu_1 p_1 & \text{if } n \equiv 1 \pmod 2, \end{cases} \quad n \in \mathbb{N}_0. \tag{3.13}$$

Remark 3.4 (Exponential function). Let $p(t_k) = p_k$, $\mu(t_k) = \mu_k$, and $p_k \in \mathbb{C}\backslash\{\frac{-1}{\mu_k}\}$ for $k = 0, 1$. By the 2-periodic nature of p given in (3.12) and (3.13), the exponential function (3.4) takes the form

$$e_p(t_n) = \begin{cases} [1 + \mu_0 p_0]^{\frac{n}{2}} [1 + \mu_1 p_1]^{\frac{n}{2}} & \text{if } n \equiv 0 \pmod 2, \\ [1 + \mu_0 p_0]^{\frac{n+1}{2}} [1 + \mu_1 p_1]^{\frac{n-1}{2}} & \text{if } n \equiv 1 \pmod 2. \end{cases} \tag{3.14}$$

Note that the conditions $p_k \in \mathbb{C}\backslash\{\frac{-1}{\mu_k}\}$ for $k \in \{0, 1\}$ are really regressive conditions, since $p(t_k) = p_k \neq \frac{-1}{\mu_k}$ is equivalent to $1 + \mu_k p_k \neq 0$, for $k = 0, 1$.

For notational convenience throughout the remainder of this section, define the key constants as

$$E_k := 1 + \mu_k p_k, \quad k = 0, 1. \tag{3.15}$$

It is left to the reader to check that e_p indeed satisfies (3.1), with $e_p(t_0) = 1$.

Theorem 3.3. *Let $\mathbb{T} = \{t_n\}_{n=0}^{\infty}$ be an isolated time scale with graininess function $\mu > 0$ such that $\lim_{n\to\infty} t_n = \infty$. Let $p(t_k) = p_k$, $\mu(t_k) = \mu_k$, and $p_k \in \mathbb{C}\backslash\{\frac{-1}{\mu_k}\}$ for $k = 0, 1$. If μ satisfies* (3.8) *and p is a 2-periodic function, then* (3.1) *is Ulam stable if and only if $|E_0 E_1| = |(1 + \mu_0 p_0)(1 + \mu_1 p_1)| \neq 1$, with Ulam constant*

$$L = \max\{|E_0| + 1, |E_1| + 1\} \frac{\mu_{\max}}{||E_0 E_1| - 1|}.$$

Proof. Recall the criteria for Ulam stability given in Definition 3.1. Since p is 2-periodic in this case, p satisfies (3.12) and has the form (3.13). Let E_k take the form (3.15) for $k = 0, 1$.

(i) First, suppose $|E_0 E_1| = 1$, and let e_p be given by (3.14). It follows in this case that $|e_p(t_n)| = |E_0|^{n \pmod 2}$, $n \in \mathbb{N}_0$. Given arbitrary $\varepsilon > 0$, set

$$m := \min\{1, |E_0|\}, \quad M := \max\{1, |E_0|\},$$

and let $\psi(t_n) = \frac{\varepsilon t_n}{M} e_p(t_n)$, $n \in \mathbb{N}_0$. Then, ψ satisfies

$$|\psi^\Delta(t_n) - p(t_n)\psi(t_n)| = \frac{\varepsilon}{M} |e_p(\sigma(t_n))| \leq \varepsilon$$

for all $t_n \in \mathbb{T}$ and $n \in \mathbb{N}_0$, so that (3.7) holds, but

$$|\psi(t_n) - z(t)| = \left|\frac{\varepsilon t_n}{M} e_p(t_n) - z_0 e_p(t_n)\right| = |e_p(t_n)| \left|\frac{\varepsilon t_n}{M} - z_0\right| \geq m \left|\frac{\varepsilon t_n}{M} - z_0\right| \to \infty$$

as $n \to \infty$, making (3.1) unstable in the Ulam sense.

(ii) Second, suppose $|E_0 E_1| > 1$, where $E_0 E_1 = (1 + \mu_0 p_0)(1 + \mu_1 p_1)$. For arbitrary $\varepsilon > 0$, suppose a function ψ fulfills (3.7) for all $t \in \mathbb{T}$. Moreover, by (3.14), the exponential function has the form

$$e_p(t_{2m}) = (E_0 E_1)^m, \quad e_p(t_{2m+1}) = E_0 (E_0 E_1)^m, \quad m \in \mathbb{N}_0,$$

and from (3.10) we have that ψ has the form

$$\psi(t_n) = \psi_0 e_p(t_n) + e_p(t_n) \sum_{j=0}^{n-1} \frac{\mu(t_j)\eta(t_j)}{e_p(t_{j+1})}, \quad |\eta(t_n)| \leq \varepsilon \tag{3.16}$$

for all $t = t_n \in \mathbb{T}$ and $n \in \mathbb{N}_0$, where $\psi(t_0) = \psi_0$. From this expression for $\psi(t_n)$, we have

$$\left|\frac{\psi(t_n)}{e_p(t_n)}\right| \leq |\psi_0| + \varepsilon \mu_{\max} \sum_{j=0}^{n-1} \frac{1}{|e_p(t_{j+1})|}, \quad n \in \mathbb{N}_0,$$

as $|\eta(t_j)| \leq \varepsilon$, and μ satisfies (3.8) for all $t = t_j \in \mathbb{T}$ and $j \in \mathbb{N}_0$ in this case. These facts allow us to rewrite ψ as

$$\psi(t_n) = \left[\psi_0 + \sum_{j=0}^{\infty} \frac{\mu(t_j)\eta(t_j)}{e_p(t_{j+1})}\right] e_p(t_n) - e_p(t_n) \sum_{j=n}^{\infty} \frac{\mu(t_j)\eta(t_j)}{e_p(t_{j+1})}, \quad n \in \mathbb{N}_0, \tag{3.17}$$

where

$$z_0 := \psi_0 + \sum_{j=0}^{\infty} \frac{\mu(t_j)\eta(t_j)}{e_p(t_{j+1})} \in \mathbb{C}$$

exists and is a finite number due to $e_p(t_n)$, $n \in \mathbb{N}_0$, given in (3.14), the assumption $|E_0 E_1| > 1$, and the absolute convergence of the infinite series. It is straightforward to see that $z(t_n) := z(t_0) e_p(t_n)$ for $t_n \in \mathbb{T}$ is a solution of (3.1), and

$$\lim_{n \to \infty} \frac{\psi(t_n)}{e_p(t_n)} = \psi_0 + \sum_{j=0}^{\infty} \frac{\mu(t_j)\eta(t_j)}{e_p(t_{j+1})} = z_0$$

exists. Consequently,

$$z(t_n) = \left(\lim_{n \to \infty} \frac{\psi(t_n)}{e_p(t_n)}\right) e_p(t_n), \quad n \in \mathbb{N}_0,$$

and

$$
\begin{aligned}
|\psi(t_n) - z(t_n)| &= \left| -e_p(t_n) \sum_{j=n}^{\infty} \frac{\mu(t_j)\eta(t_j)}{e_p(t_{j+1})} \right| \\
&\le \varepsilon\mu_{\max} |e_p(t_n)| \sum_{j=n}^{\infty} \frac{1}{|e_p(t_{j+1})|} \\
&= \frac{\varepsilon\mu_{\max}}{|E_0E_1| - 1} \begin{cases} (|E_1| + 1) & \text{if } n \equiv 0 \pmod 2, \\ (|E_0| + 1) & \text{if } n \equiv 1 \pmod 2) \end{cases}
\end{aligned}
$$

holds for all $t = t_n \in \mathbb{T}$. Therefore, (3.1) is Ulam stable with Ulam constant

$$L = \max\{|E_0| + 1, |E_1| + 1\} \frac{\mu_{\max}}{|E_0E_1| - 1}$$

for $E_0E_1 = (1 + \mu_0 p_0)(1 + \mu_1 p_1)$ and $|E_0E_1| > 1$.

(iii) Third, suppose $0 < |E_0E_1| < 1$, where $E_0E_1 = (1 + \mu_0 p_0)(1 + \mu_1 p_1)$. As before, we know that a function ψ that satisfies (3.7) takes the form (3.16) by (3.10). Let $z = z(t_n) = \psi_0 e_p(t_n)$ for $t = t_n \in \mathbb{T}$ be the solution of (3.1) such that $z(t_0) = \psi_0$. Then we have

$$\psi(t_n) - z(t_n) = e_p(t_n) \sum_{j=0}^{n-1} \frac{\mu(t_j)\eta(t_j)}{e_p(t_{j+1})},$$

and thus

$$
\begin{aligned}
|\psi(t_n) - z(t_n)| &\le \varepsilon\mu_{\max} |e_p(t_n)| \sum_{j=0}^{n-1} \frac{1}{|e_p(t_{j+1})|} \\
&\le \frac{\varepsilon\mu_{\max}}{1 - |E_0E_1|} \begin{cases} (|E_1| + 1) & \text{if } n \equiv 0 \pmod 2, \\ (|E_0| + 1) & \text{if } n \equiv 1 \pmod 2) \end{cases}
\end{aligned}
$$

for all $t = t_n \in \mathbb{T}$. Therefore, (3.1) is Ulam stable with Ulam constant

$$L = \max\{|E_0| + 1, |E_1| + 1\} \frac{\mu_{\max}}{1 - |E_0E_1|}$$

for $0 < |E_0E_1| < 1$, where $E_0E_1 = (1 + \mu_0 p_0)(1 + \mu_1 p_1)$. Overall, we have shown that (3.1) is Ulam stable if and only if $|E_0E_1| = |(1 + \mu_0 p_0)(1 + \mu_1 p_1)| \neq 1$, with Ulam constant $L = \max\{|E_0| + 1, |E_1| + 1\} \frac{\mu_{\max}}{|1 - |E_0E_1||}$, completing the proof. □

Theorem 3.4. *Let $\mathbb{T} = \{t_n\}_{n=0}^{\infty}$ be an isolated time scale. Assume p satisfies* (3.12) *and takes the form* (3.13) *for all $t = t_n \in \mathbb{T}$, i. e., p is 2-periodic. If there exist real constants $a > 0$ and $b \ge 0$ such that* (3.9) *holds, then* (3.1) *is not Ulam stable.*

Proof. Let $\varepsilon > 0$ be a fixed arbitrary constant throughout the proof. Let $p(t_k) = p_k$, $\mu(t_k) = \mu_k$, and $p_k \in \mathbb{C}\backslash\{\frac{-1}{\mu_k}\}$ for $k = 0, 1$. By the definition of Ulam stability given in Definition 3.1, we need to find a function $\psi : \mathbb{T} \to \mathbb{C}$ fulfilling (3.7), such that there is no solution z of (3.1) satisfying $|\psi(t) - z(t)| \leq L\varepsilon$ for all $t \in \mathbb{T}$, for any constant $L \in \mathbb{R}$. Suppose a function ψ fulfills (3.7) for all $t \in \mathbb{T}$, and has the form (3.10). Take $\psi(t_0) = 0$, and

$$\eta(t_j) = \frac{\varepsilon(aj + b)e_p(t_{j+1})}{\mu(t_j)|e_p(t_{j+1})|}, \qquad |\eta(t_j)| \leq \varepsilon,$$

while recalling that $(aj + b) \leq \mu(t_j)$ for $t_j \in \mathbb{T}$ from (3.9).

Suppose $n = 2m$ and $|E_0E_1| = 1$. We then have from (3.10) that

$$\begin{aligned}\psi(t_{2m}) &= \varepsilon e_p(t_{2m}) \sum_{j=0}^{2m-1} \frac{(aj + b)}{|e_p(t_{j+1})|} \\ &= \frac{\varepsilon ma(E_0E_1)^m}{|E_0|}(m|E_0| + m - 1) + \varepsilon mb(E_0E_1)^m\left(1 + \frac{1}{|E_0|}\right).\end{aligned}$$

Then, ψ satisfies (3.7) for all $t = t_{2m} \in \mathbb{T}$, but

$$\begin{aligned}&|\psi(t_{2m}) - z(t_{2m})| \\ &= \left|\frac{\varepsilon ma(E_0E_1)^m}{|E_0|}(m|E_0| + m - 1) + \varepsilon mb(E_0E_1)^m\left(1 + \frac{1}{|E_0|}\right) - z_0(E_0E_1)^m\right| \\ &= \left|\frac{\varepsilon ma}{|E_0|}(m|E_0| + m - 1) + \varepsilon mb\left(1 + \frac{1}{|E_0|}\right) - z_0\right| \\ &\to \infty\end{aligned}$$

as $m \to \infty$ for any choice of $z_0 \in \mathbb{C}$, since $|E_0E_1| = 1$, $z(t_{2m}) = z_0(E_0E_1)^m$ is the general solution of (3.1), and $a > 0$ with $b \geq 0$.

Suppose $n = 2m$, $m \in \mathbb{N}$, and $|E_0E_1| \neq 1$. We then have from (3.10) that

$$\begin{aligned}\psi(t_{2m}) &= \varepsilon e_p(t_{2m}) \sum_{j=0}^{2m-1} \frac{(aj + b)}{|e_p(t_{j+1})|} \\ &= \frac{\varepsilon a(E_0E_1)^m}{(|E_0E_1| - 1)^2|E_0E_1|^m} \\ &\quad \times (2m - 1 + |E_0E_1|^m + |E_1|(2m - 2 - |E_0|(2m|E_1| + 2m + 1)) \\ &\quad + (2 + |E_0|)|E_0E_1|^m) \\ &\quad + \frac{\varepsilon b(E_0E_1)^m(1 + |E_1|)(|E_0E_1|^m - 1)}{(|E_0E_1| - 1)|E_0E_1|^m}.\end{aligned}$$

Then, ψ satisfies (3.7) for all $t = t_{2m} \in \mathbb{T}$, but

$$\begin{aligned}&|\psi(t_{2m}) - z(t_{2m})|\\ &= \Bigg|\frac{\varepsilon a}{(|E_0E_1| - 1)^2}(2m - 1 + |E_0E_1|^m + |E_1|(2m - 2 - |E_0|(2m|E_1| + 2m + 1))\\ &\quad + (2 + |E_0|)|E_0E_1|^m) + \frac{\varepsilon b(1 + |E_1|)(|E_0E_1|^m - 1)}{(|E_0E_1| - 1)} - z_0|E_0E_1|^m\Bigg|\\ &\to \infty\end{aligned}$$

as $m \to \infty$ for any choice of $z_0 \in \mathbb{C}$, since $|E_0E_1| \neq 1$ and the coefficient of m is

$$2(1 + |E_1|)(1 - |E_0E_1|) \neq 0,$$

where $z(t_{2m}) = z_0(E_0E_1)^m$ is the general solution of (3.1), and $a > 0$ with $b \geq 0$.

Suppose $n = 2m + 1$, $m \in \mathbb{N}$, and $|E_0E_1| = 1$. We then have

$$|\psi(t_{2m+1}) - z(t_{2m+1})| = |E_0|\left|\frac{\varepsilon(am + b)(1 + m + m|E_0|)}{|E_0|} - z_0\right| \to \infty$$

as $m \to \infty$ for any choice of $z_0 \in \mathbb{C}$, since $|E_0E_1| = 1$, $z(t_{2m+1}) = z_0(E_0E_1)^m E_0$ is the general solution of (3.1), and $a > 0$ with $b \geq 0$.

Suppose $n = 2m + 1$, $m \in \mathbb{N}$, and $|E_0E_1| \neq 1$. We then have

$$\begin{aligned}&|\psi(t_{2m+1}) - z(t_{2m+1})|\\ &= \Bigg|\frac{\varepsilon}{(|E_0E_1| - 1)^2}[2am + b + |E_0|(a(2m - 1) + b - |E_1|(2a(m + 1)\\ &\quad + b + (a + b + 2am)|E_0|)\\ &\quad + (a - b + |E_1|(2a - b + |E_0|(a + b + b|E_1|)))|E_0E_1|^m)] - z_0|E_0|^{m+1}|E_1|^m\Bigg|\\ &\to \infty\end{aligned}$$

as $m \to \infty$ for any choice of $z_0 \in \mathbb{C}$, since $|E_0E_1| \neq 1$ and the coefficient of m is

$$2a(1 + |E_0|)(1 - |E_0E_1|) \neq 0,$$

where $z(t_{2m+1}) = z_0(E_0E_1)^m E_0$ is the general solution of (3.1), and $a > 0$ with $b \geq 0$.

Therefore, equation (3.1) is not Ulam stable on isolated time scales with graininess satisfying (3.9), and a 2-periodic coefficient function. □

Remark 3.5. The previous theorem can be made stronger by weakening condition (3.9) to (3.11).

3.5 Applications to two variable-coefficient equations

Let $\mathbb{S} = \{s_n\}_{n=0}^{\infty}$ be an arbitrary isolated time scale. In this section, we consider the two variable-coefficient equation

$$\alpha(s)y^{\Delta}(s) - \beta(s)y(s) = 0, \quad s \in \mathbb{S}, \tag{3.18}$$

where $\alpha : \mathbb{S} \to \mathbb{R}$ is a nonzero real-valued function, but $\beta : \mathbb{S} \to \mathbb{C}$ is a complex-valued function.

Definition 3.2 (Ulam stability). Let $\varphi : \mathbb{S} \to \mathbb{C}$ be a perturbation. Equation (3.18) has Ulam stability if and only if there exists a constant $L > 0$ such that, for arbitrary $\varepsilon > 0$, if a function $\zeta : \mathbb{S} \to \mathbb{C}$ satisfies

$$\alpha(s)\zeta^{\Delta}(s) - \beta(s)\zeta(s) = \varphi(s), \quad |\varphi(s)| \le \varepsilon, \quad s \in \mathbb{S}, \tag{3.19}$$

then there exists a solution $y : \mathbb{S} \to \mathbb{C}$ of (3.18) such that $|\zeta(s) - y(s)| \le L\varepsilon$ for all $s \in \mathbb{S}$. If (3.18) is Ulam stable, the constant L is called an Ulam stability constant for (3.18) on $\mathbb{S}$.

Theorem 3.5. *Let $\mathbb{S} = \{s_n\}_{n=0}^{\infty}$ be an isolated time scale with graininess function $\mu(s_n) > 0$. Suppose that there exists an increasing function $\phi : \mathbb{S} \to \mathbb{R}$ such that $\lim_{n\to\infty} \phi(s_n) = \infty$ and*

$$\alpha(s)\phi^{\Delta}(s) = K, \quad s \in \mathbb{S}, \tag{3.20}$$

where K is a positive constant. Suppose also that $\frac{\beta(s)}{\alpha(s)} = \frac{\beta(s_0)\mu(s_0)}{\alpha(s_0)\mu(s)}$ holds for all $s = s_n \in \mathbb{S}$, i. e., $\frac{\beta}{\alpha}$ is 1-periodic, where $\frac{\beta(s_0)}{\alpha(s_0)} \in \mathbb{C}\backslash\{\frac{-1}{\mu(s_0)}\}$. If there exists a positive constant $M \in (0, \infty)$ such that

$$\frac{\mu(s_n)}{\alpha(s_n)} \le M, \quad s_n \in \mathbb{S}, \tag{3.21}$$

then (3.18) is Ulam stable if and only if $\rho = |1 + \frac{\beta(s_0)\mu(s_0)}{\alpha(s_0)}| \neq 1$, with Ulam constant $L_s = \frac{M}{|1-\rho|}$.

Proof. Let $\mathbb{S} = \{s_n\}_{n=0}^{\infty}$ be an isolated time scale with $\mu(s_n) > 0$ for all $n \in \mathbb{N}_0$. Suppose that there exists an increasing function ϕ such that $\lim_{n\to\infty} \phi(s_n) = \infty$ and (3.20) holds. Define $t_n := \phi(s_n)$ for all $n \in \mathbb{N}_0$. Then $\mathbb{T} = \{t_n\}_{n=0}^{\infty}$ is also an isolated time scale with $\mu(t_n) > 0$, and satisfies $\lim_{n\to\infty} t_n = \infty$. Let y be a solution of (3.18). Define $z(t) := Ky(\phi^{-1}(t))$ and $p(t) := \frac{\beta(\phi^{-1}(t))}{K}$ for $t = \phi(s) \in \mathbb{T}$, where ϕ^{-1} is the inverse function of ϕ. Then, by (3.20), we have

$$z^{\Delta}(t) = K\frac{y(\phi^{-1}(t_{n+1})) - y(\phi^{-1}(t_n))}{t_{n+1} - t_n}$$

$$= K\frac{y(s_{n+1}) - y(s_n)}{\phi(s_{n+1}) - \phi(s_n)}$$
$$= \alpha(s_n)\frac{y(s_{n+1}) - y(s_n)}{s_{n+1} - s_n}$$
$$= \alpha(s)y^{\Delta}(s),$$

and so that

$$z^{\Delta}(t) - p(t)z(t) = \alpha(s)y^{\Delta}(s) - \beta(s)y(s) = 0$$

for $t \in \mathbb{T}$, that is, (3.18) is transformed into (3.1) with $p(t) = \frac{\beta(\phi^{-1}(t))}{K}$. Since $\frac{\beta(s)}{\alpha(s)} = \frac{\beta(s_0)\mu(s_0)}{\alpha(s_0)\mu(s)}$ and (3.20) hold, we see that

$$p(t) = \frac{\beta(s)}{K} = \frac{\beta(s_0)\mu(s_0)\alpha(s_n)}{K\alpha(s_0)(s_{n+1} - s_n)} = \frac{\beta(s_0)\mu(s_0)}{\alpha(s_0)(\phi(s_{n+1}) - \phi(s_n))} = \frac{p(t_0)\mu(t_0)}{\mu(t)}$$

is satisfied for all $t \in \mathbb{T}$. Moreover, by $\frac{\beta(s_0)}{\alpha(s_0)} \in \mathbb{C}\backslash\{\frac{-1}{\mu(s_0)}\}$ and (3.20), we have

$$p(t_0) = \frac{\beta(s_0)}{K} \neq \frac{-\alpha(s_0)}{K\mu(s_0)} = \frac{-1}{\mu(t_0)},$$

and thus, $p(t_0) = p_0 \in \mathbb{C}\backslash\{\frac{-1}{\mu(t_0)}\}$. If there exists a positive constant $M \in (0, \infty)$ such that (3.21) holds, then

$$\mu(t_n) = \phi(s_{n+1}) - \phi(s_n) = \frac{K\mu(s_n)}{\alpha(s_n)} \le KM = \mu_{\max}, \quad n \in \mathbb{N}_0.$$

Hence, by Theorem 3.1, we can conclude that (3.1) with $p(t) = \frac{\beta(\phi^{-1}(t))}{K}$ is Ulam stable if and only if

$$\rho = |1 + \mu(t_0)p(t_0)| = \left|1 + \frac{\beta(s_0)\mu(s_0)}{\alpha(s_0)}\right| \neq 1,$$

with Ulam constant $L = \frac{\mu_{\max}}{|1-\rho|} = \frac{KM}{|1-\rho|}$.

Next we will prove that (3.1) with $p(t) = \frac{\beta(\phi^{-1}(t))}{K}$ is Ulam stable, with an Ulam constant $\frac{KM}{|1-\rho|}$ if and only if (3.18) is Ulam stable, with an Ulam constant $\frac{M}{|1-\rho|}$. Assume that Ulam stability for (3.1), with an Ulam constant $\frac{KM}{|1-\rho|}$. Let $\varepsilon > 0$ and let $\zeta : \mathbb{S} \to \mathbb{C}$ satisfy (3.19). Define $\psi(t) := K\zeta(\phi^{-1}(t))$ for $t = \phi(s) \in \mathbb{T}$. Then, by (3.20), we obtain

$$\psi^{\Delta}(t) = K\frac{\zeta(\phi^{-1}(t_{n+1})) - \zeta(\phi^{-1}(t_n))}{t_{n+1} - t_n} = \alpha(s_n)\frac{\zeta(s_{n+1}) - \zeta(s_n)}{s_{n+1} - s_n} = \alpha(s)\zeta^{\Delta}(s),$$

and thus,

$$\eta(t) := \psi^\Delta(t) - p(t)\psi(t) = \alpha(s)\zeta^\Delta(s) - \beta(s)\zeta(s) = \varphi(s) \tag{3.22}$$

for $t \in \mathbb{T}$. Since $|\eta(t)| = |\varphi(s)| \le \varepsilon$ holds for $t = \phi(s) \in \mathbb{T}$, and (3.1) is Ulam stable on $\mathbb{T}$, there exists a solution $z(t)$ of (3.1) such that $|\psi(t)-z(t)| \le \frac{KM}{|1-\rho|}\varepsilon$ for $t \in \mathbb{T}$. Let $y(s) := \frac{z(\phi(s))}{K}$ for $s \in \mathbb{S}$. Then $y(s)$ is a solution of (3.18). Moreover, we obtain

$$K|\zeta(s) - y(s)| = |\psi(t) - z(t)| \le \frac{KM}{|1-\rho|}\varepsilon.$$

This says that (3.18) is Ulam stable on $\mathbb{S}$, with an Ulam constant $\frac{M}{|1-\rho|}$.

Conversely, we assume that Ulam stability holds for (3.18), with an Ulam constant $\frac{M}{|1-\rho|}$. Let $\varepsilon > 0$ and suppose $\psi : \mathbb{T} \to \mathbb{C}$ satisfies (3.7). Define $\zeta(s) := \frac{\psi(\phi(s))}{K}$ for $s = \phi^{-1}(t) \in \mathbb{S}$. Then, by (3.20), we obtain (3.22). Since $|\varphi(s)| = |\eta(t)| \le \varepsilon$ holds for $s = \phi^{-1}(t) \in \mathbb{S}$, and (3.18) is Ulam stable on $\mathbb{S}$, there exists a solution y of (3.18) such that $|\zeta(s) - y(s)| \le \frac{M}{|1-\rho|}\varepsilon$ for $s \in \mathbb{S}$. Let $z(t) := Ky(\phi^{-1}(t))$ for $t \in \mathbb{T}$. Then z is a solution of (3.1). Moreover, we obtain

$$\frac{1}{K}|\psi(t) - z(t)| = |\zeta(s) - y(s)| \le \frac{M}{|1-\rho|}\varepsilon, \quad s = \phi^{-1}(t) \in \mathbb{S}.$$

This says that (3.1) is Ulam stable on $\mathbb{T}$, with an Ulam constant $\frac{KM}{|1-\rho|}$.

Summarizing the above, we conclude that (3.18) is Ulam stable if and only if $\rho = |1 + \frac{\beta(s_0)\mu(s_0)}{\alpha(s_0)}| \ne 1$, with Ulam constant $L_s = \frac{M}{|1-\rho|}$. □

Theorem 3.6. *Let $\mathbb{S} = \{s_n\}_{n=0}^\infty$ be an isolated time scale with graininess function $\mu(s_n) > 0$. Suppose that there exists an increasing function $\phi : \mathbb{S} \to \mathbb{R}$ such that $\lim_{n\to\infty} \phi(s_n) = \infty$ and (3.20) holds, where K is a positive constant. Suppose also that*

$$\frac{\beta(s_n)}{\alpha(s_n)} = \frac{1}{\mu(s_n)} \begin{cases} \frac{\beta(s_0)\mu(s_0)}{\alpha(s_0)} & \text{if } n \equiv 0 \pmod 2, \\ \frac{\beta(s_1)\mu(s_1)}{\alpha(s_1)} & \text{if } n \equiv 1 \pmod 2 \end{cases} \tag{3.23}$$

holds for all $s_n \in \mathbb{S}$, i. e., $\frac{\beta}{\alpha}$ is 2-periodic, where $\frac{\beta(s_k)}{\alpha(s_k)} \in \mathbb{C}\backslash\{\frac{-1}{\mu(s_k)}\}$ for $k \in \{0,1\}$. If there exists a positive constant $M \in (0,\infty)$ such that (3.21) holds, then (3.18) is Ulam stable if and only if $|\tilde{E}_0\tilde{E}_1| \ne 1$, with Ulam constant

$$L_s = \max\{|\tilde{E}_0| + 1, |\tilde{E}_1| + 1\}\frac{M}{||\tilde{E}_0\tilde{E}_1| - 1|},$$

where

$$\tilde{E}_k := 1 + \frac{\beta(s_k)\mu(s_k)}{\alpha(s_k)}, \quad k = 0, 1.$$

Proof. Let $\mathbb{S} = \{s_n\}_{n=0}^\infty$ be an isolated time scale with $\mu(s_n) > 0$ for all $n \in \mathbb{N}_0$, and let ϕ be an increasing function such that $\lim_{n\to\infty} \phi(s_n) = \infty$ and (3.20) holds. Put $t_n := \phi(s_n)$

for all $n \in \mathbb{N}_0$. Then t_n satisfies $\lim_{n\to\infty} t_n = \infty$, and $\mathbb{T} = \{t_n\}_{n=0}^{\infty}$ is also an isolated time scale with $\mu(t_n) > 0$. Set $z(t) := Ky(\phi^{-1}(t))$ and $p(t) := \frac{\beta(\phi^{-1}(t))}{K}$ for $t = \phi(s) \in \mathbb{T}$. Then, it can be seen that (3.18) is transformed into (3.1) with $p(t) = \frac{\beta(\phi^{-1}(t))}{K}$ in the same way as in Theorem 3.5. Using (3.20) and (3.23), we see that

$$p(t_n) = \frac{\beta(s_n)}{K} = \frac{\beta(s_0)\mu(s_0)\alpha(s_n)}{K\alpha(s_0)(s_{n+1}-s_n)} = \frac{\beta(s_0)\mu(s_0)}{\alpha(s_0)(\phi(s_{n+1})-\phi(s_n))} = \frac{p(t_0)\mu(t_0)}{\mu(t_n)}$$

if $n \equiv 0 \pmod 2$, and

$$p(t_n) = \frac{\beta(s_n)}{K} = \frac{\beta(s_1)\mu(s_1)\alpha(s_n)}{K\alpha(s_1)(s_{n+1}-s_n)} = \frac{\beta(s_1)\mu(s_1)}{\alpha(s_1)(\phi(s_{n+1})-\phi(s_n))} = \frac{p(t_1)\mu(t_1)}{\mu(t_n)}$$

if $n \equiv 1 \pmod 2$. Moreover, by $\frac{\beta(s_k)}{\alpha(s_k)} \in \mathbb{C}\backslash\{\frac{-1}{\mu(s_k)}\}$ and (3.20), we have

$$p(t_k) = \frac{\beta(s_k)}{K} \neq \frac{-\alpha(s_k)}{K\mu(s_k)} = \frac{-1}{\mu(t_k)},$$

for $k \in \{0,1\}$. From these facts, $p(t)$ is 2-periodic, where $p(t_k) \in \mathbb{C}\backslash\{\frac{-1}{\mu(t_k)}\}$ for $k \in \{0,1\}$. If there exists a positive constant $M \in (0,\infty)$ such that (3.21) holds, then

$$\mu(t_n) = \phi(s_{n+1}) - \phi(s_n) = \frac{K\mu(s_n)}{\alpha(s_n)} \le KM = \mu_{\max}, \quad n \in \mathbb{N}_0.$$

Hence, by Theorem 3.3, we can conclude that (3.1) with $p(t) = \frac{\beta(\phi^{-1}(t))}{K}$ is Ulam stable if and only if

$$\begin{aligned}|E_0E_1| &= \left|(1+\mu(t_0)p(t_0))(1+\mu(t_1)p(t_1))\right| \\ &= \left|\left(1+\frac{\beta(s_0)\mu(s_0)}{\alpha(s_0)}\right)\left(1+\frac{\beta(s_1)\mu(s_1)}{\alpha(s_1)}\right)\right| \\ &= |\tilde{E}_0\tilde{E}_1| \\ &\neq 1,\end{aligned}$$

with Ulam constant

$$L = \max\{|\tilde{E}_0|+1, |\tilde{E}_1|+1\}\frac{\mu_{\max}}{||\tilde{E}_0\tilde{E}_1|-1|} = \max\{|\tilde{E}_0|+1, |\tilde{E}_1|+1\}\frac{KM}{||\tilde{E}_0\tilde{E}_1|-1|}.$$

By the same technique as Theorem 3.5, we can conclude that (3.1) with $p(t) = \frac{\beta(\phi^{-1}(t))}{K}$ is Ulam stable, with an Ulam constant

$$\max\{|\tilde{E}_0|+1, |\tilde{E}_1|+1\}\frac{KM}{||\tilde{E}_0\tilde{E}_1|-1|}$$

if and only if (3.18) is Ulam stable, with an Ulam constant

$$\max\{|\tilde{E}_0|+1, |\tilde{E}_1|+1\}\frac{M}{||\tilde{E}_0\tilde{E}_1|-1|}.$$

So, the proof is complete. □

3.6 Examples of isolated time scales with 1- or 2-periodic coefficient functions

In this section, we present several specific isolated time scales of interest, and relate their Ulam stability or lack thereof to theorems in the previous sections.

Example (h-difference equation). Let $h > 0$, and set $\mathbb{T} = \{0, h, 2h, 3h, \dots\}$. If p is a 1-periodic function on $\mathbb{T}$, then $p(t+h) = p(t)$ for all $t \in \mathbb{T}$, so that $p(t) \equiv p$ is a constant, due to the constant graininess $\mu(t) \equiv h$. It is known [2, Theorem 2.6] that (3.1) is Ulam stable for $p \in \mathbb{C}\backslash\{-\frac{1}{h}\}$ if and only if $|1+hp| \neq 1$, with best Ulam constant

$$L = \frac{h}{|1-|1+hp||} = \frac{1}{|\operatorname{Re}_h(p)|}.$$

This result also follows directly from Theorem 3.1 and (3.8), where $h = \mu_{\max} = \mu_0$. If p is 2-periodic, then $p(t+2h) = p(t)$ for all $t \in \mathbb{T}$, so

$$p(t_n) = p(nh) := \begin{cases} p_0 & \text{if } n \equiv 0 \ (\mathrm{mod}\ 2), \\ p_1 & \text{if } n \equiv 1 \ (\mathrm{mod}\ 2), \end{cases}$$

for $p_k \in \mathbb{C}\backslash\{-\frac{1}{h}\}$, $k = 0, 1$. By Theorem 3.3, (3.1) is also Ulam stable in this case, if $1 \neq |(1+hp_0)(1+hp_1)|$.

Example (Two step sizes). Let $\eta, \tau > 0$ be two step sizes, and let the isolated time scale $\mathbb{T}$ constructed with them be denoted by

$$\mathbb{T}_{\eta,\tau} := \{0, \eta, \eta+\tau, (\eta+\tau)+\eta, 2(\eta+\tau), 2(\eta+\tau)+\eta, \dots\}.$$

Then the derivative is

$$z^\Delta(t) := \begin{cases} \frac{z(t+\eta)-z(t)}{\eta} & \text{if } \frac{t}{\eta+\tau} \in \mathbb{Z}, \\ \frac{z(t+\tau)-z(t)}{\tau} & \text{if } \frac{t-\eta}{\eta+\tau} \in \mathbb{Z}, \end{cases}$$

for $t \in \mathbb{T}_{\eta,\tau}$. This time scale has been studied in [1, 4], but only for the constant coefficient case. For the periodic coefficient case, $\mu(t_n) \leq \mu_{\max} := \max\{\eta, \tau\}$, whence (3.8) holds. If p is 1-periodic, then

$$p(t_n) = p_0 \begin{cases} 1 & \text{if } n \equiv 0 \pmod 2, \\ \frac{\eta}{\tau} & \text{if } n \equiv 1 \pmod 2, \end{cases}$$

for $p_0 \in \mathbb{C} \backslash \{-\frac{1}{\eta}\}$. By Theorem 3.1, we have the new result that (3.1) is Ulam stable if and only if $\rho = |1 + \eta p_0| \neq 1$, with Ulam constant given by $L = \frac{\max\{\eta,\tau\}}{|1-\rho|}$. If p is 2-periodic, then

$$p(t_n) = \begin{cases} p_0 & \text{if } n \equiv 0 \pmod 2, \\ p_1 & \text{if } n \equiv 1 \pmod 2, \end{cases}$$

for $p_0 \in \mathbb{C} \backslash \{-\frac{1}{\eta}\}$ and $p_1 \in \mathbb{C} \backslash \{-\frac{1}{\tau}\}$. By Theorem 3.3, we have the new result that (3.1) is Ulam stable if and only if $|(1 + \eta p_0)(1 + \tau p_1)| \neq 1$, with Ulam constant given by

$$L = \max\{|1 + \eta p_0| + 1, |1 + \tau p_1| + 1\} \frac{\max\{\eta, \tau\}}{||(1 + \eta p_0)(1 + \tau p_1)| - 1|}.$$

Example (q-difference equation). Let $q > 1$ and set $\mathbb{T} = \{1, q, q^2, q^3, \ldots\}$. If p is a 1-periodic function on $\mathbb{T}$, then $qp(qt) = p(t)$ for all $t \in \mathbb{T}$. Then p takes the form

$$p(t) = \frac{p(1)}{t}, \quad t \in \mathbb{T},$$

for $p(1) \in \mathbb{C} \backslash \{\frac{-1}{q-1}\}$. The instability result follows directly from Theorem 3.2 and (3.9), where $\mu(q^n) \geq an$ for $a = (q-1)e \ln q$ for all $n \in \mathbb{N}_0$. If p is a 2-periodic function on $\mathbb{T}$, then $q^2 p(q^2 t) = p(t)$ for all $t \in \mathbb{T}$. Then p takes the form

$$p(t) := \frac{1}{t} \begin{cases} p_1 & \text{if } \log_q t \equiv 0 \pmod 2, \\ qp_q & \text{if } \log_q t \equiv 1 \pmod 2, \end{cases}$$

for $p_1 \in \mathbb{C} \backslash \{\frac{-1}{q-1}\}$ and $p_q \in \mathbb{C} \backslash \{\frac{-1}{q(q-1)}\}$. Likewise, that (3.1) is not Ulam stable for any choice of p_1 or p_q follows from Theorem 3.4.

Example (Triangular equation). Let $\mathbb{N}_0 = \mathbb{N} \cup \{0\} = \{0, 1, 2, 3, \ldots\}$, and let

$$\mathbb{T} = \left\{\frac{n(n+1)}{2}\right\}_{n=0}^{\infty} = \{0, 1, 3, 6, 10, \ldots\}$$

be the set of triangular numbers. It follows that

$$t_{n+1} - t_n = \mu(t_n) = n + 1, \quad n \in \mathbb{N}_0.$$

If p is a 1-periodic function on $\mathbb{T}$, then $(\frac{n+2}{n+1})p(t_{n+1}) = p(t_n)$ for all $t = t_n \in \mathbb{T}$. Then p takes the form

$$p(t) = \frac{p_0}{\mu(t)} \implies p(t_n) = \frac{p_0}{n+1}, \quad t \in \mathbb{T},$$

for $p_0 \in \mathbb{C}\backslash\{-1\}$. The exponential function is $e_p(t_n) = (1+p_0)^n$ for $p_0 \in \mathbb{C}\backslash\{-1\}$. Then, (3.1) is not Ulam stable over the triangular numbers by Theorem 3.2 and (3.9), since $\mu(t_n) = an + b$ for $a = 1 = b$ for all $n \in \mathbb{N}_0$.

Example (Fibonacci equation). Let

$$\mathbb{T} = \left\{ \frac{(1+\sqrt{5})^{n+1} - (1-\sqrt{5})^{n+1}}{2^{n+1}\sqrt{5}} \right\}_{n=1}^{\infty} = \{1, 2, 3, 5, 8, 13, \dots\}$$
$$= \{t_1, t_2, t_3, t_4, t_5, t_6, \dots\},$$

the set of Fibonacci numbers, where we have omitted the first 1 to avoid the redundancy of two consecutive 1s. It follows that

$$t_{n+1} - t_n = \mu(t_n) = t_{n-1}, \quad n \in \{2, 3, 4, \dots\}, \quad \mu(t_1) = 1.$$

If p is a 1-periodic function on $\mathbb{T}$, then $(\frac{t_n}{t_{n-1}})p(t_{n+1}) = p(t_n)$ for all $t = t_n \in \mathbb{T}$. This p takes the form

$$p(t) = \frac{p_1}{\mu(t)} \implies p(t_n) = \frac{p_1}{t_{n-1}}, \quad t \in \mathbb{T}, \quad t_0 = 1,$$

for $p_1 \in \mathbb{C}\backslash\{-1\}$. The exponential function is $e_p(t_n) = (1+p_1)^{n-1}$ for $p_1 \in \mathbb{C}\backslash\{-1\}$. Then, (3.1) is not Ulam stable over the Fibonacci numbers by Theorem 3.2 and (3.9), since $\mu(t_n) = t_{n-1} \geq \frac{1}{2}n$ for all $n \in \mathbb{N}_0$.

Example (Harmonic equation). Let $\mathbb{N}_0 = \mathbb{N} \cup \{0\} = \{0, 1, 2, 3, \dots\}$, and let

$$\mathbb{T} = \left\{ H_n = \sum_{j=1}^{n} \frac{1}{j} \right\}_{n=0}^{\infty} = \left\{0, 1, \frac{3}{2}, \frac{11}{6}, \frac{25}{12}, \dots\right\}, \quad n \in \mathbb{N}_0,$$

be the set of harmonic numbers. It follows that

$$H_{n+1} - H_n = \mu(H_n) = \frac{1}{n+1} \leq 1, \quad n \in \mathbb{N}_0.$$

If p is a 1-periodic function on $\mathbb{T}$, then $(\frac{n+1}{n+2})p(H_{n+1}) = p(H_n)$ for all $t = H_n \in \mathbb{T}$. This p takes the form

$$p(t) = \frac{p_0}{\mu(t)} \implies p(H_n) = (n+1)p_0, \quad t \in \mathbb{T},$$

for $p_0 \in \mathbb{C}\backslash\{-1\}$. The exponential function is $e_p(H_n) = (1+p_0)^n$ for $p_0 \in \mathbb{C}\backslash\{-1\}$. By Theorem 3.1 and condition (3.8), equation (3.1) is Ulam stable if and only if $|1 + p_0| \neq 1$. Additional work shows that (3.1) is Ulam stable with Ulam constant $L = \ln(\frac{\rho}{\rho-1})$ for $\rho > 1$, $1 + p_0 = \rho e^{i\theta}$, and $e_p(H_n) = \rho^n e^{in\theta}$, and (3.1) is Ulam stable with Ulam constant

$$L = \begin{cases} \max\{\ln(\frac{1-\rho}{\rho}), \rho + \frac{1}{2}\} & \text{if } 0 < \rho \le \frac{1}{2}, \\ \max\{\ln(\frac{\rho}{1-\rho}), \rho + \frac{1}{2}, \frac{1}{3} + \frac{\rho}{2} + \rho^2\} & \text{if } \frac{1}{2} \le \rho < 1, \end{cases}$$

for $0 < \rho < 1$, $1 + p_0 = \rho e^{i\theta}$, and $e_p(H_n) = \rho^n e^{in\theta}$.

Proof. (i) Suppose $1 + p_0 = \rho e^{i\theta}$ for any $\theta \in [0, 2\pi)$ and any real $\rho > 1$. For arbitrary $\varepsilon > 0$, suppose ψ satisfies (3.7), with $|\eta(H_n)| \le \varepsilon$ for all $t = H_n \in \mathbb{T}$. It follows that $e_p(H_n) = \rho^n e^{in\theta}$, and from (3.10) that ψ has the form

$$\psi(H_n) = \psi_0 \rho^n e^{in\theta} + \rho^n e^{in\theta} \sum_{j=0}^{n-1} \frac{\eta(H_j)}{(j+1)\rho^{j+1} e^{i(j+1)\theta}}, \quad \text{assuming } \sum_{j=0}^{-1} f(j) \equiv 0.$$

From this expression for $\psi(H_n)$, we have

$$\left| \frac{\psi(H_n)}{\rho^n e^{in\theta}} \right| \le |\psi_0| + \varepsilon \sum_{j=0}^{n-1} \frac{1}{(j+1)\rho^{j+1}},$$

as $|\eta(H_j)| \le \varepsilon$ for all $t = H_j \in \mathbb{T}$. Note that

$$\sum_{j=0}^{\infty} \frac{1}{(j+1)\rho^{j+1}} = \ln\left(\frac{\rho}{\rho - 1} \right)$$

converges, as in this case $\rho > 1$. Thus, ψ can be expressed anew as

$$\psi(H_n) = \left[\psi_0 + \sum_{j=0}^{\infty} \frac{\eta(H_j)}{(j+1)\rho^{j+1} e^{i(j+1)\theta}} \right] e_p(H_n) - e_p(H_n) \sum_{j=n}^{\infty} \frac{\eta(H_j)}{(j+1)\rho^{j+1} e^{i(j+1)\theta}}, \tag{3.24}$$

where

$$z_0 := \psi_0 + \sum_{j=0}^{\infty} \frac{\eta(H_j)}{(j+1)\rho^{j+1} e^{i(j+1)\theta}} \in \mathbb{C}$$

exists and is a finite number due to $e_p(H_n) = \rho^n e^{in\theta}$, and the assumption $\rho > 1$. It is straightforward to see that

$$z(H_n) := z(H_0) e_p(H_n) = z_0 \rho^n e^{in\theta}, \quad H_n \in \mathbb{T}$$

is a solution of (3.1), and

$$\lim_{n \to \infty} \frac{\psi(H_n)}{e_p(H_n)} = \psi_0 + \sum_{j=0}^{\infty} \frac{\eta(H_j)}{(j+1)\rho^{j+1} e^{i(j+1)\theta}} = z_0$$

exists. Consequently,

$$z(H_n) = \left(\lim_{n\to\infty} \frac{\psi(H_n)}{e_p(H_n)}\right) e_p(H_n),$$

and

$$\begin{aligned}|\psi(H_n) - z(H_n)| &= \left| -e_p(H_n) \sum_{j=n}^{\infty} \frac{\eta(H_j)}{(j+1)\rho^{j+1}e^{i(j+1)\theta}} \right| \\ &\le \varepsilon\rho^n \sum_{j=n}^{\infty} \frac{1}{(j+1)\rho^{j+1}} = \frac{\varepsilon}{\rho}\Phi\left(\frac{1}{\rho}, 1, n+1\right) \le \varepsilon \ln\left(\frac{\rho}{\rho-1}\right)\end{aligned}$$

holds for all $t \in \mathbb{T}$, where $\Phi = \Phi(\cdot,\cdot,\cdot)$ is the Lerch transcendent function. Therefore, (3.1) is Ulam stable with Ulam constant $L = \ln(\frac{\rho}{\rho-1})$ for $\rho > 1, 1+p_0 = \rho e^{i\theta}$, and $e_p(H_n) = \rho^n e^{in\theta}$.

(ii) Now assume $1 + p_0 = \rho e^{i\theta}$ for $0 < \rho < 1$. As before, we know that a function ψ that satisfies (3.7) takes the form

$$\psi(H_n) = \psi_0 e_p(H_n) + e_p(H_n) \sum_{j=0}^{n-1} \frac{\eta(H_j)}{(j+1)e_p(H_{j+1})}$$

by (3.10). Let $z = z(H_n) = \psi_0 e_p(H_n)$ for $t = H_n \in \mathbb{T}$ be the solution of (3.1) such that $z_0 = \psi_0$. Then we have

$$\psi(H_n) - z(H_n) = e_p(H_n) \sum_{j=0}^{n-1} \frac{\eta(H_j)}{(j+1)e_p(H_{j+1})},$$

and thus (using a computer algebra system and the Lerch transcendent Φ)

$$\begin{aligned}|\psi(H_n) - z(H_n)| &\le \varepsilon\rho^n \sum_{j=0}^{n-1} \frac{1}{(j+1)\rho^{j+1}} \\ &= -\frac{\varepsilon}{\rho}\Phi\left(\frac{1}{\rho}, 1, n+1\right) - \varepsilon\rho^n \ln\left(\frac{\rho-1}{\rho}\right) \\ &\le \varepsilon \begin{cases} \max\{\ln(\frac{1-\rho}{\rho}), \rho + \frac{1}{2}\} & \text{if } 0 < \rho \le \frac{1}{2}, \\ \max\{\ln(\frac{\rho}{1-\rho}), \rho + \frac{1}{2}, \frac{1}{3} + \frac{\rho}{2} + \rho^2\} & \text{if } \frac{1}{2} \le \rho < 1, \end{cases}\end{aligned}$$

for all $t = H_n \in \mathbb{T}$. Therefore, (3.1) is Ulam stable with Ulam constant

$$L = \begin{cases} \max\{\ln(\frac{1-\rho}{\rho}), \rho + \frac{1}{2}\} & \text{if } 0 < \rho \le \frac{1}{2}, \\ \max\{\ln(\frac{\rho}{1-\rho}), \rho + \frac{1}{2}, \frac{1}{3} + \frac{\rho}{2} + \rho^2\} & \text{if } \frac{1}{2} \le \rho < 1, \end{cases}$$

for $0 < \rho < 1, 1+p_0 = \rho e^{i\theta}$, and $e_p(H_n) = \rho^n e^{in\theta}$. Overall, we have shown that (3.1) is Ulam stable if $|1 + p_0| = \rho \ne 1$, completing the proof. □

Example. Let $\mathbb{N}_0 = \mathbb{N} \cup \{0\} = \{0, 1, 2, 3, \ldots\}$, and let

$$\mathbb{T} = \left\{ t_n := n + H_n = n + \sum_{j=1}^{n} \frac{1}{j} \right\}_{n=0}^{\infty} = \left\{ 0, 2, \frac{7}{2}, \frac{29}{6}, \frac{73}{12}, \ldots \right\}, \quad n \in \mathbb{N}_0,$$

where H_n is the nth harmonic number. It follows that

$$t_{n+1} - t_n = \mu(t_n) = 1 + \frac{1}{n+1} = \frac{n+2}{n+1} \leq 2, \quad n \in \mathbb{N}_0.$$

If p is a 1-periodic function on $\mathbb{T}$, then $(\frac{(n+3)(n+1)}{(n+2)^2})p(t_{n+1}) = p(t_n)$ for all $t = t_n \in \mathbb{T}$. This p takes the form

$$p(t) = \frac{2p_0}{\mu(t)} \implies p(t_n) = \frac{2(n+1)p_0}{n+2}, \quad t \in \mathbb{T},$$

for $p_0 \in \mathbb{C}\backslash\{-\frac{1}{2}\}$. The exponential function is $e_p(t_n) = (1 + 2p_0)^n$ for $p_0 \in \mathbb{C}\backslash\{-\frac{1}{2}\}$. By Theorem 3.2, equation (3.1) is Ulam stable if and only if $|1 + 2p_0| \neq 1$. Additional work shows that (3.1) is Ulam stable with Ulam constant $L = (\ln(\frac{\rho}{\rho-1}) + \frac{1}{\rho-1})$ for $\rho > 1, 1 + 2p_0 = \rho e^{i\theta}$, and $e_p(t_n) = \rho^n e^{in\theta}$, and (3.1) is Ulam stable with Ulam constant

$$L = \left(\left| \operatorname{Log}\left(\frac{\rho}{\rho - 1} \right) \right| + \frac{1}{1 - \rho} \right)$$

for $0 < \rho < 1, 1 + 2p_0 = \rho e^{i\theta}$, and $e_p(t_n) = \rho^n e^{in\theta}$.

Proof. (i) Suppose $1 + 2p_0 = \rho e^{i\theta}$ for any $\theta \in [0, 2\pi)$ and any real $\rho > 1$. For arbitrary $\varepsilon > 0$, suppose ψ satisfies (3.7), with $|\eta(t_n)| \leq \varepsilon$ for all $t = t_n \in \mathbb{T}$. It follows that $e_p(t_n) = \rho^n e^{in\theta}$, and from (3.10) that ψ has the form

$$\psi(t_n) = \psi_0 \rho^n e^{in\theta} + \rho^n e^{in\theta} \sum_{j=0}^{n-1} \frac{(j+2)\eta(t_j)}{(j+1)\rho^{j+1} e^{i(j+1)\theta}}, \quad \text{assuming } \sum_{j=0}^{-1} f(j) \equiv 0.$$

From this expression for $\psi(t_n)$, we have

$$\frac{|\psi(t_n)|}{\rho^n} = \left| \frac{\psi(t_n)}{\rho^n e^{in\theta}} \right| \leq |\psi_0| + \varepsilon \sum_{j=0}^{n-1} \frac{(j+2)}{(j+1)\rho^{j+1}},$$

as $|\eta(t_j)| \leq \varepsilon$ for all $t = t_j \in \mathbb{T}$. Note that

$$\sum_{j=0}^{\infty} \frac{(j+2)}{(j+1)\rho^{j+1}} = \ln\left(\frac{\rho}{\rho - 1} \right) + \frac{1}{\rho - 1}$$

converges, as in this case $\rho > 1$. Thus, ψ can be expressed anew as

$$\psi(t_n) = \left[\psi_0 + \sum_{j=0}^{\infty} \frac{(j+2)\eta(t_j)}{(j+1)\rho^{j+1}e^{i(j+1)\theta}}\right] e_p(t_n) - e_p(t_n) \sum_{j=n}^{\infty} \frac{(j+2)\eta(t_j)}{(j+1)\rho^{j+1}e^{i(j+1)\theta}}, \tag{3.25}$$

where

$$z_0 := \psi_0 + \sum_{j=0}^{\infty} \frac{(j+2)\eta(t_j)}{(j+1)\rho^{j+1}e^{i(j+1)\theta}} \in \mathbb{C}$$

exists and is a finite number due to $e_p(t_n) = \rho^n e^{in\theta}$, the assumption $\rho > 1$, and the absolute convergence of the infinite series. It is straightforward to see that

$$z(t_n) := z(t_0)e_p(t_n) = z_0\rho^n e^{in\theta}, \quad t_n \in \mathbb{T}$$

is a solution of (3.1), and

$$\lim_{n\to\infty} \left(\frac{\psi(t_n)}{e_p(t_n)}\right) = \psi_0 + \sum_{j=0}^{\infty} \frac{(j+2)\eta(t_j)}{(j+1)\rho^{j+1}e^{i(j+1)\theta}} = z_0$$

exists. Consequently,

$$z(t_n) = \left(\lim_{n\to\infty} \frac{\psi(t_n)}{e_p(t_n)}\right) e_p(t_n),$$

and

$$\begin{aligned}
|\psi(t_n) - z(t_n)| &= \left| -e_p(t_n) \sum_{j=n}^{\infty} \frac{(j+2)\eta(t_j)}{(j+1)\rho^{j+1}e^{i(j+1)\theta}} \right| \\
&\le \varepsilon\rho^n \sum_{j=n}^{\infty} \frac{(j+2)}{(j+1)\rho^{j+1}} \\
&= \varepsilon\rho^n B\left(\frac{1}{\rho}, n+2, -1\right) + \frac{\varepsilon(n+2)}{\rho}\Phi\left(\frac{1}{\rho}, 1, n+1\right) \\
&\le \varepsilon\left(\ln\left(\frac{\rho}{\rho-1}\right) + \frac{1}{\rho-1}\right)
\end{aligned}$$

holds for all $t \in \mathbb{T}$, where $B = B(\cdot,\cdot,\cdot)$ is the incomplete Beta function, and $\Phi = \Phi(\cdot,\cdot,\cdot)$ is the Hurwitz–Lerch transcendent function. Therefore, (3.1) is Ulam stable with Ulam constant $L = (\ln(\frac{\rho}{\rho-1}) + \frac{1}{\rho-1})$ for $\rho > 1$, $1 + 2p_0 = \rho e^{i\theta}$, and $e_p(t_n) = \rho^n e^{in\theta}$.

(ii) Now assume $1 + 2p_0 = \rho e^{i\theta}$ for $0 < \rho < 1$. As before, we know that a function ψ that satisfies (3.7) takes the form

$$\psi(t_n) = \psi_0 e_p(t_n) + e_p(t_n) \sum_{j=0}^{n-1} \frac{(j+2)\eta(t_j)}{(j+1)e_p(t_{j+1})}$$

by (3.10). Let $z = z(t_n) = \psi_0 e_p(t_n)$ for $t = t_n \in \mathbb{T}$ be the solution of (3.1) such that $z_0 = \psi_0$. Then we have

$$\psi(t_n) - z(t_n) = e_p(t_n) \sum_{j=0}^{n-1} \frac{(j+2)\eta(t_j)}{(j+1)e_p(t_{j+1})},$$

and thus (using a computer algebra system, the Hurwitz–Lerch transcendent Φ, and the principal logarithm Log)

$$\begin{aligned}
|\psi(t_n) - z(t_n)| &\le \varepsilon\rho^n \sum_{j=0}^{n-1} \frac{(j+2)}{(j+1)\rho^{j+1}} \\
&= \varepsilon\left(\frac{-1+\rho^n}{-1+\rho} - \frac{1}{\rho}\Phi\left(\frac{1}{\rho}, 1, n+1\right) - \rho^n \operatorname{Log}\left(\frac{\rho-1}{\rho}\right)\right) \\
&\le \varepsilon\left(\left|\operatorname{Log}\left(\frac{\rho}{\rho-1}\right)\right| + \frac{1}{1-\rho}\right)
\end{aligned}$$

for all $t = t_n \in \mathbb{T}$. Therefore, (3.1) is Ulam stable with Ulam constant

$$L = \left(\left|\operatorname{Log}\left(\frac{\rho}{\rho-1}\right)\right| + \frac{1}{1-\rho}\right)$$

for $0 < \rho < 1$, $1 + 2p_0 = \rho e^{i\theta}$, and $e_p(t_n) = \rho^n e^{in\theta}$. Overall, we have shown that (3.1) is Ulam stable if $|1 + 2p_0| = \rho \neq 1$, completing the proof. □

Example (Euler type q-difference equation). Let $q > 1$, and set $\mathbb{S} = q^{\mathbb{N}_0} = \{1, q, q^2, q^3, \ldots\}$. When $\alpha(s) = s$ and $\beta(s) \equiv \beta \in \mathbb{C}$, equation (3.18) becomes the first-order linear homogeneous dynamic equation of Euler type, represented by

$$sy^{\Delta}(s) - \beta y(s) = 0. \tag{3.26}$$

Note from Example 3.6 that simply dividing this equation by s is not Ulam stable. However, (3.26) is Ulam stable if and only if $\beta \in \mathbb{C}\setminus\{\frac{-1}{q-1}\}$, with Ulam constant $L_s = \frac{q-1}{|1-|1+\beta(q-1)||}$. It is known [3, Theorem 1] that the Ulam constant L_s is the best Ulam constant for (3.26) on $q^{\mathbb{Z}}$.

Proof. Let $\phi(s) = \log_q s$. Since $\mathbb{S} = \{s_n\}_{n=0}^{\infty} = \{q^n\}_{n=0}^{\infty}$ is an isolated time scale with the graininess function $\mu(s_n) = \mu(q^n) = (q-1)q^n > 0$ for $n \in \mathbb{N}_0$, we see that

$$\lim_{n\to\infty} \phi(s_n) = \lim_{n\to\infty} n = \infty,$$

and

$$\alpha(s_n)\phi^{\Delta}(s_n) = q^n \frac{\log_q q^{n+1} - \log_q q^n}{q^{n+1} - q^n} = \frac{1}{q-1} = K > 0$$

for all $s_n = s \in \mathbb{S}$; that is, (3.20) holds. Moreover,

$$\frac{\beta(s_n)}{\alpha(s_n)} = \frac{\beta}{q^n} = \frac{\beta(s_0)\mu(s_0)}{\alpha(s_0)\mu(s_n)}$$

for all $s_n = s \in \mathbb{S}$ means that $\frac{\beta(s)}{\alpha(s)}$ is 1-periodic. $\beta \in \mathbb{C}\backslash\{\frac{-1}{q-1}\}$ says that $\frac{\beta(s_0)}{\alpha(s_0)} \in \mathbb{C}\backslash\{\frac{-1}{\mu(s_0)}\}$. From $\frac{\mu(s_n)}{\alpha(s_n)} = q-1 = M$ for all $s_n = s \in \mathbb{S}$, we obtain (3.21). Hence, by Theorem 3.5, we can conclude that (3.26) is Ulam stable if and only if

$$\rho = \left|1 + \frac{\beta(s_0)\mu(s_0)}{\alpha(s_0)}\right| = |1 + \beta(q-1)| \neq 1,$$

that is, $\beta \in \mathbb{C}\backslash\{\frac{-1}{q-1}\}$, with Ulam constant $L_s = \frac{M}{|1-\rho|} = \frac{q-1}{|1-|1+\beta(q-1)||}$. □

Example (Euler type q-difference equation with variable coefficient $\beta(s)$). Let $q > 1$, and set $\mathbb{S} = q^{\mathbb{N}_0} = \{1, q, q^2, q^3, \ldots\}$. Consider the first-order linear homogeneous dynamic equation

$$sy^{\Delta}(s) - \beta(s)y(s) = 0, \tag{3.27}$$

where

$$\beta(s) := \begin{cases} \beta_0 & \text{if } \log_q s \equiv 0 \pmod 2, \\ \beta_1 & \text{if } \log_q s \equiv 1 \pmod 2. \end{cases}$$

Note from Example 3.6 that dividing this equation by s is not Ulam stable. However, (3.27) is Ulam stable if and only if $\beta_k \in \mathbb{C}\backslash\{\frac{-1}{q-1}\}$ for $k \in \{0,1\}$, with Ulam constant

$$\begin{aligned} L_s = {} & \max\{|(1+\beta_0(q-1))| + 1, |(1+\beta_1(q-1))| + 1\} \\ & \times \frac{q-1}{||(1+\beta_0(q-1))(1+\beta_1(q-1))| - 1|}. \end{aligned}$$

Proof. Let $\phi(s) = \log_q s$. Then, as in the previous example, we get the following facts: $\mathbb{S}$ is an isolated time scale with the graininess function $\mu(s_n) = (q-1)q^n > 0$ for $n \in \mathbb{N}_0$, $\lim_{n\to\infty} \phi(s_n) = \infty$, (3.20) and (3.21) hold with $K = \frac{1}{q-1}$ and $M = q-1$. Moreover, we have

$$\begin{aligned} \frac{\beta(s_n)}{\alpha(s_n)} &= \frac{1}{(q-1)q^n} \begin{cases} (q-1)\beta_0 & \text{if } n \equiv 0 \pmod 2, \\ (q-1)\beta_1 & \text{if } n \equiv 1 \pmod 2 \end{cases} \\ &= \frac{1}{\mu(s_n)} \begin{cases} \frac{\beta(s_0)\mu(s_0)}{\alpha(s_0)} & \text{if } n \equiv 0 \pmod 2, \\ \frac{\beta(s_1)\mu(s_1)}{\alpha(s_1)} & \text{if } n \equiv 1 \pmod 2. \end{cases} \end{aligned}$$

This says that (3.23) holds, and $\frac{\beta(s)}{\alpha(s)}$ is 2-periodic; $\beta_k \in \mathbb{C}\backslash\{\frac{-1}{q-1}\}$ for $k \in \{0,1\}$ implies that $\frac{\beta(s_k)}{\alpha(s_k)} \in \mathbb{C}\backslash\{\frac{-1}{\mu(s_k)}\}$ for $k \in \{0,1\}$. Using Theorem 3.6, we can conclude that (3.27) is Ulam stable if and only if

$$\left|(1+\beta_0(q-1))(1+\beta_1(q-1))\right| \neq 1;$$

that is, $\beta_k \in \mathbb{C}\backslash\{\frac{-1}{q-1}\}$ for $k \in \{0,1\}$, with Ulam constant

$$\begin{aligned} L_s = &\max\{\left|1+\beta_0(q-1)\right|+1, \left|1+\beta_1(q-1)\right|+1\} \\ &\times \frac{q-1}{\left||(1+\beta_0(q-1))(1+\beta_1(q-1))|-1\right|}. \end{aligned}$$

This completes the proof. □

Example. Set $\mathbb{S} = \{1, 2^2, 3^2, 4^2, \ldots\}$. When $\alpha(s) = 2\sqrt{s}+1$, $s \in \mathbb{S}$, and $\beta(s) \equiv \beta \in \mathbb{C}$, $s \in \mathbb{S}$, equation (3.18) becomes the first-order linear homogeneous dynamic equation

$$(2\sqrt{s}+1)y^\Delta(s) - \beta y(s) = 0, \quad s \in \mathbb{S}. \tag{3.28}$$

Then (3.28) is Ulam stable if and only if $\beta \in \mathbb{C}\backslash\{-1\}$, with Ulam constant $L_s = \frac{1}{|1-|1+\beta||}$. On the other hand, if we divide both sides of this equation by $\alpha(s)$, we get the equation

$$y^\Delta(s) - \frac{\beta}{2\sqrt{s}+1} y(s) = 0, \quad s \in \mathbb{S}. \tag{3.29}$$

Then (3.29) is not Ulam stable for any $\beta \in \mathbb{C}$.

Proof. Let $\phi(s) = \sqrt{s}$, $s \in \mathbb{S}$. Since $\mathbb{S} = \{s_n\}_{n=0}^\infty = \{(n+1)^2\}_{n=0}^\infty$ is an isolated time scale with the graininess function

$$\mu(s_n) = \mu((n+1)^2) = 2n+3 > 0$$

for $n \in \mathbb{N}_0$, we have $\lim_{n\to\infty} \phi(s_n) = \lim_{n\to\infty} n = \infty$ and $\mu(s_n) = \alpha(s_n)$. So, $\alpha(s_n)\phi^\Delta(s_n) = 1 = K > 0$ for all $s_n = s \in \mathbb{S}$. Thus, (3.20) is satisfied. In addition,

$$\frac{\beta(s_n)}{\alpha(s_n)} = \frac{\beta}{\mu(s_n)} = \frac{\beta(s_0)\mu(s_0)}{\alpha(s_0)\mu(s_n)}$$

holds for all $s_n = s \in \mathbb{S}$. Thus, $\frac{\beta(s)}{\alpha(s)}$ is 1-periodic, and $\beta \in \mathbb{C}\backslash\{-1\}$ implies $\frac{\beta(s_0)}{\alpha(s_0)} \in \mathbb{C}\backslash\{\frac{-1}{\mu(s_0)}\}$. From $\frac{\mu(s_n)}{\alpha(s_n)} = 1 = M$ for all $s_n = s \in \mathbb{S}$, we get (3.21). Hence, by Theorem 3.5, we can conclude that (3.28) is Ulam stable if and only if $\rho = |1 + \frac{\beta(s_0)\mu(s_0)}{\alpha(s_0)}| = |1+\beta| \neq 1$, that is, $\beta \in \mathbb{C}\backslash\{-1\}$, with Ulam constant $L_s = \frac{M}{|1-\rho|} = \frac{1}{|1-|1+\beta||}$.

Next we will show that (3.29) is unstable in the Ulam sense. From $\mu(s_n) = \alpha(s_n)$, it follows that

$$p(s) = \frac{\beta}{2\sqrt{s}+1} = \frac{\beta}{\alpha(s)} = \frac{\beta}{\mu(s)}, \qquad s \in \mathbb{S},$$

that is, p is 1-periodic. Since we have $\mu(s_n) = 2n+3$, (3.9) holds. By using Theorem 3.2, we see that (3.29) is not Ulam stable, completing the proof. □

Example. Set $\mathbb{S} = \{1, 2^2, 3^2, 4^2, \dots\}$. Consider the first-order linear homogeneous dynamic equations

$$(2\sqrt{s}+1)y^{\Delta}(s) - \beta(s)y(s) = 0, \qquad s \in \mathbb{S}, \tag{3.30}$$

and

$$y^{\Delta}(s) - \frac{\beta(s)}{2\sqrt{s}+1}y(s) = 0, \qquad s \in \mathbb{S}, \tag{3.31}$$

where

$$\beta(s) := \begin{cases} \beta_0 & \text{if } \sqrt{s} \equiv 0 \pmod 2, \\ \beta_1 & \text{if } \sqrt{s} \equiv 1 \pmod 2. \end{cases}$$

Then (3.30) is Ulam stable if and only if $\beta_k \in \mathbb{C}\backslash\{-1\}$ for $k \in \{0,1\}$, with Ulam constant

$$L_s = \frac{\max\{|1+\beta_0|+1, |1+\beta_1|+1\}}{||(1+\beta_0)(1+\beta_1)|-1|}.$$

On the other hand, (3.31) is not Ulam stable.

Proof. It can be easily shown by using the proofs of Examples 3.6 and 3.6, so we omit it. □

3.7 Conclusion and future directions

In this paper, we investigate the stability and instability of first-order dynamic equations on isolated time scales with 1- and 2-periodic coefficients. Additionally, we provide examples of isolated time scales with 1- or 2-periodic coefficient functions, and apply the results to these cases, including for h-difference equations, q-difference equations, triangular equations, Fibonacci equations, harmonic equations, and Euler type q-difference equations. In the future, we will extend the results to include the stability and instability of first-order linear dynamic equations on isolated time scales with n-periodic coefficients, and also to higher-order linear dynamic equations on isolated time scales.

Bibliography

[1] D. R. Anderson and M. Onitsuka, Hyers–Ulam stability for a discrete time scale with two step sizes, *Appl. Math. Comput.*, 344–345 (2019), 128–140.

[2] D. R. Anderson and M. Onitsuka, Best constant for Hyers–Ulam stability of second-order h-difference equations with constant coefficients, *Results Math.*, 74 (2019), 151. https://doi.org/10.1007/s00025-019-1077-9.

[3] D. R. Anderson and M. Onitsuka, Hyers–Ulam stability for quantum equations of Euler type, *Discrete Dyn. Nat. Soc.*, 2020 (2020), 5626481. https://doi.org/10.1155/2020/5626481.

[4] D. R. Anderson and M. Onitsuka, Best constant for Hyers–Ulam stability of two step sizes linear difference equations, *J. Math. Anal. Appl.*, 496(2) (2021), 124807. https://doi.org/10.1016/j.jmaa.2020.124807.

[5] D. R. Anderson and M. Onitsuka, Hyers–Ulam stability for Cayley quantum equations and its application to h-difference equations, *Mediterr. J. Math.*, 18 (2021), 168.

[6] D. R. Anderson and M. Onitsuka, Ulam stability for nonautonomous quantum equations, *J. Inequal. Appl.*, 2021 (2021), 161. https://doi.org/10.1186/s13660-021-02699-4.

[7] D. R. Anderson, M. Onitsuka and J. M. Rassias, Best constant for Ulam stability of first-order h-difference equations with periodic coefficient, *J. Math. Anal. Appl.*, 491 (2020), 124363. https://doi.org/10.1016/j.jmaa.2020.124363.

[8] S. András and A. R. Mészáros, Ulam–Hyers stability of dynamic equations on time scales via Picard operators, *Appl. Math. Comput.*, 219 (2013), 4853–4864.

[9] M. Bohner and R. Chieochan, Floquet theory for q-difference equations, *Sarajevo J. Math.*, 8(21)(2) (2012), 355–366.

[10] M. Bohner, J. Mesquita and S. Streipert, Periodicity on isolated time scales, *Math. Nachr.*, 295 (2022), 259–280. https://doi.org/10.1002/mana.201900360.

[11] M. Bohner and A. Peterson, Dynamic Equations on Time Scales, An Introduction with Applications, Birkhäuser, Boston, 2001.

[12] N. Brillouët-Belluot, J. Brzdek and K. Ciepliński, On some recent developments in Ulam's type stability, *Abstr. Appl. Anal.*, 2012 (2012), 716936.

[13] J. Brzdek, D. Popa, I. Rasa and B. Xu, Ulam Stability of Operators, A volume in Mathematical Analysis and its Applications, Academic Press, 2018.

[14] J. Brzdek and P. Wójcik, On approximate solutions of some difference equations, *Bull. Aust. Math. Soc.*, 95(3) (2017), 476–481.

[15] R. Fukutaka and M. Onitsuka, Best constant in Hyers–Ulam stability of first-order homogeneous linear differential equations with a periodic coefficient, *J. Math. Anal. Appl.*, 473(2) (2019), 1432–1446.

[16] L. Hua, Y. Li and J. Feng, On Hyers–Ulam stability of dynamic integral equation on time scales, *Math. Æterna*, 4(6) (2014), 559–571.

[17] D. H. Hyers, On the stability of the linear functional equation, *Proc. Natl. Acad. Sci. USA*, 27 (1941), 222–224.

[18] S.-M. Jung and Y. W. Nam, Hyers–Ulam stability of Pielou logistic difference equation, *J. Nonlinear Sci. Appl.*, 10 (2017), 3115–3122.

[19] Y. W. Nam, Hyers–Ulam stability of hyperbolic Möbius difference equation, *Filomat*, 32; 13 (2018), 4555–4575. https://doi.org/10.2298/FIL1813555N.

[20] Y. W. Nam, Hyers–Ulam stability of elliptic Möbius difference equation, *Cogent Math. & Stat.*, 5(1) (2018), 1–9.

[21] Y. W. Nam, Hyers–Ulam stability of loxodromic Möbius difference equation, *Appl. Math. Comput.*, 356(1) (2019), 119–136. https://doi.org/10.1016/j.amc.2019.03.033.

[22] M. Onitsuka, Influence of the stepsize on Hyers–Ulam stability of first-order homogeneous linear difference equations, *Int. J. Difference Equ.*, 12(2) (2017), 281–302.

[23] M. Onitsuka, Hyers–Ulam stability of second-order nonhomogeneous linear difference equations with a constant step size, *J. Comput. Anal. Appl.*, 28(1) (2020), 152–165.
[24] D. Popa, Hyers–Ulam stability of the linear recurrence with constant coefficients, *Adv. Differ. Equ.*, 2005 (2005), 407076.
[25] D. Popa, Hyers–Ulam–Rassias stability of a linear recurrence, *J. Math. Anal. Appl.*, 309 (2005), 591–597.
[26] H. Rasouli, S. Abbaszadeh and M. Eshaghi, Approximately linear recurrences, *J. Appl. Anal.*, 24(1) (2018), 81–85.
[27] Th. M. Rassias, On the stability of linear mapping in Banach spaces, *Proc. Am. Math. Soc.*, 72 (1978), 297–300.
[28] Y. H. Shen, The Ulam stability of first order linear dynamic equations on time scales, *Results Math.*, 72(4) (2017), 1881–1895.
[29] Y. H. Shen and Y. J. Li, Hyers–Ulam stability of first order nonhomogeneous linear dynamic equations on time scales, *Commun. Math. Res.*, 35(2) (2019), 139–148. https://doi.org/10.13447/j.1674-5647.2019.02.05.
[30] S. M. Ulam, A Collection of the Mathematical Problems, Interscience, New York, 1960.

Douglas R. Anderson, Martin Bohner, and Guo-Cheng Wu

4 A logarithm on time scales and its uses

Abstract: A multivalued logarithm on time scales recently introduced for delta-differentiable functions that never vanish is covered in this chapter. This is accomplished using an extended definition of the cylinder transformation from which the definition of exponential functions on time scales arose. The definition of a logarithm function on arbitrary time scales with familiar and useful properties then follows.

4.1 Prelude to the time scales logarithm

The new multivalued logarithm recapitulated here fills a gap for time scales and dynamic equations, namely how to define and represent a logarithm function on time scales with properties that extend the well-known properties of the logarithm function for the continuous case. The purpose of what follows below is to present how the novel multivalued logarithm arises naturally from the cylinder transformation employed in definitions of exponential functions for dynamic equations, and the properties that follow from this extension.

The logarithm on general time scales and its development as summarized in this chapter will proceed as follows. The definition of the traditional single-valued cylinder transformation extended to a multivalued cylinder transformation is given in Section 4.2. Useful properties across the circle plus ($\oplus$) and circle dot ($\odot$) operations are preserved by this transformation, at least for nonvanishing delta-differentiable functions. In Section 4.3, familiar and desired properties of this new logarithm are shown to hold in this more general setting. A similar logarithm for the nabla case is established in Section 4.4. The Cayley cylinder transformation is also considered, in Section 4.5, and is shown to lead to the same exact logarithm. In the literature, this is not the only approach to the logarithm question, so in Section 4.6 we give a summary of the proposed logarithm functions on time scales to date. To conclude, a numerical comparison of the various logarithms on a specific time scale is given in Section 4.7, and several examples on various time scales illustrating the properties of the new definition are also provided. These results may be found in Anderson and Bohner [1].

Douglas R. Anderson, Department of Mathematics, Concordia College, Moorhead, MN 56562, USA, e-mail: andersod@cord.edu
Martin Bohner, Department of Mathematics and Statistics, Missouri University of Science and Technology, Rolla, MO 65409-0020, USA, e-mail: bohner@mst.edu
Guo-Cheng Wu, Data Recovery Key Laboratory of Sichuan Province, College of Mathematics and Information Science, Neijiang Normal University, Neijiang 641100, PR China, e-mail: wuguocheng@gmail.com

https://doi.org/10.1515/9783111182971-004

4.2 A new logarithm on time scales

The presentation of a new definition of a logarithm for dynamic equations on general time scales starts with some motivation provided by the base definition of the cylinder transformation upon which the exponential functions for dynamic equations are built. The following definition [4, Definition 2.21] (see also Hilger [10, Section 7]) is the original cylinder transformation; the proposed modified cylinder transformation is also presented, in Section 4.5.

Any closed subset of the real line may serve as a time scale. Such a subset induces the forward jump operator $\sigma : \mathbb{T} \to \mathbb{T}$ given by $\sigma(t) = \inf\{s \in \mathbb{T} : s > t\}$, and the graininess function $\mu(t) := \sigma(t) - t$; see [2–5] for more details.

Definition 4.1 (Single-valued cylinder transformation). Fix $h > 0$, and define the cylinder transformation $\xi_h : \mathbb{C}_h \to \mathbb{Z}_h$ by

$$\xi_h(z) = \begin{cases} \frac{1}{h}\operatorname{Log}(1 + zh) & \text{for } h \neq 0, \\ z & \text{for } h = 0, \end{cases} \tag{4.1}$$

where $\mathbb{C}$ is the set of complex numbers,

$$\mathbb{C}_h = \left\{z \in \mathbb{C} : z \neq -\frac{1}{h}\right\}, \quad \mathbb{Z}_h = \left\{z \in \mathbb{C} : -\frac{\pi}{h} < \operatorname{Im}(z) \leq \frac{\pi}{h}\right\}, \tag{4.2}$$

and Log represents the principal complex logarithm function.

The following definition is [4, Definition 2.25].

Definition 4.2 (Regressive function). A function $p : \mathbb{T} \to \mathbb{R}$ is regressive granted

$$1 + \mu(\tau)p(\tau) \neq 0 \quad \text{for each } \tau \in \mathbb{T}^\kappa$$

holds. We will denote via $\mathscr{R}$ the set of all rd-continuous and regressive functions $p : \mathbb{T} \to \mathbb{R}$.

The following definition is [4, Definition 2.30]. This definition sets the foundation for offering the extended formulation of logarithms to time scales. The new definition is merely the extension to a multivalued function, which requires only a modification of the single-valued cylinder function given above in (4.1).

Definition 4.3 (Exponential function). For functions $p \in \mathscr{R}$, the time scales exponential function is formulated via

$$e_p(t, s) = \exp\left(\int_s^t \xi_{\mu(\tau)}(p(\tau))\Delta\tau\right) \quad \text{for } s, t \in \mathbb{T};$$

here, $\xi_h(z)$ is the cylinder transformation given in (4.1).

Definition 4.4 (Multivalued cylinder transformation). Fix $h > 0$, and define the multivalued cylinder transformation $\zeta_h : \mathbb{C}_h \to \mathbb{C}$ by

$$\zeta_h(z) = \begin{cases} \frac{1}{h}\log(1+zh) & \text{for } h \neq 0, \\ z & \text{for } h = 0, \end{cases} \tag{4.3}$$

where the set of complex numbers is $\mathbb{C}$, the set $\mathbb{C}_h$ is given in (4.2), and log represents the multivalued complex logarithm function.

Lemma 4.1. *Let $f, g : \mathbb{T} \to \mathbb{C}$ be Δ-differentiable functions with $f, g \neq 0$ on $\mathbb{T}$, and let the multivalued cylinder transformation ζ be given by (4.3). Then, for fixed $\tau \in \mathbb{T}^\kappa$,*

$$\zeta_{\mu(\tau)}\left(\left(\frac{f^\Delta}{f} \oplus \frac{g^\Delta}{g}\right)(\tau)\right) = \zeta_{\mu(\tau)}\left(\frac{f^\Delta(\tau)}{f(\tau)}\right) + \zeta_{\mu(\tau)}\left(\frac{g^\Delta(\tau)}{g(\tau)}\right).$$

Proof. First, note that the simple useful formula $f^\sigma = \mu f^\Delta + f$ (suppressing the variable) implies

$$\frac{(fg)^\Delta}{fg} = \frac{f^\sigma g^\Delta + f^\Delta g}{fg} = \frac{(f+\mu f^\Delta)g^\Delta}{fg} + \frac{f^\Delta}{f} = \frac{f^\Delta}{f} + \frac{g^\Delta}{g} + \mu\frac{f^\Delta g^\Delta}{fg} = \frac{f^\Delta}{f} \oplus \frac{g^\Delta}{g},$$

by the definition [4, Theorem 2.7] of $\oplus$. From this, we observe that for fixed $\tau \in \mathbb{T}^\kappa$,

$$\begin{aligned}
&\zeta_{\mu(\tau)}\left(\left(\frac{f^\Delta}{f} \oplus \frac{g^\Delta}{g}\right)(\tau)\right) \\
&= \zeta_{\mu(\tau)}\left(\frac{(fg)^\Delta(\tau)}{(fg)(\tau)}\right) \\
&= \begin{cases} \frac{1}{\mu(\tau)}\log(1+\mu(\tau)\frac{(fg)^\Delta(\tau)}{(fg)(\tau)}) & \text{for } \mu(\tau) \neq 0, \\ \frac{(fg)^\Delta(\tau)}{(fg)(\tau)} & \text{for } \mu(\tau) = 0 \end{cases} \\
&= \begin{cases} \frac{1}{\mu(\tau)}\log(\frac{(fg)^\sigma(\tau)}{(fg)(\tau)}) & \text{for } \mu(\tau) \neq 0, \\ (\frac{f^\Delta}{f} \oplus \frac{g^\Delta}{g})(\tau) & \text{for } \mu(\tau) = 0 \end{cases} \\
&= \begin{cases} \frac{1}{\mu(\tau)}\log(\frac{f^\sigma(\tau)}{f(\tau)}) + \frac{1}{\mu(\tau)}\log(\frac{g^\sigma(\tau)}{g(\tau)}) & \text{for } \mu(\tau) \neq 0, \\ (\frac{f^\Delta}{f} + \frac{g^\Delta}{g})(\tau) & \text{for } \mu(\tau) = 0 \end{cases} \\
&= \begin{cases} \frac{1}{\mu(\tau)}\log(\frac{(f+\mu f^\Delta)(\tau)}{f(\tau)}) + \frac{1}{\mu(\tau)}\log(\frac{(g+\mu g^\Delta)(\tau)}{g(\tau)}) & \text{for } \mu(\tau) \neq 0, \\ \frac{f^\Delta(\tau)}{f(\tau)} + \frac{g^\Delta(\tau)}{g(\tau)} & \text{for } \mu(\tau) = 0 \end{cases} \\
&= \zeta_{\mu(\tau)}\left(\frac{f^\Delta(\tau)}{f(\tau)}\right) + \zeta_{\mu(\tau)}\left(\frac{g^\Delta(\tau)}{g(\tau)}\right).
\end{aligned}$$

The proof is complete. □

Lemma 4.2. *Let $a \in \mathbb{R}$, and let $p : \mathbb{T} \to \mathbb{C}$ be a Δ-differentiable function with $p \neq 0$ on $\mathbb{T}$. For the multivalued cylinder transformation ζ given by* (4.3) *and for fixed $\tau \in \mathbb{T}^\kappa$,*

$$\zeta_{\mu(\tau)}\left(\left(a \odot \frac{p^\Delta}{p}\right)(\tau)\right) = a\zeta_{\mu(\tau)}\left(\frac{p^\Delta(\tau)}{p(\tau)}\right).$$

Proof. Let $a \in \mathbb{R}$, and let $p : \mathbb{T} \to \mathbb{C}$ be a Δ-differentiable function with $p \neq 0$ on $\mathbb{T}$. Then [5, Theorem 2.43] yields

$$1 + \mu(a \odot f) = (1 + \mu f)^a$$

on $\mathbb{T}^\kappa$ for $f = \frac{p^\Delta}{p}$. It follows that for fixed $\tau \in \mathbb{T}^\kappa$,

$$\begin{aligned}
&\zeta_{\mu(\tau)}\left(\left(a \odot \frac{p^\Delta}{p}\right)(\tau)\right) \\
&\quad = \begin{cases} \frac{1}{\mu(\tau)} \log(1 + \mu(\tau)(a \odot \frac{p^\Delta}{p})(\tau)) & \text{for } \mu(\tau) \neq 0, \\ (a \odot \frac{p^\Delta}{p})(\tau) & \text{for } \mu(\tau) = 0 \end{cases} \\
&\quad = \begin{cases} \frac{1}{\mu(\tau)} \log(1 + \mu(\tau)\frac{p^\Delta(\tau)}{p(\tau)})^a & \text{for } \mu(\tau) \neq 0, \\ a\frac{p^\Delta(\tau)}{p(\tau)} & \text{for } \mu(\tau) = 0 \end{cases} \\
&\quad = a \begin{cases} \frac{1}{\mu(\tau)} \log(1 + \mu(\tau)\frac{p^\Delta(\tau)}{p(\tau)}) & \text{for } \mu(\tau) \neq 0, \\ \frac{p^\Delta(\tau)}{p(\tau)} & \text{for } \mu(\tau) = 0 \end{cases} \\
&\quad = a\zeta_{\mu(\tau)}\left(\frac{p^\Delta(\tau)}{p(\tau)}\right).
\end{aligned}$$

The proof is complete. □

Definition 4.5 (Logarithm function). Given a Δ-differentiable function $p : \mathbb{T} \to \mathbb{C}$ with $p \neq 0$ on $\mathbb{T}$, the multivalued logarithm function on time scales is given by

$$\ell_p(t, s) = \int_s^t \zeta_{\mu(\tau)}\left(\frac{p^\Delta(\tau)}{p(\tau)}\right)\Delta\tau \quad \text{for } s, t \in \mathbb{T},$$

where $\zeta_h(z)$ is the multivalued cylinder transformation given in (4.3). Define the principal logarithm on time scales to be

$$L_p(t, s) = \int_s^t \xi_{\mu(\tau)}\left(\frac{p^\Delta(\tau)}{p(\tau)}\right)\Delta\tau \quad \text{for } s, t \in \mathbb{T},$$

where $\xi_h(z)$ is the single-valued cylinder transformation given in (4.1).

Remark 4.1. According to this definition, if $p \equiv$ constant, then $\ell_p(t,s) = 0$ for each $t,s \in \mathbb{T}$. Thus, this logarithm does not distinguish between either constants or constant multiples of functions. We, moreover, note here that even when we restrict the time scale to $\mathbb{T} = \mathbb{R}$, the dynamics along the negative and positive real line necessitate the existence of a logarithm with principal and multiple values, making a multivalued logarithm on general time scales both natural and expected, though heretofore unexplored.

4.3 Properties of the logarithm

Using the definition of the multivalued logarithm on time scales given above in Definition 2.7, we establish the following properties.

Theorem 4.1. *If $p : \mathbb{T} \to \mathbb{C}$ is a Δ-differentiable function with $p \neq 0$ on $\mathbb{T}$, then*

$$\exp(L_p(t,s)) = e_{\frac{p^\Delta}{p}}(t,s), \quad t,s \in \mathbb{T}.$$

In particular, if $p \in \mathcal{R}$, then

$$\exp(L_{e_p}(t,s)) = e_p(t,s), \quad t,s \in \mathbb{T}.$$

Proof. Presuming $p : \mathbb{T} \to \mathbb{C}$ is a Δ-differentiable function with $p \neq 0$ on $\mathbb{T}$,

$$L_p(t,s) = \int_s^t \xi_{\mu(\tau)}\left(\frac{p^\Delta(\tau)}{p(\tau)}\right)\Delta\tau.$$

Now, exponentiate both sides and use the definition of $e_p(t,s)$, the exponential function. □

Theorem 4.2 (Logarithm of product, quotient, and power). *Presume $f,g,p : \mathbb{T} \to \mathbb{C}$ are Δ-differentiable functions with $f,g,p \neq 0$ on $\mathbb{T}$. Then, for $s,t \in \mathbb{T}$ and $a \in \mathbb{R}$, we have the following:*

1. $\ell_{fg}(t,s) = \ell_f(t,s) + \ell_g(t,s)$,
2. $\ell_{\frac{f}{g}}(t,s) = \ell_f(t,s) - \ell_g(t,s)$,
3. $\ell_{p^a}(t,s) = a\ell_p(t,s)$.

Proof. Presume $f,g,p : \mathbb{T} \to \mathbb{C}$ are Δ-differentiable functions with $f,g,p \neq 0$ on $\mathbb{T}$. Then, for $s,t \in \mathbb{T}$, we have via Lemma 4.1 and its proof that

$$\ell_{fg}(t,s) = \int_s^t \zeta_{\mu(\tau)}\left(\frac{(fg)^\Delta(\tau)}{(fg)(\tau)}\right)\Delta\tau$$

$$
\begin{aligned}
&= \int_s^t \zeta_{\mu(\tau)}\left(\left(\frac{f^\Delta}{f} \oplus \frac{g^\Delta}{g}\right)(\tau)\right)\Delta\tau \\
&= \int_s^t \zeta_\mu\left(\frac{f^\Delta(\tau)}{f(\tau)}\right)\Delta\tau + \int_s^t \zeta_\mu\left(\frac{g^\Delta(\tau)}{g(\tau)}\right)\Delta\tau \\
&= \ell_f(t,s) + \ell_g(t,s).
\end{aligned}
$$

In a similar manner,

$$
\begin{aligned}
\ell_{\frac{f}{g}}(t,s) &= \int_s^t \zeta_{\mu(\tau)}\left(\frac{(\frac{f}{g})^\Delta(\tau)}{(\frac{f}{g})(\tau)}\right)\Delta\tau \\
&= \int_s^t \zeta_{\mu(\tau)}\left(\left(\frac{f^\Delta}{f} \ominus \frac{g^\Delta}{g}\right)(\tau)\right)\Delta\tau \\
&= \int_s^t \zeta_\mu\left(\frac{f^\Delta(\tau)}{f(\tau)}\right)\Delta\tau - \int_s^t \zeta_\mu\left(\frac{g^\Delta(\tau)}{g(\tau)}\right)\Delta\tau \\
&= \ell_f(t,s) - \ell_g(t,s).
\end{aligned}
$$

Let $\alpha \in \mathbb{R}$. For the multivalued cylinder transformation ζ given by (4.3) and for fixed $\tau \in \mathbb{T}^\kappa$,

$$
\zeta_{\mu(\tau)}\left(\left(\alpha \odot \frac{p^\Delta}{p}\right)(\tau)\right) = \alpha\zeta_{\mu(\tau)}\left(\frac{p^\Delta(\tau)}{p(\tau)}\right)
$$

using Lemma 4.2. Moreover, by [5, Theorem 2.37], we have

$$
\frac{(p^\alpha)^\Delta}{p^\alpha} = \alpha \odot \frac{p^\Delta}{p}.
$$

Consequently,

$$
\begin{aligned}
\ell_{p^\alpha}(t,s) &= \int_s^t \zeta_{\mu(\tau)}\left(\frac{(p^\alpha)^\Delta(\tau)}{p^\alpha(\tau)}\right)\Delta\tau \\
&= \int_s^t \zeta_{\mu(\tau)}\left(\left(\alpha \odot \frac{p^\Delta}{p}\right)(\tau)\right)\Delta\tau \\
&= \int_s^t \alpha\zeta_{\mu(\tau)}\left(\frac{p^\Delta(\tau)}{p(\tau)}\right)\Delta\tau \\
&= \alpha\ell_p(t,s).
\end{aligned}
$$

The proof is complete. □

Theorem 4.3. *Let $p : \mathbb{T} \to \mathbb{R}$ be a Δ-differentiable function with $p \neq 0$ on $\mathbb{T}$. Then, for $s, t \in \mathbb{T}$, we have*

$$\ell_p^{\Delta}(t,s) = \begin{cases} \frac{1}{\mu(t)} \log(\frac{p^{\sigma}(t)}{p(t)}) & \text{for } \mu(t) \neq 0, \\ \frac{p^{\Delta}(t)}{p(t)} & \text{for } \mu(t) = 0, \end{cases}$$

where Δ-differentiation is with respect to t.

Proof. Using the definition of the logarithm and Δ-differentiating with respect to t,

$$\begin{aligned} \ell_p^{\Delta}(t,s) &= \zeta_{\mu(t)}\left(\frac{p^{\Delta}(t)}{p(t)}\right) \\ &= \begin{cases} \frac{1}{\mu(t)} \log(1 + \mu(t)\frac{p^{\Delta}(t)}{p(t)}) & \text{for } \mu(t) \neq 0, \\ \frac{p^{\Delta}(t)}{p(t)} & \text{for } \mu(t) = 0. \end{cases} \end{aligned}$$

Now substitute $\mu p^{\Delta} = p^{\sigma} - p$. The argument is finished. □

4.4 The nabla case

An analogous logarithm may likewise be defined for the nabla case.

Definition 4.6 (Cylinder transformation). For $h > 0$, define the single-valued cylinder transformation $\widehat{\xi}_h : \widehat{\mathbb{C}}_h \to \mathbb{Z}_h$ by

$$\widehat{\xi}_h(z) = \begin{cases} \frac{-1}{h} \operatorname{Log}(1 - zh) & \text{for } h \neq 0, \\ z & \text{for } h = 0, \end{cases} \tag{4.4}$$

and the multivalued cylinder transformation $\widehat{\zeta}_h : \widehat{\mathbb{C}}_h \to \mathbb{C}$ by

$$\widehat{\zeta}_h(z) = \begin{cases} \frac{-1}{h} \log(1 - zh) & \text{for } h \neq 0, \\ z & \text{for } h = 0. \end{cases} \tag{4.5}$$

Here $\mathbb{C}$ is the set of complex numbers, $\mathbb{Z}_h$ is in (4.2),

$$\widehat{C}_h = \left\{ z \in \mathbb{C} : z \neq \frac{1}{h} \right\},$$

and as before Log represents the principal complex logarithm function.

The following definition is [5, Definition 3.4].

Definition 4.7 (Regressive function). A function $p : \mathbb{T} \to \mathbb{R}$ is ν-regressive granted

$$1 - \nu(t)p(t) \neq 0 \quad \text{for each } t \in \mathbb{T}_\kappa$$

holds. Let $\widehat{\mathscr{R}}$ signify the set of all ld-continuous and ν-regressive functions $p : \mathbb{T} \to \mathbb{R}$.

The following definition is [5, Definition 3.10].

Definition 4.8 (Exponential function). Let $t, s \in \mathbb{T}$. For function $p \in \widehat{\mathscr{R}}$, the time scales nabla exponential function is formulated via

$$\widehat{e}_p(t,s) = \exp\left(\int_s^t \widehat{\xi}_{\nu(\tau)}(p(\tau))\nabla\tau\right),$$

where $\widehat{\xi}_h(z)$ is the single-valued cylinder transformation given in (4.4).

We now offer a new definition of logarithms for the nabla case on time scales.

Definition 4.9 (Logarithm function). Given a ∇-differentiable function $p : \mathbb{T} \to \mathbb{R}$ with $p \neq 0$ on $\mathbb{T}$, the multivalued nabla logarithm function on time scales is given by

$$\widehat{\ell}_p(t,s) = \int_s^t \widehat{\zeta}_{\nu(\tau)}\left(\frac{p^\nabla(\tau)}{p(\tau)}\right)\nabla\tau \quad \text{for } s,t \in \mathbb{T},$$

where $\widehat{\zeta}_h(z)$ is the multivalued cylinder transformation given in (4.5), while the principal nabla logarithm is given by

$$\widehat{L}_p(t,s) = \int_s^t \widehat{\xi}_{\nu(\tau)}\left(\frac{p^\nabla(\tau)}{p(\tau)}\right)\nabla\tau \quad \text{for } s,t \in \mathbb{T},$$

where $\widehat{\xi}_h(z)$ is the single-valued nabla cylinder transformation given in (4.4).

Analogous properties to those given in the delta case earlier may be established for the nabla case in a similar manner.

4.5 Logarithms for Cayley-exponential functions

Cieśliński introduces another time scales exponential function, named the Cayley-exponential function in [7], defined by

$$E_p(t,s) = \exp\left(\int_s^t \Psi_{\mu(\tau)}(p(\tau))\Delta\tau\right), \tag{4.6}$$

where $p : \mathbb{T} \to \mathbb{C}$ is rd-continuous and satisfies the regressivity condition $\mu(\tau)p(\tau) \neq \pm 2$ for all $\tau \in \mathbb{T}^\kappa$, and the modified cylinder transformation Ψ is given by

$$\Psi_h(z) = \frac{1}{h}\operatorname{Log}\left(\frac{1+\frac{1}{2}zh}{1-\frac{1}{2}zh}\right), \quad \Psi_0(z) = z, \tag{4.7}$$

for $h > 0$. Once more, Log represents the principal complex logarithm. Now consider the multivalued function version of (4.7) denoted, i. e.,

$$\psi_h(z) = \frac{1}{h}\log\left(\frac{1+\frac{1}{2}zh}{1-\frac{1}{2}zh}\right), \quad \psi_0(z) = z, \tag{4.8}$$

where log represents the multivalued complex logarithm. The following Cayley-logarithm function on time scales may then be formulated.

Definition 4.10. For a Δ-differentiable function $p : \mathbb{T} \to \mathbb{C}$ with $p \neq 0$ on $\mathbb{T}$, the multivalued Cayley-logarithm function on time scales is given by

$$\operatorname{caylog}_p(t,s) = \int_s^t \psi_{\mu(\tau)}\left(\frac{2p^\Delta(\tau)}{p(\tau)+p^\sigma(\tau)}\right)\Delta\tau \quad \text{for } s,t \in \mathbb{T},$$

where $\psi_h(z)$ is the multivalued cylinder transformation given in (4.8). Define the principal Cayley-logarithm on time scales to be

$$\operatorname{CayLog}_p(t,s) = \int_s^t \Psi_{\mu(\tau)}\left(\frac{2p^\Delta(\tau)}{p(\tau)+p^\sigma(\tau)}\right)\Delta\tau \quad \text{for } s,t \in \mathbb{T},$$

where $\Psi_h(z)$ is the single-valued cylinder transformation given in (4.7).

Lemma 4.3. *The Cayley-logarithm functions are well-defined functions.*

Proof. For a Δ-differentiable function $p : \mathbb{T} \to \mathbb{C}$ with $p \neq 0$ on $\mathbb{T}$, we need to show that

$$\mu(\tau)\frac{2p^\Delta(\tau)}{p(\tau)+p^\sigma(\tau)} \neq \pm 2,$$

in other words, that the regressivity condition holds. The following are equivalent:

$$\frac{2\mu(\tau)p^\Delta(\tau)}{p(\tau)+p^\sigma(\tau)} = \pm 2 \iff \frac{p^\sigma(\tau)-p(\tau)}{p(\tau)+p^\sigma(\tau)} = \pm 1,$$
$$p^\sigma(\tau)-p(\tau) = \pm(p(\tau)+p^\sigma(\tau)) \iff p^\sigma(\tau) \mp p^\sigma(\tau) = p(\tau) \pm p(\tau),$$

so that we have either $0 = 2p(\tau)$ or $2p^\sigma(\tau) = 0$, both contradictions. □

Theorem 4.4. *For a Δ-differentiable function $p : \mathbb{T} \to \mathbb{C}$ with $p \neq 0$ on $\mathbb{T}$,*

$$\operatorname{caylog}_p(t,s) = \ell_p(t,s) \quad \text{and} \quad \operatorname{CayLog}_p(t,s) = L_p(t,s) \tag{4.9}$$

for all $s, t \in \mathbb{T}$.

Proof. Consider (4.8). For fixed $\tau \in \mathbb{T}^\kappa$ with $\mu(\tau) \neq 0$, observe that

$$\begin{aligned}
\psi_{\mu(\tau)}\left(\frac{2p^\Delta(\tau)}{p(\tau)+p^\sigma(\tau)}\right) &= \frac{1}{\mu(\tau)}\log\left(\frac{1+\frac{1}{2}\frac{2p^\Delta(\tau)}{p(\tau)+p^\sigma(\tau)}\mu(\tau)}{1-\frac{1}{2}\frac{2p^\Delta(\tau)}{p(\tau)+p^\sigma(\tau)}\mu(\tau)}\right)\\
&= \frac{1}{\mu(\tau)}\log\left(\frac{1+\frac{\mu(\tau)p^\Delta(\tau)}{p(\tau)+p^\sigma(\tau)}}{1-\frac{\mu(\tau)p^\Delta(\tau)}{p(\tau)+p^\sigma(\tau)}}\right)\\
&= \frac{1}{\mu(\tau)}\log\left(\frac{1+\frac{p^\sigma(\tau)-p(\tau)}{p(\tau)+p^\sigma(\tau)}}{1-\frac{p^\sigma(\tau)-p(\tau)}{p(\tau)+p^\sigma(\tau)}}\right)\\
&= \frac{1}{\mu(\tau)}\log\left(\frac{p^\sigma(\tau)}{p(\tau)}\right)\\
&= \frac{1}{\mu(\tau)}\log\left(\frac{p(\tau)+\mu(\tau)p^\Delta(\tau)}{p(\tau)}\right)\\
&= \zeta_{\mu(\tau)}\left(\frac{p^\Delta(\tau)}{p(\tau)}\right)
\end{aligned}$$

for ζ_h defined in (4.3). For fixed $\tau \in \mathbb{T}^\kappa$ with $\mu(\tau) = 0$, we have $\tau = \sigma(\tau)$ and

$$\frac{2p^\Delta(\tau)}{p(\tau)+p^\sigma(\tau)} = \frac{p^\Delta(\tau)}{p(\tau)}.$$

Consequently,

$$\psi_{\mu(\tau)}\left(\frac{2p^\Delta(\tau)}{p(\tau)+p^\sigma(\tau)}\right) = \frac{2p^\Delta(\tau)}{p(\tau)+p^\sigma(\tau)} = \frac{p^\Delta(\tau)}{p(\tau)} = \zeta_{\mu(\tau)}\left(\frac{p^\Delta(\tau)}{p(\tau)}\right).$$

Thus, in either case, we have

$$\psi_{\mu(\tau)}\left(\frac{2p^\Delta(\tau)}{p(\tau)+p^\sigma(\tau)}\right) = \zeta_{\mu(\tau)}\left(\frac{p^\Delta(\tau)}{p(\tau)}\right).$$

It follows that

$$\operatorname{caylog}_p(t,s) = \int_s^t \psi_{\mu(\tau)}\left(\frac{2p^\Delta(\tau)}{p(\tau)+p^\sigma(\tau)}\right)\Delta\tau = \int_s^t \zeta_{\mu(\tau)}\left(\frac{p^\Delta(\tau)}{p(\tau)}\right)\Delta\tau = \ell_p(t,s).$$

Similarly, we have

$$\mathrm{CayLog}_p(t,s) = L_p(t,s),$$

completing the proof. □

Remark 4.2. The previous theorem and proof may be generalized, as we will now show. Let $\theta \in [0,1]$, and set

$$\psi_h^\theta(z) = \frac{1}{h}\log\left(\frac{1+(1-\theta)zh}{1-\theta zh}\right), \quad \psi_0^\theta(z) = z. \tag{4.10}$$

Then, for a Δ-differentiable function $p : \mathbb{T} \to \mathbb{C}$ with $p \neq 0$ on $\mathbb{T}$, and for all $\tau \in \mathbb{T}^\kappa$, we have

$$\begin{aligned}
&\psi_{\mu(\tau)}^\theta\left(\frac{p^\Delta(\tau)}{(1-\theta)p(\tau)+\theta p^\sigma(\tau)}\right)\\
&= \frac{1}{\mu(\tau)}\log\left(\frac{1+(1-\theta)\mu(\tau)\frac{p^\Delta(\tau)}{(1-\theta)p(\tau)+\theta p^\sigma(\tau)}}{1-\theta\mu(\tau)\frac{p^\Delta(\tau)}{(1-\theta)p(\tau)+\theta p^\sigma(\tau)}}\right)\\
&= \frac{1}{\mu(\tau)}\log\left(\frac{(1-\theta)p(\tau)+\theta p^\sigma(\tau)+(1-\theta)\mu(\tau)p^\Delta(\tau)}{(1-\theta)p(\tau)+\theta p^\sigma(\tau)-\theta\mu(\tau)p^\Delta(\tau)}\right)\\
&= \frac{1}{\mu(\tau)}\log\left(\frac{(1-\theta)p(\tau)+\theta p^\sigma(\tau)+(1-\theta)(p^\sigma(\tau)-p(\tau))}{(1-\theta)p(\tau)+\theta p^\sigma(\tau)-\theta(p^\sigma(\tau)-p(\tau))}\right)\\
&= \frac{1}{\mu(\tau)}\log\left(\frac{p^\sigma(\tau)}{p(\tau)}\right)\\
&= \frac{1}{\mu(\tau)}\log\left(\frac{p(\tau)+\mu(\tau)p^\Delta(\tau)}{p(\tau)}\right)\\
&= \zeta_{\mu(\tau)}\left(\frac{p^\Delta(\tau)}{p(\tau)}\right)
\end{aligned}$$

for ζ_h defined in (4.3). For fixed $\tau \in \mathbb{T}^\kappa$ with $\mu(\tau) = 0$, we have $\tau = \sigma(\tau)$ and

$$\frac{p^\Delta(\tau)}{(1-\theta)p(\tau)+\theta p^\sigma(\tau)} = \frac{p^\Delta(\tau)}{p(\tau)}.$$

As a result,

$$\psi_0^\theta\left(\frac{p^\Delta(\tau)}{(1-\theta)p(\tau)+\theta p^\sigma(\tau)}\right) = \frac{p^\Delta(\tau)}{(1-\theta)p(\tau)+\theta p^\sigma(\tau)} = \frac{p^\Delta(\tau)}{p(\tau)} = \zeta_0\left(\frac{p^\Delta(\tau)}{p(\tau)}\right).$$

Thus, in either case, we have

$$\psi_{\mu(\tau)}^\theta\left(\frac{p^\Delta(\tau)}{(1-\theta)p(\tau)+\theta p^\sigma(\tau)}\right) = \zeta_{\mu(\tau)}\left(\frac{p^\Delta(\tau)}{p(\tau)}\right)$$

for all $\theta \in [0,1]$. Consequently,

$$\log_p^\theta(t,s) := \int_s^t \psi_{\mu(\tau)}^\theta\left(\frac{p^\Delta(\tau)}{(1-\theta)p(\tau)+\theta p^\sigma(\tau)}\right)\Delta\tau = \int_s^t \zeta_{\mu(\tau)}\left(\frac{p^\Delta(\tau)}{p(\tau)}\right)\Delta\tau = \ell_p(t,s).$$

This ends the remark.

4.6 Alternative logarithms on time scales

As shown in the previous sections of this chapter, the key to arriving at useful logarithm properties is to allow for a multivalued logarithm, as for the $\mathbb{T} = \mathbb{R}$ case. Here, we present the previous definitions of logarithms on time scales, noting that they are all single-valued functions. Moreover, only Definition 4.5 leads to results as given in Theorems 4.1, 4.2, 4.4, and Remark 4.2, justifying this new approach, and emphasizing the advantages of having a function satisfying familiar properties, while ensconced in the more general time scales context.

The first logarithm on time scales [11] interprets the integral

$$\int_{t_0}^t \frac{2}{\tau+\sigma(\tau)}\Delta\tau$$

as a time scales analogue of $\ln t$. This is understandable because if $\mathbb{T} = \mathbb{R}$, then $\tau = \sigma(\tau)$ and

$$\int_{t_0}^t \frac{2}{\tau+\sigma(\tau)}\Delta\tau = \int_{t_0}^t \frac{2}{2\tau}d\tau = \ln t - \ln t_0.$$

A recent paper [12] applies iterates of this logarithm to Riemann–Weber-type equations; see also [14].

A second approach [6, Section 3] is to view the slightly different integral

$$\int_{t_0}^t \frac{1}{\tau+2\mu(\tau)}\Delta\tau$$

as the time scales version of $\ln t$, due to the same fact that it reduces to $\ln t - \ln t_0$ on $\mathbb{T} = \mathbb{R}$, and as it is part of a solution form to a certain Euler–Cauchy dynamic equation whose differential equation analogue involves the natural logarithm.

A third approach [6, Section 4] could be to define a logarithm via

$$L_p(t, t_0) = \int_{t_0}^{t} \frac{p^\Delta(\tau)}{p(\tau)} \Delta\tau$$

for Δ-differentiable functions $p : \mathbb{T} \to \mathbb{R}$. Clearly, if $p(\tau) = \tau$, then this is

$$L_p(t, t_0) = \int_{t_0}^{t} \frac{p^\Delta(\tau)}{p(\tau)} \Delta\tau = \int_{t_0}^{t} \frac{1}{\tau} \Delta\tau,$$

a form that is similar to its continuous analogue for $\mathbb{T} = \mathbb{R}$.

A fourth approach [13] is to take the logarithm to be given by

$$\log_{\mathbb{T}} p(t) = \frac{p^\Delta(t)}{p(t)}$$

for Δ-differentiable functions $p : \mathbb{T} \to \mathbb{R}$, where the time scale logarithm on $\mathbb{R}$ does not play the role of the logarithm, clearly, but rather its derivative. The motivation here is to maintain some attractive algebraic properties of logarithms, and to serve in some sense as an inverse to the exponential function.

A fifth approach [15], only for time scales such that $1 \in \mathbb{T}$, is to define the natural logarithm via

$$L_{\mathbb{T}}(t) = \int_{1}^{t} \frac{1}{\tau} \Delta\tau,$$

which hearkens back to [6, Section 4]. Here the motivation is clearly that

$$L_{\mathbb{R}}(t) = \ln t, \quad L_{\mathbb{T}}(1) = 0, \quad L_{\mathbb{T}}^{\Delta}(t) = \frac{1}{t}.$$

Other possibilities may be possible but have yet to be explored.

4.7 Examples of logarithms and numerical comparisons

Each of the definitions given in the previous section has advantages and drawbacks, and each satisfies some of what one might wish for in a logarithm function. As shown earlier in this work, however, a multivalued logarithm on time scales with a definition based on cylinder transformations is a natural move that leads to nice properties, and has not been introduced until the foundational paper for this chapter. We now consider the following examples.

Example. In this example, we compare the values of the various logarithms on the time scale

$$\mathbb{T} := (-\infty, -k] \cup \{-k+1, -k+2, \ldots, -1, 0, 1, \ldots, k-2, k-1\} \cup [k, \infty), \quad k \in \mathbb{N}.$$

For $p(t) = t$ on $[1, k+3]_{\mathbb{T}}$, we have the following plot and table of comparison for the logarithms on time scales mentioned in the literature to date.

As can be seen in Table 4.1, the new definition presented in [1, Definition 2.7] and restated in Definition 2.7 of this chapter leads to a unique and accurate value for this time scale. The comparison of graphs on $[1, 8]_{\mathbb{T}} = \{1, 2, 3, 4\} \cup [5, 8]$ is given in Figure 4.1.

Table 4.1: Comparison chart of proposed logarithm functions on time scales with the new definition in Definition 2.7.

Citation	Logarithm	Value at $t = 6$	Figure 1 color
[11]	$\sum_{j=1}^{k-1} \frac{2}{2j+1} + \ln(\frac{t}{k})$	1.75692	blue
[6, Section 3]	$\sum_{j=1}^{k-1} \frac{1}{j+2} + \ln(\frac{t}{k})$	1.13232	orange
[6, Section 4]	$\sum_{j=1}^{k-1} \frac{1}{j} + \ln(\frac{t}{k})$	2.26565	green
[13]	$\sum_{j=1}^{k-1} \frac{1}{j} + \ln(\frac{t}{k})$	2.26565	green
[15]	$\sum_{j=1}^{k-1} \frac{1}{j} + \ln(\frac{t}{k})$	2.26565	green
Definition 2.7	$\sum_{j=1}^{k-1} \ln(\frac{j+1}{j}) + \ln(\frac{t}{k})$	1.79176	red

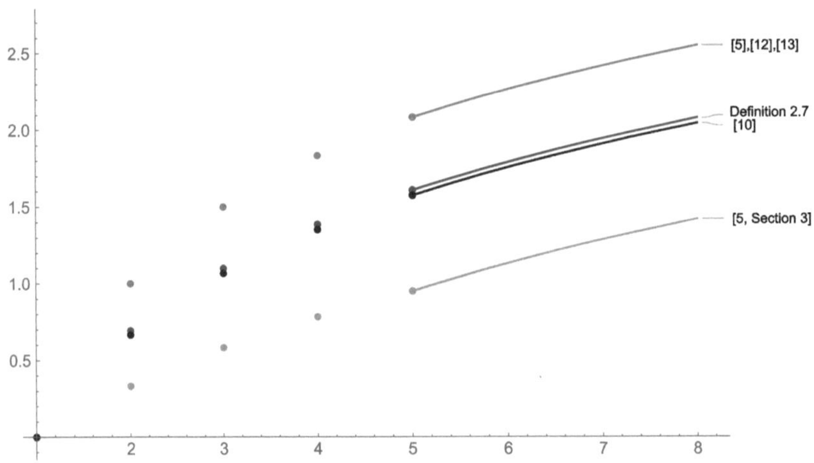

Figure 4.1: Comparison plot of various logarithms on $[1, k+3]_{\mathbb{T}}$ for $k = 5$.

In the rest of this section, we provide numerous examples of the new logarithm from Definition 2.7, for various time scales.

Example. For $\mathbb{T} = \mathbb{R}$,

$$\ell_p(t,s) = \int_s^t \zeta_{\mu(\tau)}\left(\frac{p^\Delta(\tau)}{p(\tau)}\right)\Delta\tau = \int_s^t \frac{p'(\tau)}{p(\tau)}\mathrm{d}\tau = \log\left(\frac{p(t)}{p(s)}\right),$$

where log represents the multivalued complex logarithm function. For $\mathbb{T} = h\mathbb{Z}$,

$$\Lambda^\Delta(\tau) = \Delta_h\Lambda(\tau) := \frac{\Lambda(h+\tau) - \Lambda(\tau)}{h}$$

and

$$\begin{aligned}\ell_p(t,s) &= \int_s^t \zeta_{\mu(\tau)}\left(\frac{p^\Delta(\tau)}{p(\tau)}\right)\Delta\tau = \sum_{j=\frac{s}{h}}^{\frac{t}{h}-1} \zeta_h\left(\frac{\Delta_h p(jh)}{p(jh)}\right)h \\ &= \sum_{j=\frac{s}{h}}^{\frac{t}{h}-1} \frac{1}{h}\log\left(1 + \frac{h\Delta_h p(jh)}{p(jh)}\right)h \\ &= \sum_{j=\frac{s}{h}}^{\frac{t}{h}-1} \log\left(\frac{p(jh+h)}{p(jh)}\right) = \log\left(\prod_{j=\frac{s}{h}}^{\frac{t}{h}-1} \frac{p((j+1)h)}{p(jh)}\right) = \log\left(\frac{p(t)}{p(s)}\right).\end{aligned}$$

For $\mathbb{T} = q^{\mathbb{N}_0}$,

$$f^\Delta(\tau) = D_q f(\tau) := \frac{f(q\tau) - f(\tau)}{(q-1)\tau}$$

and

$$\begin{aligned}\ell_p(t,s) &= \int_s^t \zeta_{\mu(\tau)}\left(\frac{p^\Delta(\tau)}{p(\tau)}\right)\Delta\tau = \sum_{\tau\in[s,t)} \zeta_{(q-1)\tau}\left(\frac{p^\Delta(\tau)}{p(\tau)}\right)(q-1)\tau \\ &= \sum_{\tau\in[s,t)} \frac{1}{(q-1)\tau}\log\left(1 + \frac{(q-1)\tau p^\Delta(\tau)}{p(\tau)}\right)(q-1)\tau \\ &= \sum_{\tau\in[s,t)} \log\left(\frac{p(q\tau)}{p(\tau)}\right) = \log\left(\frac{p(t)}{p(s)}\right).\end{aligned}$$

This ends the example.

Example. For real numbers a, b, c, d with $a < b < c < d$, set $\mathbb{T} = [a,b] \cup [c,d]$. Assume $p : \mathbb{T} \to \mathbb{C}$ is differentiable with $p \neq 0$ on $\mathbb{T}$. If $s, t \in [a,b)$ or $s, t \in [c,d]$, then $\mu(\tau) \equiv 0$ for $\tau \in [s,t]$, so that by the definition of the multivalued cylinder function (4.3),

$$\ell_p(t,s) = \int_s^t \frac{p'(\tau)}{p(\tau)}\mathrm{d}\tau = \log\left(\frac{p(t)}{p(s)}\right).$$

Presume without loss of generality that $s \in [a,b]$ and $t \in [c,d]$. Then $c = \sigma(b)$, and

$$\begin{aligned}
\ell_p(t,s) &= \int_s^t \zeta_{\mu(\tau)}\left(\frac{p^\Delta(\tau)}{p(\tau)}\right)\Delta\tau \\
&= \left(\int_s^b + \int_b^{\sigma(b)} + \int_{\sigma(b)}^t\right)\zeta_{\mu(\tau)}\left(\frac{p^\Delta(\tau)}{p(\tau)}\right)\Delta\tau \\
&= \log\left(\frac{p(b)}{p(s)}\right) + \log\left(\frac{p(t)}{p(\sigma(b))}\right) + \int_b^{\sigma(b)} \zeta_{\mu(\tau)}\left(\frac{p^\Delta(\tau)}{p(\tau)}\right)\Delta\tau \\
&= \log\left(\frac{p(b)}{p(s)}\right) + \log\left(\frac{p(t)}{p(c)}\right) + \mu(b)\zeta_{\mu(b)}\left(\frac{p^\Delta(b)}{p(b)}\right) \\
&= \log\left(\frac{p(b)}{p(s)}\right) + \log\left(\frac{p(t)}{p(c)}\right) + \mu(b)\left[\frac{1}{\mu(b)}\log\left(1 + \frac{\mu(b)p^\Delta(b)}{p(b)}\right)\right] \\
&= \log\left(\frac{p(b)}{p(s)}\right) + \log\left(\frac{p(t)}{p(c)}\right) + \log\left(\frac{p^\sigma(b)}{p(b)}\right) \\
&= \log\left(\frac{p(b)}{p(s)}\right) + \log\left(\frac{p(t)}{p(c)}\right) + \log\left(\frac{p(c)}{p(b)}\right) \\
&= \log\left(\frac{p(t)}{p(s)}\right).
\end{aligned}$$

Consequently, in all cases, we see that $\ell_p(t,s) = \log(\frac{p(t)}{p(s)})$ on this time scale as well.

Example. Let $\mathbb{T} = (-\infty, -4] \cup [2, \infty)$, and $p(t) = t^3$. Let $t \geq 2$ and $s = -5$. Then

$$\mu(-4) = \sigma(-4) - (-4) = 2 - (-4) = 6,$$

and the principal logarithm on this time scale is

$$\begin{aligned}
L_p(t,s) = L_p(t,-5) &= \int_{-5}^t \xi_{\mu(\tau)}\left(\frac{(\tau^3)^\Delta}{\tau^3}\right)\Delta\tau \\
&= \left(\int_{-5}^{-4} + \int_{-4}^{2} + \int_2^t\right)\xi_{\mu(\tau)}\left(\frac{\sigma(\tau)^2 + \tau\sigma(\tau) + \tau^2}{\tau^3}\right)\Delta\tau \\
&= 3\left(\int_{-5}^{-4} + \int_2^t\right)\frac{\mathrm{d}\tau}{\tau} + \mu(-4)\xi_{\mu(-4)}\left(\frac{2^2 - 4(2) + (-4)^2}{(-4)^3}\right)
\end{aligned}$$

$$= 3(\mathrm{Log}[-4] - \mathrm{Log}[-5] + \mathrm{Log}[t] - \mathrm{Log}[2]) + \mathrm{Log}\left(1 + 6\frac{12}{-64}\right)$$

$$= 3\ln\left(\frac{t}{5}\right) + i\pi,$$

where ln is the natural logarithm and Log represents the principal complex logarithm. Again for sake of comparison, the logarithms in [11] and [6, Section 3] do not apply as they are defined exclusively in terms of $p(t) = t$, and [15] does not apply as that logarithm requires $1 \in \mathbb{T}$. If we use the logarithm in [6, Section 4] or [13], we get $3\ln(\frac{2t}{5}) - \frac{9}{8}$, a real-valued function, as opposed to our principal value of $3\ln(\frac{t}{5}) + i\pi$, a complex-valued function. This example justifies our approach.

Example. Here is an example of Theorem 4.3. Let $t \in \mathbb{T}$ with $t \neq 0$, and set $p(t) = t$. For $s \in \mathbb{T}$, we have

$$\ell_p^\Delta(t,s) = \begin{cases} \frac{1}{\mu(t)} \log(\frac{\sigma(t)}{t}) & \text{for } \mu(t) \neq 0, \\ \frac{1}{t} & \text{for } \mu(t) = 0, \end{cases}$$

where Δ-differentiation is with respect to t. Thus,

$$\ell_p^\Delta(t,s) = \begin{cases} \frac{1}{t} & \text{for } \mathbb{T} = \mathbb{R}, \\ \frac{1}{h} \log(1 + \frac{h}{t}) & \text{for } \mathbb{T} = h\mathbb{Z}, \\ \frac{\log(q)}{(q-1)t} & \text{for } \mathbb{T} = q^{\mathbb{N}_0}, \end{cases}$$

where $h > 0$ and $q > 1$.

See Figure 4.2 for $h = 1$ and $\mathbb{T} = \mathbb{Z}$. This ends the example.

Example. Construct a discrete time scale with two step sizes that alternate, that is, for the two alternating step sizes $\alpha, \beta > 0$ with $\alpha \neq \beta$, let

$$\mathbb{T} := \mathbb{T}_{\alpha,\beta} = \{0, \alpha, (\alpha+\beta), (\alpha+\beta)+\alpha, 2(\alpha+\beta), 2(\alpha+\beta)+\alpha, 3(\alpha+\beta), \dots\}.$$

Then, for $t \in \mathbb{T}$ and $k \in \mathbb{N}_0 = \{0, 1, 2, \dots\}$, we have

$$\mu(t) = \begin{cases} \alpha & \text{for } t = k(\alpha+\beta), \\ \beta & \text{for } t = k(\alpha+\beta)+\alpha. \end{cases}$$

Set $p(t) = t$. We claim that for $t \in \mathbb{T}_{\alpha,\beta}$ with $t \neq 0$,

$$\ell_p^\Delta(t,s) = \begin{cases} \frac{1}{\alpha} \log(1 + \frac{\alpha}{t}) & \text{for } t = k(\alpha+\beta), \\ \frac{1}{\beta} \log(1 + \frac{\beta}{t}) & \text{for } t = k(\alpha+\beta)+\alpha. \end{cases}$$

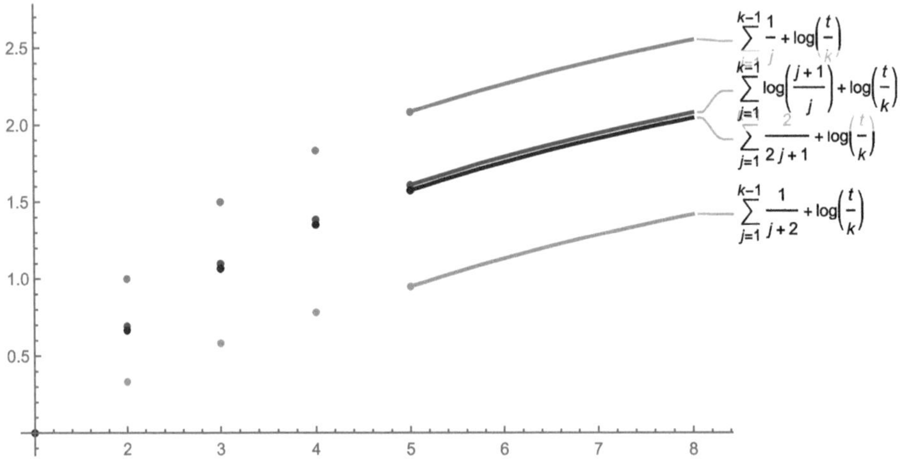

Figure 4.2: A plot of the function $\frac{1}{t}$ versus $\mathrm{Log}(1 + \frac{1}{t})$ for Example 4.7.

To verify this, note that

$$
\begin{aligned}
\ell_p^{\Delta}(t,s) &= \frac{1}{\mu(t)} \log\left(\frac{\sigma(t)}{t}\right) \\
&= \begin{cases} \frac{1}{\alpha} \log(\frac{k(\alpha+\beta)+\alpha}{k(\alpha+\beta)}) & \text{for } t = k(\alpha+\beta), \\ \frac{1}{\beta} \log(\frac{(k+1)(\alpha+\beta)}{k(\alpha+\beta)+\alpha}) & \text{for } t = k(\alpha+\beta)+\alpha \end{cases} \\
&= \begin{cases} \frac{1}{\alpha} \log(1+\frac{\alpha}{t}) & \text{for } t = k(\alpha+\beta), \\ \frac{1}{\beta} \log(1+\frac{\beta}{t}) & \text{for } t = k(\alpha+\beta)+\alpha. \end{cases}
\end{aligned}
$$

This ends the example.

Remark. As given above, the first three examples suggest that this new logarithm may be a kind of exact discretization. In other words, by definition it yields the usual logarithm function restricted to the given time scale. This remains an open question for more intricate and general time scales.

4.8 Uses of this logarithm on time scales

For trends on time scales generally, see the recent monographs [2, 3, 9] and the paper [8]. For developments associated with the logarithm introduced in [1] and summarized in the earlier sections of this chapter, we highlight the following. Song, Wu, and Wei [16] use the multivalued logarithm on time scales to formulate the Hadamard fractional calculus on time scales, and Wu, Song, and Wang [17] the Caputo–Hadamard fractional differential equations on time scales, including exploring a numerical scheme, asymptotic stability, and chaos.

The famous Hadamard derivative [18] is defined by

$$ {}_{a}^{H}I_t^{\alpha}x(t) = \frac{1}{\Gamma(\alpha)} \int_a^t (\log(t) - \log(s))^{\alpha-1} x(s)\frac{ds}{s}, \quad t \in [a, b]. \tag{4.11}$$

A "nice" $\log(t)$ is a key step to define Hadamard fractional calculus on time scales. The newly proposed logarithm function [1], summarized in this chapter, is strictly monotonous and invertible.

Consider a set $\{t_0, t_1, \dots, t_N\}$ satisfying $1 \le a = t_0 < t_1 < \cdots < t_{N-1} < t_N = b$. The logarithm function can be reduced to the standard one in $\mathbb{R}$. Motivated by the nonequidistant partition idea in numerical methods of general fractional differential equations [19, 20], we take $h = \frac{\ell_p(b,s)-\ell_p(a,s)}{N}$ and $p(t) = t$. The logarithm function and its inverse are reduced to

$$\ell_p(t,s) = \log\left(\frac{t}{s}\right) \quad \text{and} \quad \ell_p^{-1}(t,s) = se^t,$$

respectively. Then the set becomes $t_j = \ell_p^{-1}(\ell_p(a,s) + jh, s)$, $j = 0, 1, \dots, N$, and

$$t_j \in \{a, ae^h, ae^{2h}, \dots, ae^{(N-1)h}, ae^{Nh} = b\}. \tag{4.12}$$

So the backward and forward jump operators are $\rho(t) = te^{-h}$ and $\sigma(t) = te^h$. We denote the isolated time scale by $e^{(h\mathbb{N})_{\log a}}$ for $N \to \infty$.

We now use the falling factorial function

$$t^{\underline{\alpha}}_{h} := h^{\alpha}\frac{\Gamma(\frac{t}{h}+1)}{\Gamma(\frac{t}{h}+1-\alpha)}, \quad t \in (h\mathbb{N})_{\alpha h}, \quad \alpha \in \mathbb{R} \tag{4.13}$$

to give Hadamard fractional sum on the time scale.

Definition 4.11 ([16]). Suppose $f : e^{(h\mathbb{N})_{\log a}} \to \mathbb{R}$ and $\alpha > 0$. The Hadamard fractional sum of order α is defined by

$$\widetilde{\Delta}_a^{-\alpha}f(t) = \frac{1}{\Gamma(\alpha)} \int_a^{\rho^{\alpha-1}(t)} (\ell_p(t,r) - \ell_p(\sigma(s),r))_h^{\underline{\alpha-1}} \ell_p^{\Delta}(s,r) f(s)\Delta s,$$

with $t \in e^{(h\mathbb{N})_{\log a + \alpha h}}$.

Remark. Let $f : e^{(h\mathbb{N})_{\log a}} \to \mathbb{R}$ and $M - 1 < \alpha \le M$. If $t = ae^{\alpha h} \cdot e^{(N-1)h}$ and $s = ae^{jh}$, $j = 0, 1, \dots, N-1$, then the fractional sum's discrete numerical scheme reads

$$\widetilde{\Delta}_a^{-\alpha} f(t) = \frac{h^\alpha}{\Gamma(\alpha)} \sum_{j=0}^{N-1} \frac{\Gamma(N+\alpha-j-1)}{\Gamma(N-j)} f(ae^{jh}).$$

Then the Caputo and Riemann–Liouville Hadamard differences can be defined, respectively.

Definition 4.12 ([16]). Suppose $f : e^{(h\mathbb{N})_{\log a}} \to \mathbb{R}$, $M \in \mathbb{N}_1$ and $M-1 < \alpha \leq M$. The left Hadamard fractional difference of order α is given by

$$\widetilde{\Delta}_a^\alpha f(t) = \frac{1}{\Gamma(-\alpha)} \int_a^{\rho^{-\alpha-1}(t)} (\ell_p(t,r) - \ell_p(\sigma(s),r))_h^{\underline{-\alpha-1}} \ell_p^\Delta(s,r) f(s) \Delta s, \tag{4.14}$$

with $t \in e^{(h\mathbb{N})_{\log a+(M-\alpha)h}}$.

The logarithm function is used as $\ell_p(t,s) = \log(\frac{t}{s})$ from [1]. The Δ- and ∇-derivatives of $\ell_p(t,s)$ with respect to t are $\ell_p^\Delta(t,s) = \frac{1}{\mu(t)} \log(\frac{\sigma(t)}{t})$ and $\ell_p^\nabla(t,s) = \frac{1}{\nu(t)} \log(\frac{t}{\rho(t)})$, respectively.

Definition 4.13 ([17]). Let $f : e^{(h\mathbb{N})_{\log a}} \to \mathbb{R}$, $M \in \mathbb{N}_1$ and $M-1 < \alpha \leq M$. The Caputo–Hadamard fractional difference of order α is given by

$${}^C\widetilde{\Delta}_a^\alpha f(t) = \frac{1}{\Gamma(M-\alpha)} \int_a^{\rho^{M-\alpha-1}(t)} (\ell_p(t,r) - \ell_p(\sigma(s),r))_h^{\underline{M-\alpha-1}} \ell_p^\Delta(s,r)\, {}_{\mathbb{T}}\widetilde{\Delta}^M f(s) \Delta s, \tag{4.15}$$

where $t \in e^{(h\mathbb{N})_{\log a+(M-\alpha)h}}$.

The initial value problem

$${}^C\widetilde{\Delta}_a^\alpha x(t) = \lambda x(te^{\alpha h}), \quad x(a) = C, 0 < \alpha \leq 1, t \in e^{(h\mathbb{N})_{\log a+(1-\alpha)h}}, \tag{4.16}$$

has a solution

$$x(t) = C\varepsilon_\alpha(\lambda, (t-\sigma(a))^{\underline{\alpha}})$$

if we define a discrete Mittag-Leffler function of Hadamard type

$$\varepsilon_\alpha(\lambda, (t-\sigma(a))^{\underline{\alpha}}) := \sum_{k=0}^{\infty} \lambda^k \frac{(\ell_p(t,r) - \ell_p(ae^{(1-k\alpha)h}, r))_h^{\underline{k\alpha}}}{\Gamma(k\alpha+1)}, \quad t \in e^{(h\mathbb{N})_{\log a+h}}. \tag{4.17}$$

The fractional-order sum and difference satisfy many useful properties such as the composition rule and Leibniz sum laws which are similar to those of the continuous case [18]. We call them exact discretization operators. More details and the right-hand side Hadamard fractional difference can be found in [21].

Definition 4.14 ([21]). Suppose $f : e^{\log b}(h\mathbb{N}) \to \mathbb{R}$, $M \in \mathbb{N}_1$, and $M - 1 < \alpha \le M$. The right Hadamard fractional difference of order α is given by

$$ {}_b\widetilde{\Delta}^{\alpha} f(t) = \frac{1}{\Gamma(-\alpha)} \sum_{s=te^{-\alpha h}}^{b} (\ell_p(\rho(s),r) - \ell_p(t,r))_h^{\underline{-\alpha-1}} \ell_p^{\nabla}(s,r) f(s) \nu(s), \tag{4.18} $$

with $t \in e^{\log b - (M-\alpha)h}(h\mathbb{N})$.

Using Definition 4.14, we can formulate and solve the following problem. If we consider a terminal value problem of the right fractional Hadamard difference equation

$$ \begin{cases} {}_b\widetilde{\Delta}^{\alpha} x(t) = \lambda x(te^{(1-\alpha)h}), & t \in e^{\log b - (1-\alpha)h}(h\mathbb{N}), 0 < \alpha \le 1, \\ {}_b\widetilde{\Delta}^{\alpha-1} x(be^{(\alpha-1)h}) = C, \end{cases} \tag{4.19} $$

we will obtain a right Mittag-Leffler solution expressed as

$$ x(t) = C\varepsilon_{\alpha,\alpha}(\lambda, (\rho(b) - t)^{\underline{\alpha}}), \quad t \in e^{\log b - h}(h\mathbb{N}), $$

where $\varepsilon_{\alpha,\alpha}(\lambda, (\rho(b) - t)^{\underline{\alpha}})$ is defined by

$$ \varepsilon_{\alpha,\alpha}(\lambda, (\rho(b) - t)^{\underline{\alpha}}) = \sum_{m=0}^{\infty} \lambda^m \frac{(\ell_p(be^{(m+1)(\alpha-1)h}, r) - \ell_p(t,r))_h^{\underline{(m+1)\alpha-1}}}{\Gamma(m\alpha + \alpha)}, \tag{4.20} $$

with $t \in e^{\log b - h}(h\mathbb{N})$. This ends a brief introduction to one of the applications of the logarithm on time scales.

Bibliography

[1] D. R. Anderson and M. Bohner, A multivalued logarithm on time scales, *Appl. Math. Comput.*, 397 (2021), 125954.

[2] D. R. Anderson and S. G. Georgiev, Conformable Dynamic Equations on Time Scales, Chapman and Hall/CRC, Boca Raton, 2020.

[3] M. Bohner and S. G. Georgiev, Multivariable Dynamic Calculus on Time Scales, Springer, 2016.

[4] M. Bohner and A. Peterson, Dynamic Equations on Time Scales: An Introduction with Applications, Birkhäuser, Boston, 2001.

[5] M. Bohner and A. Peterson (eds.), Advances in Dynamic Equations on Time Scales, Birkhäuser, Boston, 2003.

[6] M. Bohner, The logarithm on time scales, *J. Differ. Equ. Appl.*, 11(15) (2005), 1305–1306.

[7] J. L. Cieśliński, New definitions of exponential, hyperbolic and trigonometric functions on time scales, *J. Math. Anal. Appl.*, 388 (2012), 8–22. https://doi.org/10.1016/j.jmaa.2011.11.023.

[8] T. Cuchta, D. Grow and N. Wintz, A dynamic matrix exponential via a matrix cylinder transformation, *J. Math. Anal. Appl.*, 479 (2019), 733–751. https://doi.org/10.1016/j.jmaa.2019.06.048.

[9] S. G. Georgiev, Fractional Dynamic Calculus and Fractional Dynamic Equations on Time Scales, Springer, 2018.

[10] S. Hilger, Analysis on measure chains – a unified approach to continuous and discrete calculus, *Results Math.*, 18 (1990), 18–56.

[11] S. Huff, G. Olumolode, N. Pennington and A. Peterson, Oscillation of an Euler–Cauchy dynamic equation, Proceedings of the Fourth International Conference on Dynamical Systems and Differential Equations, Discrete Contin. Dyn. Syst. 2003 (2003), 423–431. https://doi.org/10.3934/proc.2003.2003.423.

[12] B. Ito, P. Řehák and N. Yamaoka, Applications of iterated logarithm functions on time scales to Riemann–Weber-type equations, *Proc. Am. Math. Soc.*, 148 (2020), 1611–1624. https://doi.org/10.1090/proc/14812.

[13] B. Jackson, The time scale logarithm, *Appl. Math. Lett.*, 21 (2008), 215–221.

[14] J. Jekl, Closed-form solutions of second-order linear difference equations close to the self-adjoint Euler type, *Math. Methods Appl. Sci.*, 46 (2022), 5314–5327. https://doi.org/10.1002/mma.8836.

[15] D. Mozyrska and D. F. M. Torres, The natural logarithm on time scales, *J. Dyn. Syst. Geom. Theories*, 7 (2009), 41–48.

[16] T. T. Song, G. C. Wu and J. L. Wei, Hadamard fractional calculus on time scales, *Fractals*, 30 (2022), 2250145. https://doi.org/10.1142/S0218348X22501456.

[17] G. C. Wu, T. T. Song and S. Wang, Caputo–Hadamard fractional differential equations on time scales: Numerical scheme, asymptotic stability, and chaos, *Chaos*, 32 (2022), 093143. https://doi.org/10.1063/5.0098375.

[18] A. A. Kilbas, H. M. Srivastava and J. J. Trujillo, Theory and Applications of Fractional Differential Equations, North Holland Mathematical Studies, Amsterdam, 2006.

[19] S. D. Zeng, D. Baleanu, Y. R. Bai and G. C. Wu, Fractional differential equations of Caputo–Katugampola type and numerical solutions, *Appl. Math. Comput.*, 315 (2017), 549–554. https://doi.org/10.1016/j.amc.2017.07.003.

[20] G. C. Wu, H. Kong, M. Luo, H. Fu and L. L. Huang, Unified predictor–corrector method for fractional differential equations with general kernel functions, *Fract. Calc. Appl. Anal.*, 25 (2022), 648–667. https://doi.org/10.1007/s13540-022-00029-z.

[21] J. L. Wei, G. C. Wu and R. Lozi, Right Hadamard fractional differences and integration by parts, *Rocky Mt. J. Math.* (2022, accepted).

R. Vivek, S. Sivasundaram, D. Vivek, and K. Kanagarajan

5 Qualitative analysis for hybrid fuzzy differential equations involving tempered Ξ-Hilfer fractional derivative on time scales

Abstract: This chapter deals with the existence and stability results for hybrid fuzzy differential equations with tempered Ξ-Hilfer fractional derivatives on time scales. Further, we obtain sufficient conditions for the existence and uniqueness of solutions by using a hybrid fixed point theorem. In addition, it demonstrates Ulam-type stability. Finally, we give a suitable example to illustrate our main results.

5.1 Introduction

Hilger in 1990 introduced time scales to unify and extend the theory of differential equations, difference equations, and other differential systems defined over nonempty closed subset of real line. It is more realistic to model a phenomenon by a dynamical system that incorporates both continuous and discrete times, namely, as an arbitrary closed set of reals known as a time scale. In addition, the continuous and discrete processes are seen in option-pricing and stock dynamics in finance, robust 3D tracking in shape and motion estimation, frequency of markets and duration of market trading in economics, large-scale models of DNA dynamics. For a basic discussion on fuzzy-valued functions and fuzzy differential equations (FDEs), we refer to [12, 21, 27, 28, 38]. For more details on time scales, we refer to [4, 7, 16, 31, 35, 40].

The theory of fractional differential equations received considerable interest in recent years both in pure mathematics and applications, see [23, 33]. There are several kinds of fractional derivatives such as the Riemann–Liouville (RL), Caputo, Hilfer, Hadamard, and others. A new type of derivative, called the Hilfer fractional derivative (HFD), was defined by R. Hilfer [19]. In 2018, Sousa and Oliveira [24, 36] presented new HFD – the so-called Ξ-HFD was defined.

Tempered fractional calculus can be considered as the extension to traditional fractional calculus concepts, multiplying fractional derivatives and integrals by an exponential factor leads to tempered fractional derivatives and integrals. Some properties

R. Vivek, K. Kanagarajan, Department of Mathematics, Sri Ramakrishna Mission Vidyalaya College of Arts and Science, Coimbatore 641020, India, e-mails: vivekmaths1997@gmail.com, kanagarajank@gmail.com
S. Sivasundaram, College of Natural Science, Engineering and Mathematics, Daytona Beach, FL 32114, USA, e-mail: seenithi@gmail.com
D. Vivek, Department of Mathematics, PSG College of Arts & Science, Coimbatore 641014, India, e-mail: peppyvivek@gmail.com

https://doi.org/10.1515/9783111182971-005

and applications of the tempered fractional calculus are given in [25, 29, 30, 32] and the references therein.

In [5, 18, 20, 22, 26, 34], the issue of fuzzy fractional calculus and fractional FDEs has emerged as a significant subject and this new theory became very attractive to many scientists. Significant results from the theory of fuzzy differential equations and their applications have appeared. The concept of Hukuhara differentiability has been correlated with fuzzy RL-differentiability by employing the Hausdorff measure of noncompactness in [6, 10]. The stability analysis of integral and differential equations is important in many applications and for basic results and recent development on Ulam stability of integral and differential equations. Different types of stability such as Ulam–Hyers (UH), generalized UH, Ulam–Hyers–Rassias (UHR), and generalized UHR stability have been given much attention for fuzzy fractional differential equations which involve different types of operators; we refer to a series of papers [1, 14, 37, 41].

Furthermore, hybrid differential equations arise from a variety of different areas of applied mathematics and physics, for example, in the deflection of a curved beam having a constant or varying cross-section, a three-layer beam, electromagnetic waves or gravity driven flows, and so on, see [8, 15]. One more attractive class of problems involves hybrid fractional differential equations. For some works on this topic, one can refer to [2, 3, 9, 11, 13, 39].

In [17], the following first-order hybrid FDE is discussed:

$$\begin{cases} \frac{d}{dt}\left(\frac{v(t)}{f(t,v(t))}\right) = G(t, v(t)), \quad t = [0, b] \in \mathfrak{J}, \\ v(t_0) = v_0 \in E^d, \end{cases}$$

where $f \in C(\mathfrak{J} \times E^d, E^d - \{0\})$ and $G \in C(\mathfrak{J} \times E^d, E^d)$ are continuous fuzzy functions. They established some fundamental hybrid fuzzy differential inequalities that are useful for establishing the existence of extremal solutions.

Motivated by these works, we consider the following hybrid FDE involving tempered Ξ-Hilfer fractional derivative on time scale of the form:

$$\begin{cases} {}^{TH\mathbb{T}}_{0^+}\Delta_{\Xi}^{p,q,\lambda}\left(\frac{v(t)}{f(t,v(t))}\right) = G(t, v(t)), \quad 0 < \lambda < 1, \quad t \in [0, b] = \mathfrak{J} \subseteq \mathbb{T}, \\ {}^{T\mathbb{T}}_{0^+}\mathfrak{J}_{\Xi}^{1-\gamma,\lambda}\left(\frac{v(0)}{f(0,v(0))}\right) = v_0, \quad \gamma = p + q - pq, \end{cases} \tag{5.1}$$

where ${}^{TH\mathbb{T}}_{0^+}\Delta_{\Xi}^{p,q,\lambda}$ is the tempered Ξ-Hilfer fractional derivative on $\mathbb{T}$, $p \in (0, 1)$, $q \in [0, 1]$, $\lambda \in \mathbb{R}$, E^d denotes the space of fuzzy sets on $\mathbb{R}$, $v_0 \in E^d$ is given. Let $\mathbb{T}$ be a time scale and suppose that ${}^{T\mathbb{T}}_{0^+}\mathfrak{J}_{\Xi}^{1-\gamma,\lambda}$ is the tempered Ξ-RL fractional integral operator of order $1-\gamma$ defined on $\mathbb{T}$, while $f \in C(\mathbb{T} \times E^d, E^d - \{0\})$ and $G \in C(\mathbb{T} \times E^d, E^d)$ are continuous fuzzy functions.

5.2 Preliminaries

In this section, we recall some notations, definitions, and results related to time scale calculus, as well as fuzzy calculus, that are used throughout this paper.

5.2.1 Time scale calculus

Let $\mathbb{T}$ be the time scale, i. e., an arbitrary nonempty closed subset of $\mathbb{R}$. Since a time scale $\mathbb{T}$ is not connected, we need the concept of jump operators.

Definition 5.1 ([12]). The mappings $\xi, \tau : \mathbb{T} \to \mathbb{T}$ defined as

$$\xi(t) = \inf\{s \in \mathbb{T} \mid s > t\} \quad \text{and} \quad \tau(t) = \sup\{s \in \mathbb{T} \mid s < t\},$$

are called jump operators.

Definition 5.2 ([12]). A nonmaximal element $t \in \mathbb{T}$ is said to be right-scattered (rs), if $\xi(t) > t$ and right-dense (rd), if $\xi(t) = t$. A nonminimal element $t \in \mathbb{T}$ is called left-scattered (ls), if $\tau(t) < t$ and left-dense (ld), if $\tau(t) = t$. If $\mathbb{T}$ has an ls maximum m, then $\mathbb{T}^k = \mathbb{T} - \{m\}$. Otherwise $\mathbb{T}^k = \mathbb{T}$.

Definition 5.3 ([12]). The mapping $\mu : \mathbb{T} \to \mathbb{R}^+$ defined by $\mu(t) = \xi(t) - t$ is called graininess. When $\mathbb{T} = \mathbb{Z}$, $\mu(t) = 1$, and when $\mathbb{T} = \mathbb{R}$, $\mu(t) = 0$.

Definition 5.4 ([12]). The mapping $g : \mathbb{T} \to E^d$, where E^d is a fuzzy space, is called rd-continuous if

(i) it is continuous at each right-dense $(rd)t \in \mathbb{T}$,
(ii) at each left-dense point, the left-sided limit $g(t^-)$ exists.

Definition 5.5 ([12]). A fuzzy function $g : \mathbb{T} \to E^d$ is said to be differentiable at $t \in \mathbb{T}^k$, if there exists a $\delta > 0$ such that for any $\varepsilon > 0$, there exists a neighborhood $N_{\mathbb{T}}(t, \delta)$ satisfying

$$D[g(\xi(t)) \ominus_{gH} g(s), g^{\Delta}(t)(\xi(t) - s)] \le \varepsilon|\xi(t) - s|, \quad \text{for all } s \in N_{\mathbb{T}}(t, \delta),$$

where $g^{\Delta}(t) = \frac{g(\xi(t)) \ominus_{gH} g(t)}{\mu(t)}$. In this case, $g^{\Delta}(t)$ is called the delta generalized Hukuhara derivative (Δ_{gH}-derivative) of g at t. Moreover, g is said to be delta generalized Hukuhara differentiable (Δ_{gH}-differentiable) on $\mathbb{T}^k$, if $g^{\Delta}(t) \in E^d$ exists for all $t \in \mathbb{T}^k$. The fuzzy function $g^{\Delta} : \mathbb{T}^k \to E^d$ is then called the Δ_{gH}-derivative of g on $\mathbb{T}^k$.

Definition 5.6 ([12]). A fuzzy function $g : \mathbb{T} \to E^d$ is called regulated provided its right-sided limits exist (finite) at all right-dense (rd) points in $\mathbb{T}$ and its left-sided limits exist (finite) at all left-dense points in $\mathbb{T}$, and g is said to be rd-continuous if it continuous at all right-dense points in $\mathbb{T}$ and its left-sided limits exists (finite) at all left-dense points in $\mathbb{T}$. We denote the set of rd-continuous functions from $\mathbb{T}$ to E^d by $C_{rd}[\mathbb{T}, E^d]$.

Definition 5.7 ([12]). Assume that $g : \mathbb{T} \to E^d$ is a continuous fuzzy function on $\mathbb{T}$ and Δ_{gH}-differentiable on $\mathbb{T}^k$. If $g^{\Delta}(t) \geq 0$ for all $t \in \mathbb{T}^k$, then g is called nondecreasing on $\mathbb{T}$. If $g^{\Delta_{gH}}(t_0) \leq 0$ for all $t \in \mathbb{T}^k$ then g is called nonincreasing on $\mathbb{T}$.

Theorem 5.1 ([21]). *Let $g : \mathbb{T} \to E^d$ be a fuzzy function and $t \in \mathbb{T}^k$. Then we have the following:*

(i) *If g is Δ_{gH}-differentiable at t, then g is continuous at t.*

(ii) *If g is continuous at t and t is rs, then g is Δ_{gH}-differentiable at t with*

$$g^{\Delta}(t) = \frac{g(\xi(t)) \ominus_{gH} g(t)}{\mu(t)}.$$

(iii) *If t is rd, then g is Δ_{gH}-differentiable at t if and only if the limits*

$$\lim_{h \to 0^+} \frac{g(t+h) \ominus_{gH} g(t)}{h}, \quad \lim_{h \to 0^+} \frac{g(t) \ominus_{gH} g(t-h)}{h}$$

exist. In this case

$$\lim_{h \to 0^+} \frac{g(t+h) \ominus_{gH} g(t)}{h} = \lim_{h \to 0^+} \frac{g(t) \ominus_{gH} g(t-h)}{h} = g^{\Delta}(t).$$

(iv) *If g is Δ_{gH}-differentiable at t, then*

$$g(\xi(t)) \ominus_{gH} g(t) = \mu(t) g^{\Delta}(t).$$

Theorem 5.2. *Let $g : [a,b]_{\mathbb{T}} \to E^d$ be a continuous fuzzy function on $[a,b]_{\mathbb{T}}$. If g is Δ_{gH}-differentiable on $[a,b)_{\mathbb{T}}$ such that g^{Δ} is continuous on $[a,b)_{\mathbb{T}}$, then*

$$\int_a^b g^{\Delta}(t)\Delta(t) = g(b) \ominus_{gH} g(a). \tag{5.2}$$

Remark 5.1. If g is nondecreasing on $[a,b]_{\mathbb{T}}$, then Eq. (5.2) is equivalent to

$$g(b) = g(a) + \int_a^b g^{\Delta}(t)\Delta t,$$

and if g is nonincreasing, then Eq. (5.2) is equivalent to

$$g(b) = g(a) \ominus_{gH} (-1) \int_a^b g^{\Delta}(t)\Delta t.$$

Theorem 5.3. *Suppose $a, b \in \mathbb{T}$, $a < b$, and $g(t)$ is continuous on $[a, b]$. Then*

$$\int_a^b g^{\Delta}(t)\Delta t = [\xi(a) - a]g(a) + \int_{\xi(a)}^b g(t)\Delta t.$$

Theorem 5.4. *Suppose $\mathbb{T}$ is a time scale and g is an increasing continuous function on $[a, b]_{\mathbb{T}}$. If g is the extension of g_1 to the real interval $[a, b]$ given by*

$$g(s) = \begin{cases} g_1(s), & s \in \mathbb{T}, \\ g_1(t), & s \in (t, \xi(t)) \notin \mathbb{T}, \end{cases}$$

then

$$\int_a^b g_1(t)\Delta t \leq \int_a^b g(t)dt. \tag{5.3}$$

5.2.2 Fuzzy calculus

Definition 5.8 ([34]). Let $\mathfrak{P}_k(\mathbb{R})$ denote the family of all nonempty compact convex subsets of $\mathbb{R}$, then the Hausdorff metric d_H is defined by

$$d_H(\mathfrak{A}, \mathfrak{B}) = \max\left\{\sup_{y\in\mathfrak{A}} \inf_{z\in\mathfrak{B}} \|y - z\|, \sup_{z\in\mathfrak{B}} \inf_{y\in\mathfrak{A}} \|y - z\|\right\}, \quad \mathfrak{A}, \mathfrak{B} \in \mathfrak{P}_k(\mathbb{R}),$$

where $\|\cdot\|$ denotes the usual Euclidean norm on $\mathbb{R}$.

Definition 5.9 ([34]). Let E^d denote the class of fuzzy subsets of the real axis $v : \mathbb{R} \to [0, 1]$, which satisfy the following conditions:

(1) v is normal, i. e., there exists $y_0 \in \mathbb{R}$ with $v(y_0) = 1$;
(2) v is a convex fuzzy set, i. e., $v(\omega y + (1 - \omega)z) \geq \min\{v(y), v(z)\}$, for all $\omega \in [0, 1]$ and $y, z \in \mathbb{R}$;
(3) v is upper semicontinuous on $\mathbb{R}$;
(4) the closure of $\{y \in \mathbb{R} : v(y) > 0\}$, denoted by $[v]^0$, is compact.

We call E^d the space of fuzzy numbers or a fuzzy set; obviously, $E^d \subset \mathfrak{P}_k(\mathbb{R})$.

Definition 5.10 ([34]). Let E^d represent the fuzzy set $v : \mathbb{R} \to [0, 1]$ such that $[v]^\alpha \in \mathfrak{P}_k(\mathbb{R})$ for all $\alpha \in [0, 1]$, where $[v]^\alpha = \{y \in \mathbb{R} \mid v(y) \geq \alpha\}$ for $0 < \alpha \leq 1$, and $[v]^0 = \mathrm{cl}\{y \in \mathbb{R} : v(y) > 0\}$.

The notation $[v]^\alpha = [v_l^\alpha, v_r^\alpha]$ denotes the α-level set of v. We refer to v_l, v_r as the left and right branches of v.

Definition 5.11 ([26]). Let v_1 and v_2 be two fuzzy numbers defined on E^d and $\omega \in \mathbb{R}$. Due to Zadeh's extension principle, $v_1 + v_2$ and ωv_1 are in E^d and are defined as

$$\begin{aligned}[v_1 + v_2]^\alpha &= [v_1]^\alpha + [v_2]^\alpha, \\ [\omega v_1]^\alpha &= \omega[v_1]^\alpha, \quad \text{for all } \alpha \in [0,1],\end{aligned}$$

where $[v_1]^\alpha + [v_2]^\alpha$ is the usual addition of two intervals of $\mathbb{R}$ and $\omega[v_1]^\alpha$ is the usual scalar product of a number and a subset of $\mathbb{R}$.

Definition 5.12 ([26]). The Hausdorff distance between fuzzy numbers $v_1, v_2 \in E^d$ is given as

$$D[v_1, v_2] = \sup_{\alpha\in[0,1]} \{d_H([v_1]^\alpha, [v_2]^\alpha)\},$$

where $d_H([v_1]^\alpha, [v_2]^\alpha)$ is the Hausdorff distance between two sets $[v_1]^\alpha, [v_2]^\alpha$ of E^d.

Definition 5.13 ([22]). Let $g : [a, b] \to E^d$ be measurable and bounded. The integral of g over $[a, b]$, denoted by $\int_a^b g(t)dt$, is defined level-wise by the expression

$$\begin{aligned}\left[\int_a^b g(t)dt\right]^\alpha &= \int_a^b [g(t)]^\alpha dt \\ &= \left\{\int_a^b \bar{g}(t)dt \mid \bar{g} : [a, b] \to \mathbb{R} \text{ is a measurable selection of } [g(\cdot)]^\alpha\right\}.\end{aligned}$$

Definition 5.14 ([10]). The generalized Hukuhara difference (gH-difference) of two fuzzy numbers $v_1, v_2 \in E^d$ is $v_3 \in E^d$, defined as

$$v_1 \ominus_{gH} v_2 = v_3 \iff \begin{cases} v_1 = v_2 + v_3, & \text{or} \\ v_2 = v_1 + (-1)v_3. \end{cases} \tag{5.4}$$

Definition 5.15. The delta generalized Hukuhara derivative (Δ_{gH}-derivative) of a fuzzy function $g : [a, b] \to E^d$ at t is defined as

$$g^\Delta(t) = \lim_{h\to 0} \frac{g(t + h) \ominus_{gH} g(t)}{h}.$$

If $g^\Delta(t) \in E^d$, then g is said to be Δ_{gH}-differentiable at t. Also, g is $[(i)\text{-}\Delta_{gH}]$-differentiable at t if (i) $g^\Delta(t; \alpha) = [(g_l^\alpha)^\Delta(t, \alpha), (g_r^\alpha)^\Delta(t, \alpha)]$ and g is $[(ii)\text{-}\Delta_{gH}]$-differentiable at t if (ii) $g^\Delta(t; \alpha) = [(g_r^\alpha)^\Delta(t, \alpha), (g_l^\alpha)^\Delta(t, \alpha)]$, where $\alpha \in (0, 1)$.

Let $[0, b] \subset \mathbb{R}$ be an interval and suppose $\Xi \in C^1([0, b], \mathbb{R}^+)$ is such that Ξ is increasing and positive-defined, as well as $\Xi^\Delta(t) \neq 0$ for every $t \in [0, b]$. We consider the weighted space $C_{1-\gamma,\Xi}([0, b], E^d)$ of fuzzy functions g on $(0, b]$ defined by

$$C_{1-\gamma,\Xi}([0,b],E^d) = \{g : (0,b] \to E^d : (\Xi(t) - \Xi(0))^{1-\gamma} g(t) \in C([0,b],E^d)\},$$

with $0 < \gamma \leq 1$ and the norm

$$\|g\|_{C_{1-\gamma,\Xi}([0,b],E^d)} = D[g(t),\widehat{0}] = \max_{t\in[0,b]} D[(\Xi(t) - \Xi(0))^{1-\gamma} g(t),\widehat{0}].$$

We denote by $C([0,b],E^d)$ the space of all continuous fuzzy functions on $[0,b]$.

Definition 5.16. Suppose $\mathbb{T}$ is a time scale and $[0,b]$ an interval of $\mathbb{T}$. Let g be an integrable fuzzy-valued function defined on $[0,b]$. Then the tempered Ξ-RL fuzzy fractional integral on time scales of order p of the fuzzy-valued function g with respect to Ξ is defined by

$$\left({}_{0^+}^{T\mathbb{T}}\mathfrak{I}_{\Xi}^{p,\lambda} g\right)(t) = \frac{1}{\Gamma(p)} \int_0^t \Xi^{\Delta}(s)(\Xi(t) - \Xi(s))^{p-1} e^{-\lambda(\Xi(t)-\Xi(s))} g(s)\Delta s. \tag{5.5}$$

Definition 5.17. Suppose $\mathbb{T}$ is a time scale and $[0,b]$ an interval of $\mathbb{T}$. Then, the tempered Ξ-Hilfer fuzzy fractional derivative on time scales of order $p \in (0,1)$ and type $q \in [0,1]$ of the fuzzy-valued function $g \in C^1([0,b],E^d)$ is defined by

$$\left({}_{0^+}^{TH\mathbb{T}}\Delta_{\Xi}^{p,q,\lambda} g\right)(t) = \left({}_{0^+}^{T}\mathfrak{I}_{\Xi}^{q(1-p),\Xi}\right)\left(\frac{\Delta}{\Xi^{\Delta}(t)} + \lambda\right)\left({}_{0^+}^{T\mathbb{T}}\mathfrak{I}_{\Xi}^{(1-q)(1-p),\lambda} g\right)(t),$$

where ${}^{\mathbb{T}}\Delta = \frac{d}{dt}$.

Lemma 5.1. *Let $\mathbb{T}$ be a time scale. Let also $p \in (0,\infty)$, $\lambda \in \mathbb{R}$, and $\nu \in (-1,\infty)$. Then, we have*

$$\begin{aligned}
{}_{0^+}^{T\mathbb{T}}\mathfrak{I}_{\Xi}^{p,\lambda}(\Xi(t) - \Xi(0))^{\nu} &= e^{-\lambda(\Xi(t)-\Xi(0))}(\Xi(t) - \Xi(0))^{\nu+p}\frac{\Gamma(\nu+1)}{\Gamma(\nu+p+1)} \\
&\quad \times {}_1G_1(\nu+1;\nu+p+1;\lambda(\Xi(t) - \Xi(0))), \\
{}_{0^+}^{TR\mathbb{T}}\Delta_{\Xi}^{p,\lambda}(\Xi(t) - \Xi(0))^{\nu} &= e^{-\lambda(\Xi(t)-\Xi(0))}(\Xi(t) - \Xi(0))^{\nu-p}\frac{\Gamma(\nu+1)}{\Gamma(\nu-p+1)} \\
&\quad \times {}_1G_1(\nu+1;\nu-p+1;\lambda(\Xi(t) - \Xi(0))).
\end{aligned}$$

Proof. Using the expression for the Ξ-tempered integral as a conjugation of the Ξ-RL integral, together with the functions of the form $\frac{(\Xi(t)-\Xi(0))^{\nu}}{\Gamma(\nu+1)}$, we have

$$\begin{aligned}
&{}_{0^+}^{T\mathbb{T}}\mathfrak{I}_{\Xi}^{p,\lambda}(\Xi(t) - \Xi(0))^{\nu} \\
&\quad = e^{-\lambda(\Xi(t)-\Xi(0))}\,{}_{0^+}^{\mathbb{T}}\mathfrak{I}_{\Xi}^{p,\lambda}\left(e^{\lambda(\Xi(t)-\Xi(0))}(\Xi(t) - \Xi(0))^{\nu}\right) \\
&\quad = e^{-\lambda(\Xi(t)-\Xi(0))}\,{}_{0^+}^{\mathbb{T}}\mathfrak{I}_{\Xi}^{p,\lambda}\left(\sum_{l=0}^{\infty}\frac{(\lambda(\Xi(t)) - \Xi(0))^{l}}{l!}(\Xi(t) - \Xi(0))^{\nu}\right)
\end{aligned}$$

$$
\begin{aligned}
&= e^{-\lambda(\Xi(t)-\Xi(0))} \sum_{l=0}^{\infty} \frac{\lambda^l}{l!} {}^{\mathbb{T}}_{0^+}\mathfrak{I}_{\Xi}^{p,\lambda} (\Xi(t) - \Xi(0))^{\nu+l} \\
&= e^{-\lambda(\Xi(t)-\Xi(0))} \sum_{l=0}^{\infty} \frac{\lambda^l}{l!} \frac{\Gamma(\nu+l+1)}{\Gamma(\nu+l+p+1)} (\Xi(t) - \Xi(0))^{\nu+p+l} \\
&= e^{-\lambda(\Xi(t)-\Xi(0))} (\Xi(t) - \Xi(0))^{\nu+p} \frac{\Gamma(\nu+1)}{\Gamma(\nu+p+1)} \\
&\quad \times \left\{ \frac{\Gamma(\nu+p+1)}{\Gamma(\nu+1)} \sum_{l=0}^{\infty} \frac{\Gamma(\nu+1+l)}{\Gamma(\nu+p+1+l)} \frac{(\lambda(\Xi(t) - \Xi(0))^l)}{l!} \right\} \\
&= \frac{\Gamma(\nu+1)}{\Gamma(\nu+p+1)} (\Xi(t) - \Xi(0))^{\nu+p} e^{-\lambda(\Xi(t)-\Xi(0))} \\
&\quad \times {}_1G_1(\nu+1; \nu+p+1; \lambda(\Xi(t) - \Xi(0))).
\end{aligned}
$$

This proves the first of the stated identities. The identity for the Ξ-tempered derivative of RL type can be proved similarly by direct calculation. □

Lemma 5.2. *Let $g \in C^1([0,b], E^d)$, $0 < p < 1$, and $0 \le q \le 1$. Then, we have:*

(i) ${}^{TT}_{0^+}\mathfrak{I}_{\Xi}^{p,\lambda} \, {}^{THT}_{0^+}\Delta_{\Xi}^{p,q,\lambda} g(t)$

$$
= g(t) - e^{-\lambda(\Xi(t))} \frac{(\Xi(t) - \Xi(0))^{\gamma-1}}{\Gamma(\gamma)} \left[{}^{TT}_{0^+}\mathfrak{I}_{\Xi}^{1-\gamma,\lambda} (e^{\lambda \Xi(t)} g(t)) \right]_{t=0},
$$

with $\gamma = p + q(1-p)$.

(ii) ${}^{THT}_{0^+}\Delta_{\Xi}^{p,q,\lambda} \, {}^{TT}_{0^+}\mathfrak{I}_{\Xi}^{p,\lambda} g(t) = g(t)$.

Theorem 5.5 (Gronwall's inequality, [30]). *Let $p > 0$, $\lambda \in \mathbb{R}$ and let $\mathscr{X}(t)$, $\mathscr{Y}(t)$ be two integrable functions and $\mathscr{Z}(t)$ be a continuous function on $[0,b]$. Assume that $\mathscr{X}(t)$, $\mathscr{Y}(t)$ are nonnegative, and $\mathscr{Z}(t)$ is nonnegative and nondecreasing. If*

$$
\mathscr{X}(t) \le \mathscr{Y}(t) + \mathscr{Z}(t) \int_0^t \Xi'(s)(\Xi(t) - \Xi(s))^{p-1} e^{-\lambda(\Xi(t)-\Xi(s))} \mathscr{X}(s) ds, \quad t \in [0,b],
$$

then

$$
\mathscr{X}(t) \le \mathscr{Y}(t) + \int_0^t \sum_{k=1}^{\infty} \frac{[Z(t)\Gamma(p)]^k}{\Gamma(pk)} \Xi'(s)(\Xi(t) - \Xi(s))^{pk-1} e^{-\lambda(\Xi(t)-\Xi(s))} \mathscr{Y}(s) ds.
$$

Let $\mathscr{Y}$ be a nondecreasing function. Then

$$
\mathscr{X}(t) \le \mathscr{Y}(t) E_p \{ \mathscr{Z}(t) \Gamma(p) (\Xi(t) - \Xi(0))^p \}.
$$

Lemma 5.3 (Hybrid fixed point theorem, [15]). *Let S be a nonempty closed convex and bounded subset of a Banach algebra Y and let $P : Y \to Y$, $Q : S \to Y$ be two operators* *such that*

(i) *P is Lipschitzian with Lipschitz constant β;*
(ii) *Q is completely continuous;*
(iii) *$v = PvQ\bar{v} \leftrightarrow v \in S$ for every $\bar{v} \in S$;*
(iv) *$L\beta < 1$, where $L = \sup\{\|Q(s)\| : \bar{v} \in S\}$.*

Then the operator equation $v = PvQv$ has a solution in S.

5.3 Main results

Lemma 5.4. *Assume that $\mathfrak{J} = [0, b] \subseteq \mathbb{T}$. Let $\gamma = p + q - pq$, where $p \in (0, 1)$ and $q \in [0, 1]$. Let also $G : \mathfrak{J} \times E^d \to E^d$ be a function such that $G(\cdot, v(\cdot)) \in C_{1-\gamma,\Xi}(\mathfrak{J}, E^d)$ for any $v \in C_{1-\gamma,\Xi}(\mathfrak{J}, E^d)$. A fuzzy function v is a solution of the problem (5.1) if and only if v satisfies the following integral equation:*

$$v(t) = f(t, v(t))\left(\frac{v_0 e^{-\lambda(\Xi(t)-\Xi(0))}}{\Gamma(\gamma)}(\Xi(t) - \Xi(0))^{\gamma-1}\right.$$
$$\left.\ominus (-1)\frac{1}{\Gamma(p)}\int_0^t \Xi^\Delta(s)(\Xi(t) - \Xi(s))^{p-1} e^{-\lambda(\Xi(t)-\Xi(s))} G(s, v(s))\Delta s\right). \tag{5.6}$$

Proof. Let $v \in C_{1-\gamma,\Xi}(\mathfrak{J}, E^d)$ be a solution of the problem (5.1). We will prove that v is also a solution of the integral equation (5.6). Applying ${}^{\mathbb{TT}}_{0^+}\mathfrak{J}^{p,\lambda}_{\Xi}$ on both sides of the first equation of the problem and using Lemma 5.2, we obtain

$$\frac{v(t)}{f(t, v(t))} - \frac{e^{-\lambda(\Xi(t)-\Xi(0))}(\Xi(t) - \Xi(0))^{\gamma-1}}{\Gamma(\gamma)}\left[{}^{\mathbb{TT}}_{0^+}\mathfrak{J}^{1-\gamma,\lambda}_{\Xi}\left(\frac{v(t)}{f(t, v(t))}\right)\right]_{t=0}$$
$$= {}^{\mathbb{TT}}_{0^+}\mathfrak{J}^{p,\lambda}_{\Xi} G(t, v(t)). \tag{5.7}$$

Hence, we get the integral equation (5.6). In the latter equation we have used that

$${}^{\mathbb{TT}}_{0^+}\mathfrak{J}^{1-\gamma,\lambda}_{\Xi}\left(\frac{v(0)}{f(0, v(0))}\right) = v_0. \tag{5.8}$$

Thus v is a solution of the integral equation (5.6).

Conversely, assume that $v \in C_{1-\gamma,\Xi}(\mathfrak{J}, E^d)$ is a solution of Eq. (5.6). Then Eq. (5.7) becomes

$$\frac{v(t)}{f(t, v(t))} = \left(\frac{v_0 e^{-\lambda(\Xi(t)-\Xi(0))}}{\Gamma(\gamma)}(\Xi(t) - \Xi(0))^{\gamma-1}\right.$$
$$\left.\ominus (-1)\frac{1}{\Gamma(p)}\int_0^t \Xi^\Delta(s)(\Xi(t) - \Xi(s))^{p-1} e^{-\lambda(\Xi(t)-\Xi(s))} G(s, v(s))\Delta s\right). \tag{5.9}$$

Applying ${}^{THT}_{0^+}\Delta^{p,q,\lambda}_{\Xi}$ on both sides of the above equation and using ${}^{THT}_{0^+}\Delta^{p,q,\lambda}_{\Xi}(\Xi(t) - \Xi(0))^{\gamma-1} = 0$, together with Lemma 5.2, we obtain

$$
\begin{aligned}
{}^{THT}_{0^+}\Delta^{p,q,\lambda}_{\Xi}\left[\frac{v(t)}{f(t,v(t))}\right] &= \left(\frac{v_0 e^{-\lambda(\Xi(t)-\Xi(0))}}{\Gamma(\gamma)}\,{}^{THT}_{0^+}\Delta^{p,q,\lambda}_{\Xi}(\Xi(t)-\Xi(0))^{\gamma-1}\right.\\
&\quad \left.\ominus (-1)^{THT}_{0^+}\Delta^{p,q,\lambda}_{\Xi}\,{}^{TT}_{0^+}\mathfrak{J}^{p,\lambda}_{\Xi}G(t,v(t))\right)\\
&= G(t,v(t)).
\end{aligned}
$$

Finally, we need to verify that the initial condition ${}^{TT}_{0^+}\mathfrak{J}^{1-\gamma,\lambda}_{\Xi}(\frac{v(0)}{f(0,v(0))}) = v_0$ in the problem (5.1) also is satisfied. From Eq. (5.9), we have

$$
{}^{TT}_{0^+}\mathfrak{J}^{1-\gamma,\lambda}_{\Xi}\left[\frac{v(t)}{f(t,v(t))}\right] = \left(\frac{v_0}{\Gamma(\gamma)} \ominus (-1)^{TT}_{0^+}\mathfrak{J}^{1-\gamma+p,\lambda}_{\Xi}G(t,v(t))\right).
$$

For this purpose, we put $t = 0$ and then the above equation reduces to

$$
{}^{TT}_{0^+}\mathfrak{J}^{1-\gamma,\lambda}_{\Xi}\left[\frac{v(0)}{f(t,v(0))}\right] = v_0,
$$

which is the initial condition of (5.1). This completes the proof. □

For establishing the main results, we will impose the following hypotheses on f and G:

$(H1)$ The functions $f : \mathfrak{J} \times E^d \to E^d - \{0\}$ and $G : \mathfrak{J} \times E^d \to E^d$ are continuous and there exists a bounded function $p_1 : \mathfrak{J} \to \mathbb{R}^+$ such that

$$
D[f(t,v(t)), f(t,\bar{v}(t))] \le p_1(t)D[v(t),\bar{v}(t)],
$$

with $p_1^* = \sup_{t\in\mathfrak{J}} D[p_1(t),\widehat{0}]$, for $v,\bar{v} \in E^d$ for every $t \in \mathfrak{J}$.

$(H2)$ There exist a function $q_1 \in C(\mathfrak{J},\mathbb{R}^+)$ and a continuous nondecreasing function $U : [0,\infty) \to [0,\infty)$ such that

$$
D[G(t,v(t)),\widehat{0}] \le q_1(t)U(D[v,\widehat{0}]),
$$

with $q_1^* = \sup_{t\in\mathfrak{J}} D[q_1(t),\widehat{0}]$, for $v \in E^d$, for every $t \in \mathfrak{J}$.

$(H3)$ There exists a constant $\mathscr{L}_1 > 0$ such that

$$
D[f(t,v(t)), f(t,\bar{v}(t))] \le \mathscr{L}_1 D[v,\bar{v}], \quad \text{for every } v,\bar{v} \in E^d,
$$

and we put $D[f(t,v(t)),\widehat{0}] \le \mathscr{M}_1$.

$(H4)$ There exists a constant $\mathscr{L}_2 > 0$ such that

$$
D[G(t,v(t)), G(t,\bar{v}(t))] \le \mathscr{L}_2 D[v,\bar{v}], \quad \text{for every } v,\bar{v} \in E^d,
$$

and we put $D[G(t, v(t)), \widehat{0}] \leq \mathscr{M}_2$.

(H5) Let $\varphi \in C_{1-\gamma,\Xi}(\mathfrak{J}, \mathbb{R}^+)$ be a nondecreasing function. Then there exists $\mathfrak{K} > 0$ such that

$$ {}_{0^+}^{T\mathbb{T}}\mathfrak{J}_{\Xi}^{p,\lambda}\varphi(t) \leq \mathfrak{K}\varphi(t). $$

Theorem 5.6. *Assume that* (H1) *and* (H2) *are satisfied. Then, the problem* (5.1) *has a solution* $v \in C_{1-\gamma,\Xi}(\mathfrak{J}, E^d)$ *provided*

$$ p_1^*\left(\frac{v_0}{\Gamma(\gamma)} + \frac{(\Xi(b) - \Xi(0))^{1-\gamma+p}}{\Gamma(p+1)} e^{-\lambda(\Xi(b)-\Xi(0))} q_1^* U(r)\right) < 1. \tag{5.10} $$

Proof. Define a subset S of $C_{1-\gamma,\Xi}(\mathfrak{J}, E^d)$ by

$$ S = \{v \in C_{1-\gamma,\Xi}(\mathfrak{J}, E^d) \mid \|v\| \leq r\}, \tag{5.11} $$

where

$$ r = \mathscr{K}\left(\frac{v_0}{\Gamma(\gamma)} + \frac{(\Xi(b) - \Xi(0))^{1-\gamma+p}}{\Gamma(p+1)} e^{-\lambda(\Xi(b)-\Xi(0))} q_1^* U(r)\right), $$

and $\mathscr{K}$ is bounded on f. Obviously, S is a closed, convex, and bounded subset of $C_{1-\gamma,\Xi}(\mathfrak{J}, E^d)$. By Lemma 5.4, the problem (5.1) is equivalent to the integral equation (5.6). Let us consider two operators $P : C_{1-\gamma,\Xi}(\mathfrak{J}, E^d) \to C_{1-\gamma,\Xi}(\mathfrak{J}, E^d)$ defined by

$$ Pv(t) = f(t, v(t)), \quad t \in \mathfrak{J}, $$

and $Q : S \to C_{1-\gamma,\Xi}(\mathfrak{J}, E^d)$ by

$$ \begin{aligned} Qv(t) = {} & \frac{v_0 e^{-\lambda(\Xi(t)-\Xi(0))}}{\Gamma(\gamma)}(\Xi(t) - \Xi(0))^{\gamma-1} \\ & \ominus (-1)\frac{1}{\Gamma(p)} \int_0^t \Xi^{\Delta}(s)(\Xi(t) - \Xi(s))^{p-1} e^{-\lambda(\Xi(t)-\Xi(s))} G(s, v(s))\Delta s. \end{aligned} $$

Then, $v = PvQv$. We shall show that the operators P and Q fulfill all the assumptions of Lemma 5.3. We divide our proof into several steps.

Step 1: P is Lipschitzian on $C_{1-\gamma,\Xi}(\mathfrak{J}, E^d)$.

Let $v, \bar{v} \in C_{1-\gamma,\Xi}(\mathfrak{J}, E^d)$. Then by (H1) we have

$$ \begin{aligned} & D[(\Xi(t) - \Xi(0))^{1-\gamma} Pv(t), (\Xi(t) - \Xi(0))^{1-\gamma} P\bar{v}(t)] \\ & \leq D[(\Xi(t) - \Xi(0))^{1-\gamma} f(t, v(t)), (\Xi(t) - \Xi(0))^{1-\gamma} f(t, \bar{v}(t))] \\ & \leq p_1(t) D[(\Xi(t) - \Xi(0))^{1-\gamma} v(t), (\Xi(t) - \Xi(0))^{1-\gamma} \bar{v}(t)]. \end{aligned} $$

This gives

$$D[Pv - P\bar{v}] \leq p_1^* D[v - \bar{v}]. \tag{5.12}$$

Thus, the operator P is Lipschitzian on $C_{1-\gamma,\Xi}(\mathfrak{J}, E^d)$ with Lipschitz constant p_1^*.
Step 2: Q is completely continuous on S.
Let us take a sequence $\{v_n\} \subset S$ and $v \in S$ such that $D[v_n, v] \to 0$ as $n \to \infty$. Then the Lebesgue dominated convergence theorem gives that

$$\begin{aligned}
&\lim_{n\to\infty} (\Xi(t) - \Xi(0))^{1-\gamma} Qv_n(t) \\
&= \lim_{n\to\infty} \left(\frac{v_0 e^{-\lambda(\Xi(t)-\Xi(0))}}{\Gamma(\gamma)} \ominus (-1) \frac{(\Xi(t) - \Xi(0))^{1-\gamma}}{\Gamma(p)} \right. \\
&\quad \left. \times \int_0^t \Xi^\Delta(s)(\Xi(t) - \Xi(s))^{p-1} e^{-\lambda(\Xi(t)-\Xi(s))} D[G(s, v_n(s)), \hat{0}]\Delta s \right) \\
&= \frac{v_0 e^{-\lambda(\Xi(t)-\Xi(0))}}{\Gamma(\gamma)} \ominus (-1) \frac{(\Xi(t) - \Xi(0))^{1-\gamma}}{\Gamma(p)} \\
&\quad \times \int_0^t \Xi^\Delta(s)(\Xi(t) - \Xi(s))^{p-1} e^{-\lambda(\Xi(t)-\Xi(s))} \lim_{n\to\infty} D[G(s, v_n(s)), \hat{0}]\Delta s \\
&= \frac{v_0 e^{-\lambda(\Xi(t)-\Xi(0))}}{\Gamma(\gamma)} \ominus (-1) \frac{(\Xi(t) - \Xi(0))^{1-\gamma}}{\Gamma(p)} \\
&\quad \times \int_0^t \Xi^\Delta(s)(\Xi(t) - \Xi(s))^{p-1} e^{-\lambda(\Xi(t)-\Xi(s))} D[G(s, v(s)), \hat{0}]ds \\
&= (\Xi(t) - \Xi(0))^{1-\gamma} Qv(t), \quad \text{for every } t \in \mathfrak{J}.
\end{aligned}$$

This proves that Q is continuous on S.
Step 3: Q is uniformly bounded in S.
For each $t \in \mathfrak{J}$, $v \in S$ and by $(H2)$, one has

$$\begin{aligned}
&D[(\Xi(t) - \Xi(0))^{1-\gamma} Qv(t), \hat{0}] \\
&\leq \frac{v_0 e^{-\lambda(\Xi(t)-\Xi(0))}}{\Gamma(\gamma)} \ominus (-1) \frac{(\Xi(t) - \Xi(0))^{1-\gamma}}{\Gamma(p)} \\
&\quad \times \int_0^t \Xi^\Delta(s)(\Xi(t) - \Xi(s))^{p-1} e^{-\lambda(\Xi(t)-\Xi(s))} D[G(s, v(s)), \hat{0}]\Delta s \\
&\leq \frac{v_0 e^{-\lambda(\Xi(t)-\Xi(0))}}{\Gamma(\gamma)} + q_1(t)U(r) \frac{(\Xi(b) - \Xi(0))^{1-\gamma+p}}{\Gamma(p+1)} e^{-\lambda(\Xi(b)-\Xi(0))}.
\end{aligned}$$

Taking the supremum over the interval $[0, b]$, the above inequality becomes

$$\|Qv\| \leq \frac{v_0}{\Gamma(\gamma)} + q_1^* U(r) \frac{(\Xi(b) - \Xi(0))^{1-\gamma+p}}{\Gamma(p+1)} e^{-\lambda(\Xi(b)-\Xi(0))}, \tag{5.13}$$

for every $v \in S$. This proves that Q is uniformly bounded on S.

Step 4: $Q(s)$ is an equicontinuous set in $C_{1-\gamma,\Xi}(\mathfrak{J}, E^d)$.

Indeed, let $v \in S$ and $0 \leq t_1 < t_2 \leq b$ with $t_1 < t_2$. Then one has

$$\begin{aligned}
&D[(\Xi(t_2) - \Xi(0))^{1-\gamma} Qv(t_2), (\Xi(t_1) - \Xi(0))^{1-\gamma} Qv(t_1)] \\
&\leq \mathscr{K}\Bigg(\frac{v_0}{\Gamma(\gamma)} \big(e^{-\lambda(\Xi(t_1)-\Xi(0))} - e^{-\lambda(\Xi(t_2)-\Xi(0))}\big) + \frac{(\Xi(t_2) - \Xi(0))^{1-\gamma}}{\Gamma(p)} e^{-\lambda\Xi(t_2)} \\
&\quad \times \int_0^{t_2} \Xi^\Delta(s)(\Xi(t_2) - \Xi(0))^{p-1} e^{\lambda\Xi(s)} D[G(s, v(s)), \widehat{0}]\Delta s \\
&\quad + \frac{(\Xi(t_1) - \Xi(0))^{1-\gamma}}{\Gamma(p)} e^{-\lambda\Xi(t_1)} \\
&\quad \times \int_0^{t_1} \Xi^\Delta(s)(\Xi(t_1) - \Xi(0))^{p-1} e^{\lambda\Xi(s)} D[G(s, v(s)), \widehat{0}]\Delta s \Bigg).
\end{aligned}$$

Thus, as $t_2 \to t_1$, the right-hand side of the above inequality tends to zero. As a consequence, the Arzela–Ascoli theorem gives that Q is a completely continuous operator on S.

Step 5: For $v \in C_{1-\gamma,\Xi}(\mathfrak{J}, E^d)$, $v = Pv Q\bar{v} \leftrightarrow v \in S$ for every $\bar{v} \in S$.

Let $v \in C_{1-\gamma,\Xi}(\mathfrak{J}, E^d)$ and $\bar{v} \in S$ be such that $v = PvQ\bar{v}$. Then, one obtains that

$$\begin{aligned}
&D[(\Xi(t) - \Xi(0))^{1-\gamma} v(t), \widehat{0}] \\
&= D[(\Xi(t) - \Xi(0))^{1-\gamma} Pv(t) Q\bar{v}(t), \widehat{0}] \\
&\leq D[g(t, v(t)), \widehat{0}] \Bigg\{ \frac{v_0 e^{-\lambda(\Xi(t)-\Xi(0))}}{\Gamma(\gamma)} \ominus (-1) \frac{(\Xi(t) - \Xi(0))^{1-\gamma}}{\Gamma(p)} \\
&\quad \times \int_0^t \Xi^\Delta(s)(\Xi(t) - \Xi(s))^{p-1} e^{-\lambda(\Xi(t)-\Xi(s))} D[G(s, v(s)), \widehat{0}]\Delta s \Bigg\} \\
&\leq \mathscr{K} \Bigg\{ \frac{v_0 e^{-\lambda(\Xi(t)-\Xi(0))}}{\Gamma(\gamma)} + q_1(t) U(r) \frac{(\Xi(b) - \Xi(0))^{1-\gamma+p}}{\Gamma(p+1)} e^{-\lambda(\Xi(b)-\Xi(0))} \Bigg\}.
\end{aligned}$$

Taking the supremum for $t \in \mathfrak{J}$, we obtain

$$\|v\| \leq \mathscr{K} \left\{ \frac{v_0}{\Gamma(\gamma)} + q_1^* U(r) \frac{(\Xi(b) - \Xi(0))^{1-\gamma+p}}{\Gamma(p+1)} e^{-\lambda(\Xi(b)-\Xi(0))} \right\} = r.$$

This implies that $v \in S$.

Step 6: We will prove that $\beta L < 1$, where $\beta = p_1^*$ and $L = \sup\{\|Qv\| : v \in S\}$.

From inequality (5.13), we have

$$L = \sup\{\|Qv\| : v \in S\} \leq \frac{v_0}{\Gamma(\gamma)} + q_1^* U(r) \frac{(\Xi(b) - \Xi(0))^{1-\gamma+p}}{\Gamma(p+1)} e^{-\lambda(\Xi(b)-\Xi(0))}.$$

From the inequality (5.12), we have $\beta = L$. Hence, by using Eq. (5.10), we have

$$\beta L \leq p_1^* \left(\frac{v_0}{\Gamma(\gamma)} + q_1^* U(r) \frac{(\Xi(b) - \Xi(0))^{1-\gamma+p}}{\Gamma(p+1)} e^{-\lambda(\Xi(b)-\Xi(0))} \right) < 1.$$

Thus, all the conditions of Lemma 5.3 hold true and hence the operator equation $v = PvQv$ has a solution in S. Consequently, the problem (5.1) has a solution $\mathfrak{J}$. This completes the proof. □

Theorem 5.7. *Assume that* (H3)–(H4) *are satisfied. If*

$$\frac{(\Xi(b) - \Xi(0))^p}{\Gamma(p+1)} e^{-\lambda(\Xi(b)-\Xi(0))} (\mathscr{L}_1 \mathscr{M}_2 + \mathscr{M}_1 \mathscr{L}_2) < 1, \tag{5.14}$$

then the problem (5.1) *has a unique solution on* $\mathfrak{J}$.

Proof. Consider the operator $\Pi : C_{1-\gamma,\Xi}(\mathfrak{J}, E^d) \to C_{1-\gamma,\Xi}(\mathfrak{J}, E^d)$ defined as follows:

$$(\Pi v)(t) = f(t, v(t)) \left(\frac{v_0 e^{-\lambda(\Xi(t)-\Xi(0))}}{\Gamma(\gamma)} (\Xi(t) - \Xi(0))^{\gamma-1} \right.$$
$$\left. \ominus (-1) \frac{1}{\Gamma(p)} \int_0^t \Xi^\Delta(s)(\Xi(t) - \Xi(s))^{p-1} e^{-\lambda(\Xi(t)-\Xi(s))} G(s, v(s)) \Delta s \right). \tag{5.15}$$

By Eq. (5.15), finding a solution of Eq. (5.1) in $C_{1-\gamma,\Xi}(\mathfrak{J}, E^d)$ is equivalent to finding a fixed point of the operator Π. For any $v, \bar{v} \in C_{1-\gamma,\Xi}(\mathfrak{J}, E^d)$ and each $t \in \mathfrak{J}$, we have

$$D[(\Pi v)(t), (\Pi \bar{v})(t)]$$
$$= \frac{1}{\Gamma(p)} D\left[f(t, v(t)) \int_0^t \Xi^\Delta(s)(\Xi(t) - \Xi(s))^{p-1} e^{-\lambda(\Xi(t)-\Xi(s))} G(s, v(s)) \Delta s, \right.$$
$$\left. f(t, \bar{v}(t)) \int_0^t \Xi^\Delta(s)(\Xi(t) - \Xi(s))^{p-1} e^{-\lambda(\Xi(t)-\Xi(s))} G(s, \bar{v}(s)) \Delta s \right]$$
$$= \frac{1}{\Gamma(p)} \left[D[f(t, v(t)), f(t, \bar{v}(t))] \right.$$
$$\times \int_0^t \Xi^\Delta(s)(\Xi(t) - \Xi(s))^{p-1} e^{-\lambda(\Xi(t)-\Xi(s))} D[G(s, v(s)), \widehat{0}] \Delta s$$

$$+ D[f(t,\bar{v}(t)),\hat{0}]$$
$$\times \int_0^t \Xi^\Delta(s)(\Xi(t)-\Xi(s))^{p-1} e^{-\lambda(\Xi(t)-\Xi(s))} D[G(s,\bar{v}(s)), G(s,\bar{v}(s))]\Delta s\Bigg]$$
$$\le \max_{t\in\mathfrak{J}}\left[\frac{1}{\Gamma(p)}\left\{(\mathscr{L}_1\mathscr{M}_2+\mathscr{M}_1\mathscr{L}_2)\int_0^t \Xi^\Delta(s)(\Xi(t)-\Xi(s))^{p-1} e^{-\lambda(\Xi(t)-\Xi(s))}\Delta s\right\}\right]$$
$$\times D[v,\bar{v}]$$
$$\le \frac{(\Xi(b)-\Xi(0))^p}{\Gamma(p+1)} e^{-\lambda(\Xi(b)-\Xi(0))}(\mathscr{L}_1\mathscr{M}_2+\mathscr{M}_1\mathscr{L}_2)D[v,\bar{v}].$$

This further implies that

$$\frac{(\Xi(b)-\Xi(0))^p}{\Gamma(p+1)} e^{-\lambda(\Xi(b)-\Xi(0))}(\mathscr{L}_1\mathscr{M}_2+\mathscr{M}_1\mathscr{L}_2)D[v,\bar{v}] \ge 0. \tag{5.16}$$

By the condition (5.14), the relation (5.16) is true whenever $D[(\Pi v)(t),(\Pi\bar{v})(t)] = 0$, which further implies that $\Pi v(t) = \Pi\bar{v}(t)$. Ultimately, the problem (5.1) has a unique solution on $\mathfrak{J}$. This completes the proof. □

5.4 Stability theory

Our aim in this section is to prove that the problem (5.1) is UH stable and UHR stable. In the fuzzy setting, we have the following definitions.

Definition 5.18. Equation (5.1) is called UH stable if there exists a constant $C_G > 0$ such that for each $\varepsilon > 0$ and for each solution $\bar{v} \in C_{1-\gamma,\Xi}(\mathfrak{J},E^d)$ of the inequality

$$D\left[{}^{THT}_{0^+}\Delta_\Xi^{p,q,\lambda}\left(\frac{\bar{v}(t)}{f(t,\bar{v}(t))}\right), G(t,\bar{v}(t))\right] \le \varepsilon, \quad t\in\mathfrak{J}, \tag{5.17}$$

there exists a solution $v \in C_{1-\gamma,\Xi}(\mathfrak{J},E^d)$ satisfying (5.1) with

$$D[\bar{v}(t),v(t)] \le C_G\varepsilon, \quad t\in\mathfrak{J}.$$

Definition 5.19. Equation (5.1) is called generalized UH stable if there exists $\chi_G \in C(\mathbb{R}^+,\mathbb{R}^+)$, $\chi_G(0) = 0$ such that, for each solution $\bar{v} \in C_{1-\gamma,\Xi}(\mathfrak{J},E^d)$ of the inequality (5.17), there exists a solution $v \in C_{1-\gamma,\Xi}(\mathfrak{J},E^d)$ satisfying (5.1) with

$$D[\bar{v}(t),v(t)] \le \chi_G\varepsilon, \quad t\in\mathfrak{J}.$$

Definition 5.20. Equation (5.1) is UHR stable with respect to $\varphi \in C_{1-\gamma,\Xi}(\mathfrak{J}, E^d)$ if there exists a constant $C_{G,\varphi} > 0$ such that, for each $\varepsilon > 0$ and for each solution $\bar{v} \in C_{1-\gamma,\Xi}(\mathfrak{J}, E^d)$ of the inequality

$$D\left[{}^{THT}_{0^+}\Delta_{\Xi}^{p,q,\lambda}\left(\frac{\bar{v}(t)}{f(t,\bar{v}(t))}\right), G(t,\bar{v}(t))\right] \le \varepsilon\varphi(t), \quad t \in \mathfrak{J}, \tag{5.18}$$

there exists a solution $v \in C_{1-\gamma,\Xi}(\mathfrak{J}, E^d)$ satisfying (5.1) with

$$D[(\Xi(t)-\Xi(0))^{1-\gamma}\bar{v}(t), (\Xi(t)-\Xi(0))^{1-\gamma}v(t)] \le \varepsilon C_{G,\varphi}\varphi(t), \quad t \in \mathfrak{J}.$$

Definition 5.21. Equation (5.1) is called generalized UHR stable with respect to $\varphi \in C_{1-\gamma,\Xi}(\mathfrak{J}, E^d)$ if there exists a constant $C_{G,\varphi} > 0$ such that for each solution $\bar{v} \in C_{1-\gamma,\Xi}(\mathfrak{J}, E^d)$ of the inequality

$$D\left[{}^{THT}_{0^+}\Delta_{\Xi}^{p,q,\lambda}\left(\frac{\bar{v}(t)}{f(t,\bar{v}(t))}\right), G(t,\bar{v}(t))\right] \le \varphi(t), \quad t \in \mathfrak{J}, \tag{5.19}$$

there exists a solution $v \in C_{1-\gamma,\Xi}(\mathfrak{J}, E^d)$ satisfying Eq. (5.1) with

$$D[(\Xi(t)-\Xi(0))^{1-\gamma}\bar{v}(t), (\Xi(t)-\Xi(0))^{1-\gamma}v(t)] \le C_{G,\varphi}\varphi(t), \quad t \in \mathfrak{J}.$$

Remark 5.2. A fuzzy function $\bar{v} \in C_{1-\gamma,\Xi}(\mathfrak{J}, E^d)$ is a solution of the inequality

$$D\left[{}^{THT}_{0^+}\Delta_{\Xi}^{p,q,\lambda}\left(\frac{\bar{v}(t)}{f(t,\bar{v}(t))}\right), G(t,\bar{v}(t))\right] \le \varepsilon, \quad t \in \mathfrak{J},$$

if and only if there exists a $h \in C_{1-\gamma,\Xi}(\mathfrak{J}, E^d)$ such that

$$\begin{cases} \text{(i)} \quad D[h(t),\hat{0}] \le \varepsilon\varphi(t), \quad t \in \mathfrak{J}, \\ \text{(ii)} \quad {}^{THT}_{0^+}\Delta_{\Xi}^{p,q,\lambda}\left(\frac{\bar{v}(t)}{f(t,\bar{v}(t))}\right) = G(t,\bar{v}(t)) + h(t), \quad t \in \mathfrak{J}. \end{cases} \tag{5.20}$$

Theorem 5.8. *Assume that (H1), (H2), (H4) and (5.14) are satisfied. Then, the problem (5.1) is UHR stable.*

Proof. Fix $\varepsilon > 0$ and let $\bar{v} \in C_{1-\gamma,\Xi}(\mathfrak{J}, E^d)$ be a solution of the inequality

$$D\left[{}^{THT}_{0^+}\Delta_{\Xi}^{p,q,\lambda}\left(\frac{\bar{v}(t)}{f(t,\bar{v}(t))}\right), G(t,\bar{v}(t))\right] \le \varepsilon\varphi(t), \quad t \in \mathfrak{J}. \tag{5.21}$$

□

Let $v \in C_{1-\gamma,\Xi}(\mathfrak{J}, E^d)$ be the unique solution of Eq. (5.1). By Lemma 5.4,

$$v(t) = f(t,v(t))\left(\frac{v_0 e^{-\lambda(\Xi(t)-\Xi(0))}}{\Gamma(\gamma)}(\Xi(t)-\Xi(0))^{\gamma-1}\right.$$

$$\ominus (-1)\frac{1}{\Gamma(p)}\int_0^t \Xi^{\Delta}(s)(\Xi(t)-\Xi(s))^{p-1}e^{-\lambda(\Xi(t)-\Xi(s))}G(s,v(s))\Delta s\Bigg),$$

while the solution of Eq. (5.20) is given by

$$\bar{v}(t)=\begin{cases} f(t,\bar{v}(t))\big(\frac{\bar{v}_0 e^{-\lambda(\Xi(t)-\Xi(0))}}{\Gamma(\gamma)}(\Xi(t)-\Xi(0))^{\gamma-1} \\ \quad \ominus (-1)\frac{1}{\Gamma(p)}\int_0^t \Xi^{\Delta}(s)(\Xi(t)-\Xi(s))^{p-1}e^{-\lambda(\Xi(t)-\Xi(s))}G(s,\bar{v}(s))\Delta s\big) \\ \quad + {}_{0^+}^{\mathcal{T}\mathbb{T}}\mathfrak{I}_{\Xi}^{p,\lambda}h(t). \end{cases} \tag{5.22}$$

From Eq. (5.22) we obtain

$$\begin{aligned} D\Bigg[&\bar{v}(t), f(t,\bar{v}(t))\Bigg(\frac{\bar{v}_0 e^{-\lambda(\Xi(t)-\Xi(0))}}{\Gamma(\gamma)}(\Xi(t)-\Xi(0))^{\gamma-1} \\ &\ominus (-1)\frac{1}{\Gamma(p)}\int_0^t \Xi^{\Delta}(s)(\Xi(t)-\Xi(s))^{p-1}e^{-\lambda(\Xi(t)-\Xi(s))}G(s,\bar{v}(s))\Delta s\Bigg)\Bigg] \\ &\le {}_{0^+}^{\mathcal{T}\mathbb{T}}\mathfrak{I}_{\Xi}^{p,\lambda}D[h(t),\widehat{0}] \le {}_{0^+}^{\mathcal{T}\mathbb{T}}\mathfrak{I}_{\Xi}^{p,\lambda}\varphi(t) \le \varepsilon\mathfrak{K}\varphi(t). \end{aligned}$$

On the other hand, we have

$$\begin{aligned} D[\bar{v}(t),v(t)] &\le \varepsilon\mathfrak{K}\varphi(t) + \frac{1}{\Gamma(p)}D\Bigg[f(t,\bar{v}(t))\int_0^t \Xi^{\Delta}(s)(\Xi(t)-\Xi(s))^{p-1} \\ &\quad \times e^{-\lambda(\Xi(t)-\Xi(s))}G(s,\bar{v}(s))\Delta s, f(t,v(t))\int_0^t \Xi^{\Delta}(s)(\Xi(t)-\Xi(s))^{p-1} \\ &\quad \times e^{-\lambda(\Xi(t)-\Xi(s))}G(s,v(s))\Delta s\Bigg] \\ &\le \varepsilon\mathfrak{K}\varphi(t) + \frac{1}{\Gamma(p)}\Bigg[D[f(t,\bar{v}(t)),f(t,v(t))]\int_0^t \Xi^{\Delta}(s)(\Xi(t)-\Xi(s))^{p-1} \\ &\quad \times e^{-\lambda(\Xi(t)-\Xi(s))}D[G(s,\bar{v}(s)),\widehat{0}]\Delta s + D[f(t,v(t)),\widehat{0}]\int_0^t \Xi^{\Delta}(s)(\Xi(t)-\Xi(s))^{p-1} \\ &\quad \times e^{-\lambda(\Xi(t)-\Xi(s))}D[G(s,\bar{v}(s)),G(s,v(s))]\Delta s\Bigg] \\ &\le \varepsilon\mathfrak{K}\varphi(t) + \frac{1}{\Gamma(p)}\Bigg\{(\mathscr{L}_1\mathscr{M}_2+\mathscr{M}_1\mathscr{L}_2)\int_0^t \Xi^{\Delta}(s)(\Xi(t)-\Xi(s))^{p-1} \\ &\quad \times e^{-\lambda(\Xi(t)-\Xi(s))}D[\bar{v}(s),v(s)]\Delta s\Bigg\}. \end{aligned}$$

Applying Gronwall's inequality given that $\mathscr{X}(t) = D[\bar{v}(t), v(t)]$, $\mathscr{Y}(t) = \varepsilon\mathfrak{K}\varphi(t)$, and $\mathscr{Z}(t) = \frac{(\mathscr{L}_1\mathscr{M}_2+\mathscr{M}_1\mathscr{L}_2)}{\Gamma(p)}$, we obtain

$$D[\bar{v}(t), v(t)] \leq \varepsilon\mathfrak{K}\varphi(t)E_p((\mathscr{L}_1\mathscr{M}_2 + \mathscr{M}_1\mathscr{L}_2)(\Xi(t) - \Xi(0))^p).$$

Therefore,

$$\begin{aligned} &D[(\Xi(t) - \Xi(0))^{1-\gamma}\bar{v}(t), (\Xi(t) - \Xi(0))^{1-\gamma}v(t)] \\ &\quad \leq \varepsilon\mathfrak{K}\varphi(t)(\Xi(t) - \Xi(0))^{1-\gamma}E_p((\mathscr{L}_1\mathscr{M}_2 + \mathscr{M}_1\mathscr{L}_2)(\Xi(t) - \Xi(0))^p) \\ &\quad \leq \varepsilon C_{G,\varphi}\varphi(t), \end{aligned} \tag{5.23}$$

where $C_{G,\varphi} = \mathfrak{K}(\Xi(t) - \Xi(0))^{1-\gamma}E_p((\mathscr{L}_1\mathscr{M}_2 + \mathscr{M}_1\mathscr{L}_2)(\Xi(t) - \Xi(0))^p)$. This proves that Eq. (5.1) is UHR stable. Further, in the same fashion, it is easy to check that the solution of the problem (5.1) is generalized UHR stable, which follows by taking $\varepsilon = 1$ in the inequality (5.23).

Corollary 5.1.
(i) *The proof that Eq.* (5.1) *is UH stable follows by taking $\varphi(t) = 1$ in the inequality* (5.23).
(ii) *The proof that Eq.* (5.1) *is generalized UH stable follows by taking $\varphi(t) = 1$ and $\chi_G(\varepsilon) = \varepsilon C_{G,1}$ in the inequality* (5.23).

5.5 An example

Consider the following hybrid FDE involving tempered Ξ-Hilfer fractional derivative on time scales

$$\begin{cases} {}^{THT}_{0^+}\Delta^{\frac{1}{2},\frac{1}{3},1}_{t^2}\left(\frac{v(t)}{f(t,v(t))}\right) = G(t, v(t)), \quad t \in [0,1] \subseteq \mathbb{T}, \\ {}^{TT}_{0^+}\mathfrak{I}^{1-\frac{2}{3},\lambda}_{t^2}\left(\frac{v(0)}{f(0,v(0))}\right) = 1, \quad \gamma = \frac{1}{2} + \frac{1}{3} - \frac{1}{6} = \frac{2}{3}. \end{cases} \tag{5.24}$$

Here, $p = \frac{1}{2}$, $q = \frac{1}{3}$, $\lambda = 1$, $b = 1$, $v_0 = 1$ and $\Xi(t) = t^2$. Set $f(t, v(t)) = \frac{e^{-t}v(t)}{18+e^t}$ and $G(t, v(t)) = \frac{t}{4}\cos v(t)$.

It is observed that the functions f and G are continuous. Let $v, \bar{v} \in E^d$ and $t \in [0,1]$. Then, we get

$$D[f(t, v(t)), f(t, \bar{v}(t))] = \frac{e^{-t}}{18 + e^t}D[v(t), \bar{v}(t)] \leq \frac{e^{-t}}{18 + e^t}D[v, \bar{v}].$$

Thus, hypothesis $(H1)$ holds true with $p_1(t) = \frac{e^{-t}}{18+e^t}$ and $p^* = \frac{1}{19}$.

Moreover, for $v \in E^d$ and $t \in [0,1]$, we get

$$D[G(t,v(t)),\widehat{0}] = D\left[\frac{t}{4}\cos v(t),\widehat{0}\right].$$

This shows that hypothesis ($H2$) holds true with $q_1(t) = \frac{t}{4}$, $q_1^* = \frac{1}{4}$, and $U(r) = 1$.

Hence, all the conditions of Theorem 5.6 are satisfied, along with condition that

$$\begin{aligned} &p_1^*\left(\frac{v_0}{\Gamma(\gamma)} + \frac{(\Xi(b)-\Xi(0))^{1-\gamma+p}}{\Gamma(p+1)}e^{-\lambda(\Xi(b)-\Xi(0))}q_1^* U(r)\right) \\ &= \frac{1}{19}\left(\frac{1}{1.35412} + \frac{1}{0.88623}\left(\frac{1}{4}\right)(0.36787)\right) \\ &= 0.04432 < 1. \end{aligned}$$

It follows from Theorem 5.7 that the problem (5.24) has a unique solution on $[0,1]$. Moreover, Theorem 5.8 ensures that the problem (5.24) is UHR stable.

Bibliography

[1] S. Abbas, M. Benchohra, J. E. Lazreg, A. Alsaedi and Y. Zhou, Existence and Ulam stability for fractional differential equations of Hilfer–Hadamard type, *Adv. Differ. Equ.*, 2017 (2017), 14.

[2] M. I. Abbas and M. A. Ragusa, On the hybrid fractional differential equations with fractional proportional derivatives of a function with respect to a certain function, *Symmetry*, 13(2) (2021), 264.

[3] M. I. Abbas, Existence results and the Ulam stability for fractional differential equations with hybrid proportional-Caputo derivatives, *J. Nonlinear Anal.*, 2020 (2020), 1–14.

[4] P. Agarwal and I. H. Ibrahim, A new type of hybrid multi step multi derivative formula for solving stiff IVPs, *Adv. Differ. Equ.*, 286(2019) (2019), 1–14.

[5] R. P. Agarwal, V. Lakshmikantham and J. J. Nieto, On the concept of solution for fractional differential equations with uncertainty, *Nonlinear Anal.*, 72 (2010), 2859–2862.

[6] R. P. Agarwal, S. Arshad, D. O'Regan and V. Lupulescu, Fuzzy fractional integral equations under compactness type conditions, *Fract. Calc. Appl. Anal.*, 15 (2012), 572–590.

[7] R. P. Agarwal and M. Bohner, Basic calculus on time scales and some of its applications, *Results Math.*, 35 (1999), 3–22.

[8] B. Ahmad and S. K. Ntouyas, Initial-value problems for hybrid Hadamard fractional differential equations, *Electron. J. Differ. Equ.*, 2014 (2014), 1–8.

[9] B. Ahmad and S. Sivasundaram, Existence results for nonlinear impulsive hybrid boundary value problems involving fractional differential equations, *Nonlinear Anal. Hybrid Syst.*, 3 (2009), 251–258.

[10] B. Bede and S. G. Gal, Generalizations of the differentiability of fuzzy-number-valued functions with applications to fuzzy differential equations, *Fuzzy Sets Syst.*, 151 (2005), 581–599.

[11] M. Benchohra and J. E. Lazreg, On stability for nonlinear implicit fractional differential equations, *Matematiche*, 70 (2015), 49–61.

[12] M. Bohner and A. Peterson, Dynamic equations on time scales, An introduction with Applications, Birkhäuser, Boston, 2001.

[13] M. M. E. Borai, W. G. E. Sayed, A. A. Badr and A. Tarek, Initial value problem for stochastic hybrid Hadamard fractional differential equation, *J. Adv. Math.*, 16 (2019), 8288–8296.

[14] R. Chaharpashlou, R. Saadati and A. Atangana, Ulam–Hyers–Rassias stability for nonlinear ψ-Hilfer stochastic fractional differential equation with uncertainty, *Adv. Differ. Equ.*, 10 (2020), 339.
[15] B. C. Dhage, A fixed point theorem in Banach algebras with applications to functional integral equations, *Kyungpook Math. J.*, 44 (2004), 145–155.
[16] M. Hannabou and K. Hilal, Existence results for a system of coupled hybrid differential equations with fractional order, *Int. J. Differ. Equ.*, 2020 (2020), 1–8.
[17] A. Harir, S. Melliani and L. S. Chadli, Hybrid fuzzy differential equations, *AIMS Math.*, 5(1) (2020), 273–285.
[18] A. Harir, S. Melliani and L. S. Chadli, An algorithm for the solution of fuzzy fractional differential equations, *J. Univers. Math.*, 3(1) (2020), 11–20.
[19] R. Hilfer, Applications of fractional Calculus in Physics, World Scientific, Singapore, 1999.
[20] N. V. Hoa, V. Lupulescu and D. O'Regan, A note on initial value problems for fractional fuzzy differential equations, *Fuzzy Sets Syst.*, 347 (2018), 54–69.
[21] S. Hong, Differentiability of multivalued functions on time scales and applications to multivalued dynamic equations, *Nonlinear Anal.*, 71 (2009), 3622–3637.
[22] O. Kaleva, Fuzzy differential equations, *Fuzzy Sets Syst.*, 24 (1987), 301–317.
[23] A. A. Kilbas, H. M. Srivastava and J. J. Trujillo, Theory and applications of the fractional differential equations, North-Holland Mathematics Studies, Amsterdam, Netherlands, 2006.
[24] K. D. Kucche, A. D. Mali and J. V. D. C. Sousa, On the nonlinear ψ-Hilfer fractional differential equations, *Comput. Appl. Math.*, 38(2) (2019), 73.
[25] K. K. Kucche, A. D. Mali, A. Fernandez and H. M. Fahad, On tempered Hilfer fractional derivatives with respect to functions and the associated fractional differential equations, *Chaos Solitons Fractals*, 163 (2022), 112547.
[26] V. Lakshmikantham and R. N. Mohapatra, Theory of Fuzzy Differential Equations and Applications, London, 2003.
[27] V. Lakshmikantham, S. Sivasundaram and B. Kaymakcalan, Dynamic systems on measure chains, Kluwer Academic Publishers, Dordrecht, 1996.
[28] V. Lakshmikantham and A. S. Vatsala, Hybrid systems on time scales, *J. Comput. Appl. Math.*, 141 (2002), 227–335.
[29] C. Li, W. Deng and L. Zhao, Well-posedness and numerical approximation of tempered fractional ordinary differential equations, *Discrete Contin. Dyn. Syst.*, 20(5) (2017), 1239–1262.
[30] A. D. Mali, K. D. Kucche, A. Fernandez and H. M. Fahad, On tempered fractional calculus with respect to functions and associated fractional differential equations, *Math. Methods Appl. Sci.*, 45(17) (2022), 11134–11157.
[31] S. Manikandan, S. Sivasundaram, D. Vivek and K. Kanagarajan, Controllability and qualitative property results for Ambartsumian equation via Ξ-Hilfer generalized proportional fractional derivative on time scales, *Nonlinear Stud.*, 29(4) (2022), 1–23.
[32] M. D. Obeidat and D. E. Bentil, New theories and applications of tempered fractional differential equations, *Nonlinear Dyn.*, 105 (2021), 1689–1702.
[33] I. Podlubny, Fractional differential equations, Academic Press, New York, 1999.
[34] A. Rivaz, O. S. Fard and T. A. Bidgoli, On the existence and uniqueness of solutions for fuzzy fractional differential equations, *Tbil. Math. J.*, 10 (2017), 197–205.
[35] J. V. C. Sousa, D. S. Oliveira, G. S. F. Frederico and D. F. M. Torres, Existence, uniqueness and controllability for g-Hilfer differential equations on time scales, *hal-03726777*, 2022.
[36] J. V. C. Sousa and E. C. Oliveira, On the ψ-Hilfer fractional derivative, *Commun. Nonlinear Sci. Numer. Simul.*, 60 (2018), 72–91.
[37] J. V. C. Sousa and E. C. Oliveira, On the Ulam–Hyers–Rassias stability for nonlinear fractional differential equations using the ψ-Hilfer operator, *J. Fixed Point Theory Appl.*, 20 (2018), 1–21.
[38] Ch. Vasavi, G. Suresh Kumar and M. S. N. Murty, Fuzzy Hukuhara delta differential and applications to fuzzy dynamic equations on time scales, *J. Uncertain Syst.*, 10(3) (2016), 163–180.

[39] D. Vivek, O. Baghani and K. Kanagarajan, Existence results for hybrid fractional differential equations with Hilfer fractional derivative, *Casp. J. Math. Sci.*, 9(2) (2020), 294–304.
[40] D. Vivek, K. Kanagarajan and S. Sivasundaram, On the behavior of solutions of fractional differential equations on time scale via Hilfer fractional derivatives, *Fract. Calc. Appl. Anal.*, 21(4) (2018), 1120–1138.
[41] H. Vu, J. M. Rassias and N. V. Hoa, Ulam–Hyers–Rassias stability for fuzzy fractional integral equations, *Iran. J. Fuzzy Syst.*, 38(2) (2020), 2245–2259.

S. Manikandan, S. Sivasundaram, K. Kanagarajan, and D. Vivek

6 The dynamical analysis of nonlinear Ambartsumian equation via tempered Ξ-Hilfer fractional derivative on time scales

Abstract: In this chapter, we examine a new class of Ambartsumian equations of the fractional type with tempered Ξ-Hilfer fractional derivative with boundary conditions. The provided problem is transformed into an equivalent fixed point problem, which is then solved by using the Banach and Krasnosel'skii fixed point theorems. Ulam stability is investigated. An example is included to verify the theoretical results.

6.1 Introduction

In comparison to differential equations of classical order, fractional-order differential equations more correctly model a variety of real-world phenomena. Recently, fractional differential equations (FDEs) have been used in a wide range of engineering, mathematics, physics, bioengineering, and applied sciences fields; we refer to [27–29] for some key findings in the theory of fractional calculus and its applications.

The literature contains numerous definitions of fractional integrals and derivatives. We refer to the references given therein for more information on the various uses of FDEs employing a Hilfer derivative. There are actual events in the real world that have unusual dynamics, such as the atmospheric diffusion of pollution, the transmission of signals through powerful magnetic fields, the impact of stock market profitability theory, the theoretical simulation of dielectric relaxation in glass-forming materials, network traffic, and so on. The presence of initial and boundary value problem solutions for FDEs with the Hilfer fractional derivative has received a lot of attention recently, see [8, 10–12, 20, 21, 24].

Several authors [1, 7, 9, 17, 23] used particular versions of the proportional derivatives, called modified conformable derivatives, to present the fractional counterpart proportional derivatives and integrals. Later, authors [25] generalized proportional derivatives and used them to generate more general classes of nonlocal fractional integrals and derivatives.

S. Manikandan, K. Kanagarajan, Department of Mathematics, Sri Ramakrishna Mission Vidyalaya College of Arts and Science, Coimbatore 641020, India, e-mails: p.s.manikandanshanmugam@gmail.com, kanagarajank@gmail.com
S. Sivasundaram, Department of Mathematics, College of Natural Science, Engineering and Mathematics, Daytona Beach, FL 32114, USA, e-mail: seenithi@gmail.com
D. Vivek, Department of Mathematics, PSG College of Arts & Science, Coimbatore 641014, India, e-mail: peppyvivek@gmail.com

https://doi.org/10.1515/9783111182971-006

In recent years, many fractional models with tempered fractional derivatives have been widely applied in many fields of science and technology, and a lot of research results have been obtained. For more details, see [5, 14, 16, 19]. In [3, 6, 18, 26], the authors worked on time scales.

Ulam brought up the topic of functional equation stability first, and Hyers followed suit. Ulam–Hyers stability is the name given to this sort of stability nowadays. By taking variables into account, Rassias accomplished a surprising expansion of the Ulam–Hyers stability of maps. When we swap out the functional equation with an inequality that perturbs the equation, the idea of stability for a functional equation emerges. The study of Ulam–Hyers and Ulam–Hyers–Rassias stability for various functional equations has received significant attention.

The prime target of this work is to investigate a fractional type of the Ambartsumian equation. This equation is very useful to describe the surface brightness of the Milky Way. In [2, 15, 22], the authors discussed about Ambartsumian equations in different aspects.

In this work, we discuss the existence theory and stability result for the following problem:

$$\begin{cases} ({}_{a^+}^{TH\mathbb{T}}\Delta_{\Xi}^{p,q,\lambda}\mathscr{A})(t) = \mathbb{Q}\Big(t, \mathscr{A}(t), \mathscr{A}\Big(\frac{t}{\eta}\Big)\Big), & t \in J = [a,b], \\ \mathscr{A}(a) = 0, \\ \mathscr{A}(b) = \sum_{j=1}^{m} \alpha_j \mathscr{A}(\zeta_j) + \sum_{i=1}^{n} \beta_i {}_{a^+}^{TH\mathbb{T}}\mathscr{I}_{\Xi}^{\phi_i,\lambda}\mathscr{A}(\theta_i) + \sum_{k=1}^{r} \varrho_k {}_{a^+}^{TH\mathbb{T}}\Delta_{\Xi}^{\omega_k,\lambda}\mathscr{A}(\mu_k), \end{cases} \tag{6.1}$$

where $\mathbb{Q}(t, \mathscr{A}(t), \mathscr{A}(\frac{t}{\eta})) = \frac{1}{\eta}\mathscr{A}(\frac{t}{\eta}) - \mathscr{A}(t), \eta > 1$, ${}_{a^+}^{TH\mathbb{T}}\Delta_{\Xi}^{p,q,\lambda}$ is the tempered Ξ-Hilfer fractional derivative on time scales of order $p > 0$ and type $q > 0$ and $\alpha_j, \beta_i, \varrho_k \in \mathbb{R}$, $\lambda \in \mathbb{R}$ are given constants, $\mathbb{Q} : J \times \mathbb{R} \times \mathbb{R} \to \mathbb{R}$ is a given continuous function, $\phi_i, \omega_k > 0$ and $\zeta_j, \theta_i, \mu_k \in J, j = 1, 2, \ldots, m, i = 1, 2, \ldots, n,\ k = 1, 2, \ldots, r$.

6.2 Prerequisites

Let $0 \leq a < b < \infty$, J be a finite interval, and ϑ a parameter such that $n - 1 \leq \vartheta < n$.

Let $X = C(J, \mathbb{R})$ be the Banach space of the continuous functions $\mathbb{Q}$ on J with the norm defined by

$$\|\mathbb{Q}\|_{C(J,\mathbb{R})} = \max_{t \in J} |\mathbb{Q}(t)|.$$

Definition 6.1 ([26]). A time scale $\mathbb{T}$ is an arbitrary nonempty closed subset of the real numbers. For $t \in \mathbb{T}$, one defines the forward jump operator $\sigma : \mathbb{T} \to \mathbb{T}$ by

$$\sigma(t) = \inf\{s \in \mathbb{T} : s > t\},$$

while the backward jump operator $\rho : \mathbb{T} \to \mathbb{T}$ is defined by

$$\rho(t) = \sup\{s \in \mathbb{T} : s > t\}.$$

If $\max \mathbb{T}$ is finite and there exists a finite $\min \mathbb{T}$ in addition, we put $\sigma(\max \mathbb{T}) = \max \mathbb{T}$ and $\rho(\min \mathbb{T}) = \min \mathbb{T}$.

If $\sigma(t) > t$, then we say that t is right-scattered, while if $\rho(t) < t$, then we say that t is left-scattered and also if $t < \max \mathbb{T}$, and $\sigma(t) = t$ then t is called right-dense, and if $t > \min \mathbb{T}$ and $\rho(t) = t$, then t is called right-dense.

The derivative makes use of the set $\mathbb{T}^{\kappa}$, which is derived from the time scale $\mathbb{T}$ as follows: if $\mathbb{T}$ has left-scattered maximum M, then $\mathbb{T}^{\kappa} := \mathbb{T}\backslash\{M\}$; otherwise, $\mathbb{T}^{\kappa} = \mathbb{T}$.

Definition 6.2 (Delta derivative, [26]). Suppose that $\mathbb{Q} : \mathbb{T} \to \mathbb{R}$ and let $t \in \mathbb{T}^{\kappa}$. The delta derivative is defined by

$$\mathbb{Q}^{\triangle}(t) = \lim_{s \to t} \frac{\mathbb{Q}(\sigma(s)) - \mathbb{Q}(t)}{\sigma(s) - t}, \quad t \neq \sigma(s).$$

Definition 6.3 ([26]). Let J denote a closed bounded interval in $\mathbb{T}$. A function $F : J \to \mathbb{R}$ is called a delta antiderivative of a function $f : [a, b) \to \mathbb{R}$ provided F is continuous on J, delta differentiable on $[a, b)$, and $F^{\triangle}(t) = f(t)$ for all $t \in [a, b)$. Then we define the $\triangle$-integral by

$$\int_a^b f(t) \triangle (t) = F(b) - F(a).$$

Proposition 6.1 ([26]). *Suppose $a, b \in \mathbb{T}, a < b$ and $\mathbb{Q}$ is continuous on J. Then,*

$$\int_a^b \mathbb{Q}(t) \triangle t = [\sigma(a) - a]\mathbb{Q}(a) + \int_{\sigma(a)}^b \mathbb{Q}(t) \triangle t.$$

Proposition 6.2 ([26]). *Let $\mathbb{T}$ be a time scale and $\mathbb{Q}$ be an increasing continuous function on J. If $\mathbb{Q}$ is the extension of $\mathbb{Q}$ to the real interval J given by*

$$\mathbb{Q}(s) = \begin{cases} \mathbb{Q}(s), & s \in \mathbb{T}, \\ \mathbb{Q}(t), & s \in (t, \sigma(t)) \not\subset \mathbb{T}, \end{cases}$$

then

$$\int_a^b \mathbb{Q}(t) \triangle t \leq \int_a^b \mathbb{Q}(t) dt.$$

Definition 6.4. Suppose $\mathbb{T}$ is a time scale, $\mathbb{Q}$ is an absolutely Ξ integrable function defined on $[a,b]$. Then the Ξ-tempered fractional integral of order $p \in (0,2)$ and the index $\lambda \in \mathbb{R}$ of the function $\mathbb{Q}$ is defined by

$$ {}_{a^+}^{T\mathbb{T}}\mathscr{I}_{\Xi}^{p,\lambda}\mathbb{Q}(t) = \frac{1}{\Gamma(p)} \int_{a^+}^{t} \Xi^{\triangle}(s)(\Xi(t) - \Xi(s))^{p-1} e^{-\lambda(\Xi(t)) - \Xi(s))}\mathbb{Q}(s) \triangle s. $$

Definition 6.5. Suppose $\mathbb{T}$ is a time scale, $\lambda \in \mathbb{R}$, Ξ is an increasing real C^n function on J such that $\Xi^{\triangle} > 0$ on J. Then the tempered (nonfractional) tempered derivative with respect to Ξ is defined by

$$ {}_{a^+}^{T\mathbb{T}}\triangle_{\Xi}^{1,\lambda} = \frac{{}^{\mathbb{T}}\triangle}{\Xi^{\triangle}(t)} + \lambda, $$

which can be taken to the nth power ($n \in \mathbb{N}$) to get

$$ {}_{a^+}^{T\mathbb{T}}\triangle_{\Xi}^{n,\lambda} = \left(\frac{{}^{\mathbb{T}}\triangle}{\Xi^{\triangle}(t)} + \lambda\right)^n, $$

where ${}^{\mathbb{T}}\triangle = \frac{d}{dt}$.

Definition 6.6. Let $p, \lambda \in \mathbb{R}$ with $p > 0$ and Ξ be an increasing real C^n function on J such that $\Xi^{\triangle} > 0$ on J. Then, the tempered fractional integrals of order p and index λ with respect to Ξ, of Riemann–Liouville and Caputo type, respectively, are defined as follows, applied to a function $\mathbb{Q} \in AC_{\Xi}^{n}J$:

$$ {}_{a^+}^{TR\mathbb{T}}\triangle_{\Xi}^{p,\lambda}\mathbb{Q}(t) = {}_{a^+}^{T\mathbb{T}}\triangle_{\Xi}^{n,\lambda}\,{}_{a^+}^{T\mathbb{T}}\mathscr{I}_{\Xi}^{n-p,\lambda}\mathbb{Q}(t), $$
$$ {}_{a^+}^{TC\mathbb{T}}\triangle_{\Xi}^{p,\lambda}\mathbb{Q}(t) = {}_{a^+}^{T\mathbb{T}}\mathscr{I}_{\Xi}^{n-p,\lambda}\,{}_{a^+}^{T\mathbb{T}}\triangle_{\Xi}^{n,\lambda}\mathbb{Q}(t). $$

Definition 6.7. Let $n-1 < p < n$ with $n \in \mathbb{N}$, J be the interval such that $-\infty \le a \le b \le \infty$ and $\mathbb{Q}, \Xi \in C^n(J, \mathbb{R})$ two functions such that Ξ is an increasing function and $\Xi^{\triangle}(t) \neq 0$, for all $t \in J$. The tempered Ξ-Hilfer fractional derivative of a function $\mathbb{Q}$ of order p and type q is defined by

$$ {}_{a^+}^{TH\mathbb{T}}\triangle_{\Xi}^{p,q,\lambda}\mathbb{Q}(t) = {}_{a^+}^{T\mathbb{T}}\mathscr{I}_{\Xi}^{q(n-p),\lambda}\left(\frac{{}^{\mathbb{T}}\triangle}{\Xi^{\triangle}(t)} + \lambda\right){}_{a^+}^{T\mathbb{T}}\triangle_{\Xi}^{(1-q)(n-p),\lambda}\mathbb{Q}(t). $$

Lemma 6.1. *Let $p, q > 0$. Then we have the following semigroup property:*

$$ {}_{a^+}^{T\mathbb{T}}\mathscr{I}_{\Xi}^{p,\lambda}\,{}_{a^+}^{T\mathbb{T}}\mathscr{I}_{\Xi}^{q,\lambda}\mathbb{Q}(t) = {}_{a^+}^{T\mathbb{T}}\mathscr{I}_{\Xi}^{p+q,\lambda}\mathbb{Q}(t), \quad t > a. $$

Lemma 6.2. *Let $\mathbb{Q} \in C^n(J, \mathbb{R})$, $0 \le a \le b \le \infty$, $p, q > 0$, and $\vartheta = p + q(n-p)$. If $\mathbb{Q} \in L^1(J)$, $({}_{a^+}^{T\mathbb{T}}\triangle_{\Xi}^{p,q,\lambda}\mathbb{Q})(t) \in AC_{\Xi}^{n}J$, then*

$$
\begin{aligned}
&{}^{T\mathbb{T}}_{a^+}\mathscr{I}^{p,\lambda}_{\Xi}\left({}^{T\mathbb{T}}_{a^+}\Delta^{p,q,\lambda}_{\Xi}\mathbb{Q}\right)(t)\\
&\quad = \mathbb{Q}(t)\\
&\quad - \sum_{k=1}^{n}\frac{e^{-\lambda(\Xi(t)-\Xi(a))}(\Xi(t)-\Xi(a))^{\vartheta-k}}{\Gamma(\vartheta-k+1)}\left(\frac{{}^{\mathbb{T}}\Delta}{\Xi^{\Delta}(t)}\right)_{\Xi}^{[n-k]}{}^{T\mathbb{T}}_{a^+}\mathscr{I}^{(1-q)(n-p),\lambda}_{\Xi}\mathbb{Q}(a).
\end{aligned} \tag{6.2}
$$

Lemma 6.3 (Banach fixed point theorem, [4]). *Let X be a Banach space, $D \subset X$ closed and $F : D \to D$ a strict contraction, i. e., $|Fx - Fy| \le k|x - y|$ for some $k \in (0,1)$ and all $x, y \in D$. Then F has a fixed point in D.*

Lemma 6.4 (Krasnoselskii's fixed point theorem, [13]). *Let M be a closed, bounded, convex, and nonempty subset of a Banach space X. Let A, B be operators such that*
- *$Ax + By \in M$ whenever $x, y \in M$,*
- *A is compact and continuous,*
- *B is a contraction mapping.*

Then there exists $z \in M$ such that $z = Az + Bz$.

Theorem 6.1 (Gronwall's inequality, [16]). *Let $p \in \mathbb{R}^+$, $\lambda \in \mathbb{R}$, $u, v \in L^1(J, d\Xi)$ be nonnegative functions, and let $w : J \to [0,\infty)$ be a continuous, nonnegative and nondecreasing function. If for all $t \in J$ we have*

$$
u(t) = v(t) + w(t)\int_{a^+}^{t}\Xi'(s)(\Xi(t)-\Xi(s))^{p-1}e^{-\lambda(\Xi(t)-\Xi(s))}u(s)ds,
$$

then for all $t \in J$ we have

$$
u(t) \le v(t) + \int_{a^+}^{t}\sum_{k=1}^{\infty}\frac{|w(t)\Gamma(p)|^k}{\Gamma(pk)}\Xi'(s)(\Xi(t)-\Xi(s))^{pk-1}e^{-\lambda(\Xi(t)-\Xi(s))}v(s)ds.
$$

Further, if v is a nondecreasing function, then for all $t \in J$ we have

$$
u(t) \le v(t)E_p\big(w(t)\Gamma(p)(\Xi(t)-\Xi(a))^p\big).
$$

6.3 Existence theory

The problem (6.1) will be converted into an equivalent fixed point problem using the auxiliary lemma in this part. Standard fixed point theorems are used to study the main results.

Lemma 6.5. *Let* $\mathbb{Q} \in C(J, \mathbb{R})$ *and*

$$\begin{aligned}\wedge = {} & e^{-\lambda(\Xi(b)-\Xi(a))}(\Xi(b) - \Xi(a))^{\vartheta-1} - \sum_{j=1}^{m} a_j e^{-\lambda(\Xi(\zeta_j)-\Xi(a))}(\Xi(\zeta_j) - \Xi(a))^{\vartheta-1} \\ & - \sum_{i=1}^{n} \beta_i \frac{\vartheta}{\Gamma(\vartheta + \phi_i)} e^{-\lambda(\Xi(\theta_i)-\Xi(a))}(\Xi(\theta_i) - \Xi(a))^{\vartheta+\phi_i-1} \\ & - \sum_{k=1}^{r} \varrho_k \frac{\vartheta}{\Gamma(\vartheta - \omega_k)} e^{-\lambda(\Xi(\mu_k)-\Xi(a))}(\Xi(\mu_k) - \Xi(a))^{\vartheta-\omega_k-1} \neq 0. \end{aligned} \tag{6.3}$$

Then, $\mathscr{A}$ *is a solution of our proposed problem* (6.1) *if and only if it satisfies the integral equation*

$$\begin{aligned}\mathscr{A}(t) = {} & {}_{a^+}^{\mathbb{TT}}\mathscr{I}_{\Xi}^{p,\lambda}\mathbb{Q}\left(t, \mathscr{A}(t), \mathscr{A}\left(\frac{t}{\eta}\right)\right) + \frac{e^{-\lambda(\Xi(t)-\Xi(a))}(\Xi(t) - \Xi(a))^{\vartheta-1}}{\wedge} \\ & \times \left(\sum_{j=1}^{m} a_j\, {}_{a^+}^{\mathbb{TT}}\mathscr{I}_{\Xi}^{p,\lambda}\mathbb{Q}\left(\zeta_j, \mathscr{A}(\zeta_j), \mathscr{A}\left(\frac{\zeta_j}{\eta}\right)\right)\right. \\ & - \sum_{i=1}^{n} \beta_i\, {}_{a^+}^{\mathbb{TT}}\mathscr{I}_{\Xi}^{p,\lambda}\mathbb{Q}\left(\theta_i, \mathscr{A}(\theta_i), \mathscr{A}\left(\frac{\theta_i}{\eta}\right)\right) \\ & - \sum_{k=1}^{r} \varrho_k\, {}_{a^+}^{\mathbb{TT}}\mathscr{I}_{\Xi}^{p-\omega_k,\lambda}\mathbb{Q}\left(\mu_k, \mathscr{A}(\mu_k), \mathscr{A}\left(\frac{\mu_k}{\eta}\right)\right) \\ & \left. - {}_{a^+}^{\mathbb{TT}}\mathscr{I}_{\Xi}^{p,\lambda}\mathbb{Q}\left(b, \mathscr{A}(b), \mathscr{A}\left(\frac{b}{\eta}\right)\right)\right). \end{aligned} \tag{6.4}$$

Proof. Applying the tempered fractional integral operator of order p on both sides of Eq. (6.4) and using Lemma 6.2, we find that

$$\begin{aligned}\mathscr{A}(t) = {} & \frac{\delta({}_{a^+}^{\mathbb{TT}}\mathscr{I}_{\Xi}^{2-\vartheta,\lambda}\mathscr{A})(a)}{\Gamma(\vartheta)} e^{-\lambda(\Xi(t)-\Xi(a))}(\Xi(t) - \Xi(a))^{\vartheta-1} \\ & - \frac{\delta({}_{a^+}^{\mathbb{TT}}\mathscr{I}_{\Xi}^{2-\vartheta,\lambda}\mathscr{A})(a)}{\Gamma(\vartheta - 1)} e^{-\lambda(\Xi(t)-\Xi(a))}(\Xi(t) - \Xi(a))^{\vartheta-2} \\ = {} & {}_{a}^{\mathbb{TT}}\mathscr{I}_{\Xi}^{p,\lambda}\mathbb{Q}\left(t, \mathscr{A}(t), \mathscr{A}\left(\frac{t}{\eta}\right)\right), \end{aligned} \tag{6.5}$$

which can be rewritten as follows:

$$\begin{aligned}\mathscr{A}(t) = {} & c_0 e^{-\lambda(\Xi(t)-\Xi(a))}(\Xi(t) - \Xi(a))^{\vartheta-1} + c_1 e^{-\lambda(\Xi(t)-\Xi(a))}(\Xi(t) - \Xi(a))^{\vartheta-2} \\ & + \frac{1}{\Gamma(p)} \int_a^t \Xi^{\triangle}(s) e^{-\lambda(\Xi(t)-\Xi(s))}(\Xi(t) - \Xi(s))^{p-1}\mathbb{Q}\left(s, \mathscr{A}(s), \mathscr{A}\left(\frac{s}{\eta}\right)\right) \triangle s, \end{aligned} \tag{6.6}$$

where c_0, c_1 are arbitrary constants. The first boundary condition $\mathscr{A}(a) = 0$ in Eq. (6.6) gives $c_1 = 0$, since $\vartheta \in [p, 2]$. In consequence, Eq. (6.6) takes the following form:

$$\begin{aligned}\mathscr{A}(t) &= c_0 e^{-\lambda(\Xi(t)-\Xi(a))}(\Xi(t)-\Xi(a))^{\vartheta-1}\\ &+ \frac{1}{\Gamma(p)}\int_a^t \Xi^{\triangle}(s)e^{-\lambda(\Xi(t)-\Xi(s))}(\Xi(t)-\Xi(s))^{p-1}\mathbb{Q}\left(s,\mathscr{A}(s),\mathscr{A}\left(\frac{s}{\eta}\right)\right)\triangle s. \qquad (6.7)\end{aligned}$$

Now, employing the second boundary condition,

$$\mathscr{A}(b) = \sum_{j=1}^{m} \alpha_j \mathscr{A}(\zeta_j) + \sum_{i=1}^{n} \beta_i {}_{a^+}^{THT}\mathscr{I}_{\Xi}^{\phi_i,\lambda}\mathscr{A}(\theta_i) + \sum_{k=1}^{r} \varrho_k {}_{a^+}^{THT}\triangle_{\Xi}^{\omega_k,\lambda}\mathscr{A}(\mu_k),$$

and using Eq. (6.5), we obtain the following:

$$\begin{aligned}c_0 = \frac{1}{\wedge}\Bigg\{&\sum_{j=1}^{m} \alpha_j {}_{a^+}^{TT}\mathscr{I}_{\Xi}^{p,\lambda}\mathbb{Q}\left(\zeta_j,\mathscr{A}(\zeta_j),\mathscr{A}\left(\frac{\zeta_j}{\eta}\right)\right)\\ &- \sum_{i=1}^{n} \beta_i {}_{a^+}^{TT}\mathscr{I}_{\Xi}^{p,\lambda}\mathbb{Q}\left(\theta_i,\mathscr{A}(\theta_i),\mathscr{A}\left(\frac{\theta_i}{\eta}\right)\right)\\ &- \sum_{k=1}^{r} \varrho_k {}_{a^+}^{TT}\mathscr{I}_{\Xi}^{p-\omega_k,\lambda}\mathbb{Q}\left(\mu_k,\mathscr{A}(\mu_k),\mathscr{A}\left(\frac{\mu_k}{\eta}\right)\right)\\ &- {}_{a^+}^{TT}\mathscr{I}_{\Xi}^{p,\lambda}\mathbb{Q}\left(b,\mathscr{A}(b),\mathscr{A}\left(\frac{b}{\eta}\right)\right)\Bigg\}.\end{aligned}$$

After substituting the value of c_0 into Eq. (6.7), we get the solution.

Hence $\mathscr{A}$ satisfies our proposed problem (6.1). By direct computation, one can obtain the converse of the lemma. The proof is completed. □

Next, we define the operator $\mathscr{T}: X \to X$ associated with the problem (6.1) as follows:

$$\begin{aligned}\mathscr{T}\mathscr{A}(t) &= \frac{1}{\Gamma(p)}\int_a^t \Xi^{\triangle}(z)e^{-\lambda(\Xi(t)-\Xi(z))}(\Xi(t)-\Xi(z))^{p-1}\mathbb{Q}\left(z,\mathscr{A}(z),\mathscr{A}\left(\frac{z}{\eta}\right)\right)\triangle z\\ &+ \frac{e^{-\lambda(\Xi(t)-\Xi(z))}(\Xi(t)-\Xi(z))^{\vartheta-1}}{\wedge}\\ &\times\left(\sum_{j=1}^{m}\frac{\alpha_j}{\Gamma(p)}\int_a^{\zeta_j}\Xi^{\triangle}(z)e^{-\lambda(\Xi(t)-\Xi(z))}(\Xi(t)-\Xi(z))^{p-1}\right.\\ &\times\mathbb{Q}\left(z,\mathscr{A}(z),\mathscr{A}\left(\frac{z}{\eta}\right)\right)\triangle z\\ &-\sum_{i=1}^{n}\frac{\beta_i}{\Gamma(p+\phi_i)}\int_a^{\theta_i}\Xi^{\triangle}(z)e^{-\lambda(\Xi(t)-\Xi(z))}(\Xi(t)-\Xi(z))^{p+\phi_i-1}\\ &\times\mathbb{Q}\left(z,\mathscr{A}(z),\mathscr{A}\left(\frac{z}{\eta}\right)\right)\triangle z\end{aligned}$$

$$- \sum_{k=1}^{r} \frac{\varrho_k}{\Gamma(p-\omega_k)} \int_a^{\mu_k} \Xi^{\triangle}(z) e^{-\lambda(\Xi(t)-\Xi(z))} (\Xi(t)-\Xi(z))^{p-\omega_k-1}$$
$$\times \mathbb{Q}\left(z, \mathscr{A}(z), \mathscr{A}\left(\frac{z}{\eta}\right)\right) \triangle z$$
$$- \frac{1}{\Gamma(p)} \int_a^{b} \Xi^{\triangle}(z) e^{-\lambda(\Xi(t)-\Xi(z))} (\Xi(t)-\Xi(z))^{p-1}$$
$$\times \mathbb{Q}\left(z, \mathscr{A}(z), \mathscr{A}\left(\frac{z}{\eta}\right)\right) \triangle z\Bigg), \quad t \in J. \tag{6.8}$$

In the sequel, we use the following notation:

$$\Omega = \frac{e^{-\lambda(\Xi(b)-\Xi(a))}(\Xi(b)-\Xi(a))^p}{\Gamma(p+1)} + \frac{e^{-\lambda(\Xi(b)-\Xi(a))}(\Xi(b)-\Xi(a))^{\vartheta-1}}{|\wedge|}$$
$$\times \left\{ \sum_{j=1}^{m} \frac{|\alpha_j| e^{-\lambda(\Xi(\zeta_j)-\Xi(a))}(\Xi(\zeta_j)-\Xi(a))^p}{\Gamma(p+1)} \right.$$
$$+ \sum_{i=1}^{n} \frac{|\beta_i| e^{-\lambda(\Xi(\theta_i)-\Xi(a))}(\Xi(\theta_i)-\Xi(a))^{p+\phi_i}}{\Gamma(p+\phi_i+1)}$$
$$+ \sum_{k=1}^{r} \frac{|\varrho_k| e^{-\lambda(\Xi(\mu_k)-\Xi(a))}(\Xi(\mu_k)-\Xi(a))^{p-\omega_k}}{\Gamma(p-\omega_k+1)}$$
$$\left. + \frac{e^{-\lambda(\Xi(b)-\Xi(a))}(\Xi(b)-\Xi(a))^p}{\Gamma(p+1)} \right\}. \tag{6.9}$$

The following result is discussed by using the Banach fixed point theorem.

Theorem 6.2. *Suppose that the following condition holds:*
(H_1) There exits a constant $l > 0$ such that for all $t \in J$ and $u, v, \overline{u}, \overline{v} \in \mathbb{R}$,

$$|\mathbb{Q}(t,u,v) - \mathbb{Q}(t,\overline{u},\overline{v})| \le l\{|u-\overline{u}| + |v-\overline{v}|\}.$$

Then, the problem (6.1) *has a unique solution on J, if $\Omega < 1$, where Ω is defined by* (6.9).

Proof. We will verify that the operator $\mathscr{T}$ satisfies the hypotheses of the Banach contraction mapping principle. Fixing $N = \max_{t\in J} |\mathbb{Q}(t,0,0)| < \infty$ and using assumption (H_1), we obtain the following:

$$\left|\mathbb{Q}\left(t, \mathscr{A}(t), \mathscr{A}\left(\frac{t}{\eta}\right)\right)\right| \le l|\mathscr{A}(t)| + |\mathbb{Q}(t,0,0)| \le l\|\mathscr{A}\| + N. \tag{6.10}$$

The proof is divided into two steps.

Step I. We will show that $\mathscr{T}(B_r) \subset B_r$, where

$$(B_r) = \{\mathscr{A} \in X : \|\mathscr{A}\| < r\} \quad \text{with } r \geq \frac{N\Omega}{(1 - l\Omega)}.$$

Let $\mathscr{A} \in B_r$. Then, we have the following:

$$\begin{aligned}
|\mathscr{T}\mathscr{A}(t)| &= \frac{1}{\Gamma(p)} \int_a^t \Xi^{\triangle}(z) e^{-\lambda(\Xi(t)-\Xi(z))} (\Xi(t) - \Xi(z))^{p-1} \left|\mathbb{Q}\left(z, \mathscr{A}(z), \mathscr{A}\left(\frac{z}{\eta}\right)\right)\right| \triangle z \\
&\quad + \frac{e^{-\lambda(\Xi(t)-\Xi(z))} (\Xi(t) - \Xi(z))^{\vartheta-1}}{\wedge} \\
&\quad \times \left\{ \sum_{j=1}^{m} \frac{|a_j|}{\Gamma(p)} \int_a^{\zeta_j} \Xi^{\triangle}(z) e^{-\lambda(\Xi(t)-\Xi(z))} (\Xi(t) - \Xi(z))^{p-1} \right. \\
&\quad \times \left|\mathbb{Q}\left(z, \mathscr{A}(z), \mathscr{A}\left(\frac{z}{\eta}\right)\right)\right| \triangle z \\
&\quad - \sum_{i=1}^{n} \frac{|\beta_i|}{\Gamma(p + \phi_i)} \int_a^{\theta_i} \Xi^{\triangle}(z) e^{-\lambda(\Xi(t)-\Xi(z))} (\Xi(t) - \Xi(z))^{p+\phi_i-1} \\
&\quad \times \left|\mathbb{Q}\left(z, \mathscr{A}(z), \mathscr{A}\left(\frac{z}{\eta}\right)\right)\right| \triangle z \\
&\quad - \sum_{k=1}^{r} \frac{|\varrho_k|}{\Gamma(p - \omega_k)} \int_a^{\mu_k} \Xi^{\triangle}(z) e^{-\lambda(\Xi(t)-\Xi(z))} (\Xi(t) - \Xi(z))^{p-\omega_k-1} \\
&\quad \times \left|\mathbb{Q}\left(z, \mathscr{A}(z), \mathscr{A}\left(\frac{z}{\eta}\right)\right)\right| \triangle z \\
&\quad - \frac{1}{\Gamma(p)} \int_a^{b} \Xi^{\triangle}(z) e^{-\lambda(\Xi(t)-\Xi(z))} (\Xi(t) - \Xi(z))^{p-1} \\
&\quad \left. \times \left|\mathbb{Q}\left(z, \mathscr{A}(z), \mathscr{A}\left(\frac{z}{\eta}\right)\right)\right| \triangle z \right\} \\
&= \frac{e^{-\lambda(\Xi(a)-\Xi(b))} (\Xi(a) - \Xi(b))^{p}}{\Gamma(p+1)} (l\|\mathscr{A}\| + N) + \frac{e^{-\lambda(\Xi(b)-\Xi(a))} (\Xi(b) - \Xi(a))^{\vartheta-1}}{|\wedge|} \\
&\quad \times \left\{ \sum_{j=1}^{m} \frac{|a_j| e^{-\lambda(\Xi(\zeta_j)-\Xi(a))} (\Xi(\zeta_j) - \Xi(a))^{p}}{\Gamma(p+1)} \right. \\
&\quad + \sum_{i=1}^{n} \frac{|\beta_i| e^{-\lambda(\Xi(\theta_i)-\Xi(a))} (\Xi(\theta_i) - \Xi(a))^{p+\phi_i}}{\Gamma(p + \phi_i + 1)} \\
&\quad + \sum_{k=1}^{r} \frac{|\varrho_k| e^{-\lambda(\Xi(\mu_k)-\Xi(a))} (\Xi(\mu_k) - \Xi(a))^{p-\omega_k}}{\Gamma(p - \omega_k + 1)}
\end{aligned}$$

$$
\begin{aligned}
&\left.+\frac{e^{-\lambda(\Xi(b)-\Xi(a))}(\Xi(b)-\Xi(a))^{p}}{\Gamma(p+1)}\right\}(l\|\mathscr{A}\|+N)\\
&\leq\left[\frac{e^{-\lambda(\Xi(a)-\Xi(b))}(\Xi(a)-\Xi(b))^{p}}{\Gamma(p+1)}+\frac{e^{-\lambda(\Xi(b)-\Xi(a))}(\Xi(b)-\Xi(a))^{\vartheta-1}}{|\wedge|}\right.\\
&\times\left\{\sum_{j=1}^{m}\frac{|a_j|e^{-\lambda(\Xi(\zeta_j)-\Xi(a))}(\Xi(\zeta_j)-\Xi(a))^{p}}{\Gamma(p+1)}\right.\\
&+\sum_{i=1}^{n}\frac{|\beta_i|e^{-\lambda(\Xi(\theta_i)-\Xi(a))}(\Xi(\theta_i)-\Xi(a))^{p+\phi_i}}{\Gamma(p+\phi_i+1)}\\
&+\sum_{k=1}^{r}\frac{|\varrho_k|e^{-\lambda(\Xi(\mu_k)-\Xi(a))}(\Xi(\mu_k)-\Xi(a))^{p-\omega_k}}{\Gamma(p-\omega_k+1)}\\
&\left.\left.+\frac{e^{-\lambda(\Xi(b)-\Xi(a))}(\Xi(b)-\Xi(a))^{p}}{\Gamma(p+1)}\right\}\right](lr+N)\\
&=\Omega(lr+N)\leq r.
\end{aligned}
$$

Thus the following is the case:

$$
\|\mathscr{T}(\mathscr{A})\|=\max_{t\in J}|\mathscr{T}u(t)|\leq r,
$$

which means that $\mathscr{T}(B_r)\subset B_r$.

Step II. We will show that the operator $\mathscr{A}$ is a contraction.

Let $\mathscr{A},\overline{\mathscr{A}}\in X$. Then for any $t\in J$, we have the following:

$$
\begin{aligned}
|\mathscr{T}\overline{\mathscr{A}}-\mathscr{T}\mathscr{A}|\leq{}&\frac{1}{\Gamma(p)}\int_a^t\Xi^{\triangle}(z)e^{-\lambda(\Xi(t)-\Xi(z))}(\Xi(t)-\Xi(z))^{p-1}\\
&\times\left|\mathbb{Q}\left(z,\mathscr{A}(z),\mathscr{A}\left(\frac{z}{\eta}\right)\right)-\mathbb{Q}\left(z,\overline{\mathscr{A}}(z),\overline{\mathscr{A}}\left(\frac{z}{\eta}\right)\right)\right|\triangle z\\
&+\frac{e^{-\lambda(\Xi(t)-\Xi(z))}(\Xi(t)-\Xi(z))^{\vartheta-1}}{\wedge}\\
&\times\left\{\sum_{j=1}^{m}\frac{|a_j|}{\Gamma(p)}\int_a^{\zeta_j}\Xi^{\triangle}(z)e^{-\lambda(\Xi(t)-\Xi(z))}(\Xi(t)-\Xi(z))^{p-1}\right.\\
&\times\left|\mathbb{Q}\left(z,\mathscr{A}(z),\mathscr{A}\left(\frac{z}{\eta}\right)\right)-\mathbb{Q}\left(z,\overline{\mathscr{A}}(z),\overline{\mathscr{A}}\left(\frac{z}{\eta}\right)\right)\right|\triangle z\\
&-\sum_{i=1}^{n}\frac{|\beta_i|}{\Gamma(p+\phi_i)}\int_a^{\theta_i}\Xi^{\triangle}(z)e^{-\lambda(\Xi(t)-\Xi(z))}(\Xi(t)-\Xi(z))^{p+\phi_i-1}\\
&\times\left|\mathbb{Q}\left(z,\mathscr{A}(z),\mathscr{A}\left(\frac{z}{\eta}\right)\right)-\mathbb{Q}\left(z,\overline{\mathscr{A}}(z),\overline{\mathscr{A}}\left(\frac{z}{\eta}\right)\right)\right|\triangle z
\end{aligned}
$$

$$
\begin{aligned}
&- \sum_{k=1}^{r} \frac{|\varrho_k|}{\Gamma(p-\omega_k)} \int_a^{\mu_k} \Xi^{\triangle}(z) e^{-\lambda(\Xi(t)-\Xi(z))} (\Xi(t)-\Xi(z))^{p-\omega_k-1} \\
&\times \left| \mathbb{Q}\left(z, \mathscr{A}(z), \mathscr{A}\left(\frac{z}{\eta}\right)\right) - \mathbb{Q}\left(z, \overline{\mathscr{A}}(z), \overline{\mathscr{A}}\left(\frac{z}{\eta}\right)\right) \right| \triangle z \\
&- \frac{1}{\Gamma(p)} \int_a^{b} \Xi^{\triangle}(z) e^{-\lambda(\Xi(t)-\Xi(z))} (\Xi(t)-\Xi(z))^{p-1} \\
&\times \left| \mathbb{Q}\left(z, \mathscr{A}(z), \mathscr{A}\left(\frac{z}{\eta}\right)\right) - \mathbb{Q}\left(z, \overline{\mathscr{A}}(z), \overline{\mathscr{A}}\left(\frac{z}{\eta}\right)\right) \right| \triangle z \Bigg\} \\
\le & \left[\frac{e^{-\lambda(\Xi(a)-\Xi(b))}(\Xi(a)-\Xi(b))^p}{\Gamma(p+1)} + \frac{e^{-\lambda(\Xi(b)-\Xi(a))}(\Xi(b)-\Xi(a))^{\vartheta-1}}{|\wedge|} \right. \\
&\times \left\{ \sum_{j=1}^{m} \frac{|\alpha_j| e^{-\lambda(\Xi(\zeta_j)-\Xi(a))}(\Xi(\zeta_j)-\Xi(a))^p}{\Gamma(p+1)} \right. \\
&+ \sum_{i=1}^{n} \frac{|\beta_i| e^{-\lambda(\Xi(\theta_i)-\Xi(a))}(\Xi(\theta_i)-\Xi(a))^{p+\phi_i}}{\Gamma(p+\phi_i+1)} \\
&+ \sum_{k=1}^{r} \frac{|\varrho_k| e^{-\lambda(\Xi(\mu_k)-\Xi(a))}(\Xi(\mu_k)-\Xi(a))^{p-\omega_k}}{\Gamma(p-\omega_k+1)} \\
&\left.\left. + \frac{e^{-\lambda(\Xi(b)-\Xi(a))}(\Xi(b)-\Xi(a))^p}{\Gamma(p+1)} \right\} \right] l \|\overline{\mathscr{A}} - \mathscr{A}\|.
\end{aligned}
$$

Thus, the following is the case:

$$
\|\mathscr{T}\overline{\mathscr{A}} - \mathscr{T}\mathscr{A}\| = \max_{t\in J} |\mathscr{T}\overline{\mathscr{A}} - \mathscr{T}\mathscr{A}| \le l\Omega \|\overline{\mathscr{A}} - \mathscr{A}\|,
$$

which, in view of $l\Omega < 1$, shows that the operator $\mathscr{T}$ is a contraction. Hence, the operator $\mathscr{T}$ has a unique fixed point by the Banach contraction mapping principle. Therefore, our proposed problem (6.1) has a unique solution on J. The proof is completed. □

Now the existence of solution for our proposed problem is discussed by using Krasnoselskii's fixed point theorem.

Theorem 6.3. *Let* $\mathbb{Q} : J \times \mathbb{R} \times \mathbb{R} \to \mathbb{R}$ *be a continuous function satisfying* (H_1). *In addition, we assume that the following condition is satisfied:*

(H_2) *There exists a continuous function* $\phi \in X$ *such that*

$$
\left| \mathbb{Q}\left(t, \mathscr{A}(t), \mathscr{A}\left(\frac{t}{\eta}\right)\right) \right| \le \phi(t).
$$

Then, our proposed problem (6.1) *has at least one solution on* J, *provided the following condition holds:*

$$\frac{e^{-\lambda(\Xi(b)-\Xi(a))}(\Xi(b)-\Xi(a))^p}{\Gamma(p+1)} + \frac{e^{-\lambda(\Xi(b)-\Xi(a))}(\Xi(b)-\Xi(a))^{\vartheta-1}}{|\wedge|}$$
$$\times\left\{\sum_{j=1}^{m}\frac{|\alpha_j|e^{-\lambda(\Xi(\zeta_j)-\Xi(a))}(\Xi(\zeta_j)-\Xi(a))^p}{\Gamma(p+1)}\right.$$
$$+\sum_{i=1}^{n}\frac{|\beta_i|e^{-\lambda(\Xi(\theta_i)-\Xi(a))}(\Xi(\theta_i)-\Xi(a))^{p+\phi_i}}{\Gamma(p+\phi_i+1)}$$
$$+\sum_{k=1}^{r}\frac{|\varrho_k|e^{-\lambda(\Xi(\mu_k)-\Xi(a))}(\Xi(\mu_k)-\Xi(a))^{p-\omega_k}}{\Gamma(p-\omega_k+1)}$$
$$\left.+\frac{e^{-\lambda(\Xi(b)-\Xi(a))}(\Xi(b)-\Xi(a))^p}{\Gamma(p+1)}\right\}l < 1. \tag{6.11}$$

Proof. By assumption (H_2), we can fix $\rho \geq \Omega\|\phi\|$ and consider a closed ball $B_\rho = \{\mathscr{A} \in C(J,\mathbb{R}) : \|\mathscr{A}\| \leq \rho\}$, where $\|\phi\| = \sup_{t\in J}|\phi(t)|$. We verify the hypotheses of Krasnoselskii's fixed point theorem by splitting the operator $\mathscr{T}$ as $\mathscr{T} = \mathscr{G} + \mathscr{H}$, where $\mathscr{G}, \mathscr{H}$ are defined by the following:

$$\mathscr{G}\mathscr{A}(t) = \frac{1}{\Gamma(p)}\int_a^t \Xi^{\triangle}(z)e^{-\lambda(\Xi(t)-\Xi(z))}(\Xi(t)-\Xi(z))^{p-1}\left|\mathbb{Q}\left(z,\mathscr{A}(z),\mathscr{A}\left(\frac{z}{\eta}\right)\right)\right|\triangle z,$$
$$\mathscr{H}\mathscr{A}(t) = \frac{e^{-\lambda(\Xi(t)-\Xi(z))}(\Xi(t)-\Xi(z))^{\vartheta-1}}{\wedge}$$
$$\times\left\{\sum_{j=1}^{m}\frac{|\alpha_j|}{\Gamma(p)}\int_a^{\zeta_j}\Xi^{\triangle}(z)e^{-\lambda(\Xi(t)-\Xi(z))}(\Xi(t)-\Xi(z))^{p-1}\right.$$
$$\times\left|\mathbb{Q}\left(z,\mathscr{A}(z),\mathscr{A}\left(\frac{z}{\eta}\right)\right)\right|\triangle z$$
$$-\sum_{i=1}^{n}\frac{|\beta_i|}{\Gamma(p+\phi_i)}\int_a^{\theta_i}\Xi^{\triangle}(z)e^{-\lambda(\Xi(t)-\Xi(z))}(\Xi(t)-\Xi(z))^{p+\phi_i-1}$$
$$\times\left|\mathbb{Q}\left(z,\mathscr{A}(z),\mathscr{A}\left(\frac{z}{\eta}\right)\right)\right|\triangle z$$
$$-\sum_{k=1}^{r}\frac{|\varrho_k|}{\Gamma(p-\omega_k)}\int_a^{\mu_k}\Xi^{\triangle}(z)e^{-\lambda(\Xi(t)-\Xi(z))}(\Xi(t)-\Xi(z))^{p-\omega_k-1}$$
$$\times\left|\mathbb{Q}\left(z,\mathscr{A}(z),\mathscr{A}\left(\frac{z}{\eta}\right)\right)\right|\triangle z$$
$$-\frac{1}{\Gamma(p)}\int_a^{b}\Xi^{\triangle}(z)e^{-\lambda(\Xi(t)-\Xi(z))}(\Xi(t)-\Xi(z))^{p-1}$$
$$\left.\times\left|\mathbb{Q}\left(z,\mathscr{A}(z),\mathscr{A}\left(\frac{z}{\eta}\right)\right)\right|\triangle z\right\}.$$

For any $\mathscr{A}, \overline{\mathscr{A}} \in B_\rho$, we have the following:

$$
\begin{aligned}
&\left|\mathscr{G}\mathscr{A}(t) + \mathscr{H}\overline{\mathscr{A}}(t)\right| \\
&= \frac{1}{\Gamma(p)} \int_a^t \Xi^{\triangle}(z) e^{-\lambda(\Xi(t)-\Xi(z))} (\Xi(t) - \Xi(z))^{p-1} \left|\mathbb{Q}\left(z, \mathscr{A}(z), \mathscr{A}\left(\frac{z}{\eta}\right)\right)\right| \triangle z \\
&\quad + \frac{e^{-\lambda(\Xi(t)-\Xi(z))}(\Xi(t) - \Xi(z))^{\vartheta-1}}{\wedge} \\
&\quad \times \left\{ \sum_{j=1}^{m} \frac{|\alpha_j|}{\Gamma(p)} \int_a^{\zeta_j} \Xi^{\triangle}(z) e^{-\lambda(\Xi(t)-\Xi(z))} (\Xi(t) - \Xi(z))^{p-1} \right. \\
&\quad \times \left|\mathbb{Q}\left(z, \overline{\mathscr{A}}(z), \overline{\mathscr{A}}\left(\frac{z}{\eta}\right)\right)\right| \triangle z \\
&\quad - \sum_{i=1}^{n} \frac{|\beta_i|}{\Gamma(p+\phi_i)} \int_a^{\theta_i} \Xi^{\triangle}(z) e^{-\lambda(\Xi(t)-\Xi(z))} (\Xi(t) - \Xi(z))^{p+\phi_i-1} \\
&\quad \times \left|\mathbb{Q}\left(z, \overline{\mathscr{A}}(z), \overline{\mathscr{A}}\left(\frac{z}{\eta}\right)\right)\right| \triangle z \\
&\quad - \sum_{k=1}^{r} \frac{|\varrho_k|}{\Gamma(p-\omega_k)} \int_a^{\mu_k} \Xi^{\triangle}(z) e^{-\lambda(\Xi(t)-\Xi(z))} (\Xi(t) - \Xi(z))^{p-\omega_k-1} \\
&\quad \times \left|\mathbb{Q}\left(z, \overline{\mathscr{A}}(z), \overline{\mathscr{A}}\left(\frac{z}{\eta}\right)\right)\right| \triangle z \\
&\quad - \frac{1}{\Gamma(p)} \int_a^b \Xi^{\triangle}(z) e^{-\lambda(\Xi(t)-\Xi(z))} (\Xi(t) - \Xi(z))^{p-1} \\
&\quad \left. \times \left|\mathbb{Q}\left(z, \overline{\mathscr{A}}(z), \overline{\mathscr{A}}\left(\frac{z}{\eta}\right)\right)\right| \triangle z \right\} \\
&\le \left[\frac{e^{-\lambda(\Xi(a)-\Xi(b))}(\Xi(a) - \Xi(b))^{p}}{\Gamma(p+1)} + \frac{e^{-\lambda(\Xi(b)-\Xi(a))}(\Xi(b) - \Xi(a))^{\vartheta-1}}{|\wedge|} \right. \\
&\quad \times \left\{ \sum_{j=1}^{m} \frac{|\alpha_j| e^{-\lambda(\Xi(\zeta_j)-\Xi(a))}(\Xi(\zeta_j) - \Xi(a))^{p}}{\Gamma(p+1)} \right. \\
&\quad + \sum_{i=1}^{n} \frac{|\beta_i| e^{-\lambda(\Xi(\theta_i)-\Xi(a))}(\Xi(\theta_i) - \Xi(a))^{p+\phi_i}}{\Gamma(p+\phi_i+1)} \\
&\quad + \sum_{k=1}^{r} \frac{|\varrho_k| e^{-\lambda(\Xi(\mu_k)-\Xi(a))}(\Xi(\mu_k) - \Xi(a))^{p-\omega_k}}{\Gamma(p-\omega_k+1)} \\
&\quad \left.\left. + \frac{e^{-\lambda(\Xi(b)-\Xi(a))}(\Xi(b) - \Xi(a))^{p}}{\Gamma(p+1)} \right\} \right] \|\phi\| \\
&= \Omega\|\phi\| \le \rho.
\end{aligned}
$$

Hence $\|\mathcal{G}\mathcal{A}(t)+\mathcal{H}\overline{\mathcal{A}}(t)\| \leq \rho$, which shows that $\mathcal{G}\mathcal{A}(t)+\mathcal{H}\overline{\mathcal{A}}(t) \in B_\rho$. Now, it is easy to prove that the operator $\mathcal{H}$ is a contraction mapping. The operator $\mathcal{G}$ is continuous by the continuity of $\mathbb{Q}$. Moreover, $\mathcal{G}$ is uniformly bounded on B_ρ since

$$\|\mathcal{G}\mathcal{A}\| \leq \frac{e^{-\lambda(\Xi(b)-\Xi(a))}(\Xi(b)-\Xi(a))^p}{\Gamma(p+1)}\|\phi\|.$$

Finally, we prove the compactness of the operator $\mathcal{G}$. For $t_1, t_2 \in J$, $t_1 < t_2$, we have the following:

$$\begin{aligned}
&|\mathcal{G}\mathcal{A}(t_2)-\mathcal{G}\mathcal{A}(t_1)| \\
&\leq \frac{1}{\Gamma(p)}\int_a^{t_1}\Xi^{\triangle}(z)\Big[e^{-\lambda(\Xi(t_2)-\Xi(z))}(\Xi(t_2)-\Xi(z))^{p-1} \\
&\quad - e^{-\lambda(\Xi(t_1)-\Xi(z))}(\Xi(t_1)-\Xi(z))^{p-1}\Big]\left|\mathbb{Q}\left(z,\mathcal{A}(z),\mathcal{A}\left(\frac{z}{\eta}\right)\right)\right|\triangle z \\
&\quad + \frac{1}{\Gamma(p)}\int_{t_1}^{t_2}\Xi^{\triangle}(z)e^{-\lambda(\Xi(t_2)-\Xi(z))}(\Xi(t_2)-\Xi(z))^{p-1}\left|\mathbb{Q}\left(z,\mathcal{A}(z),\mathcal{A}\left(\frac{z}{\eta}\right)\right)\right|\triangle z \\
&\leq \frac{\|\phi\|}{\Gamma(p+1)}[2(e^{-\lambda(\Xi(t_2)-\Xi(t_1))}(\Xi(t_2)-\Xi(t_1))^p) \\
&\quad + |e^{-\lambda(\Xi(t_2)-\Xi(t_1))}(\Xi(t_2)-\Xi(t_1))^p|],
\end{aligned}$$

so that $|\mathcal{G}\mathcal{A}(t_2)-\mathcal{G}\mathcal{A}(t_1)| \to 0$, as $t_2 \to t_1$. Thus, $\mathcal{G}$ is equicontinuous, By applying the Arzela–Ascoli theorem, we deduce that the operator $\mathcal{G}$ is compact on B_ρ. Thus, the hypotheses of Krasnoselskii's fixed point theorem hold true. In consequence, there exists at least one solution for our proposed problem (6.1) on J. This completes the proof. □

6.4 Stability theory

Definition 6.8. Let $p > 0, \lambda \in \mathbb{R}$, and $\mathbb{Q} \in C(J,\mathbb{R})$. Equation (6.1) is said to be Ulam–Hyers (UH) stable if there exists a constant $c_{\mathbb{Q}} > 0$ such that for each $\epsilon > 0$ and for each $z \in AC^n(J,\mathbb{R})$ the following inequality holds true:

$$\left|\left({}^{THT}_{a^+}\triangle_{\Xi}^{p,q,\lambda} z\right)(t) - \mathbb{Q}\left(t, z(t), z\left(\frac{t}{\eta}\right)\right)\right| \leq \epsilon, \quad t \in [a,b]. \tag{6.12}$$

Then there exists a solution $\mathcal{A} \in AC^n(J,\mathbb{R})$ of Eq. (6.1) with

$$|z-\mathcal{A}|_{AC^n(J,\mathbb{R})} \leq c_{\mathbb{Q}}\epsilon.$$

Equation (6.1) is said to be generalized UH stable if there exits a function $c_{\mathbb{Q}} \in C[\mathbb{R}_0^+, \mathbb{R}_0^+]$ with $c_{\mathbb{Q}}(0) = 0$ such that for each $\epsilon > 0$ and for each $z_{AC^n(J,\mathbb{R})}$ satisfying the inequality (6.12), there exits a solution $\mathscr{A} \in AC^n[J, \mathbb{R}]$ of Eq. (6.1) with

$$|z - \mathscr{A}|_{AC^n([a,b],\mathbb{R})} \leq c_{\mathbb{Q}}\epsilon.$$

Definition 6.9. Let $p > 0, \lambda \in \mathbb{R}$, and $\mathbb{Q} \in C(J \times \mathbb{R} \times \mathbb{R}, \mathbb{R})$. Equation (6.1) is said to be Ulam–Hyers–Rassias (UHR) stable with respect to a given function $\nu \in C(J, \mathbb{R}_+)$ if there exits a constant $\phi_{\mathbb{Q},\nu} > 0$ such that, for each $\epsilon > 0$ and for each $z \in AC^n(J, \mathbb{R})$ satisfying the inequality

$$\left|\left({}^{THT}_{a^+} \triangle_{\Xi}^{p,q,\lambda} \mathscr{A}\right)(t) - \mathbb{Q}\left(t, z(t), z\left(\frac{t}{\eta}\right)\right)\right| \leq \epsilon\nu(t), \quad t \in J, \tag{6.13}$$

there exists a solution $\mathscr{A} \in AC^n(J, \mathbb{R})$ of Eq. (6.1) with

$$\|z - \mathscr{A}\|_{AC^n(J,\mathbb{R})} \leq \epsilon\phi_{\mathbb{Q},\nu}\nu(t), \quad t \in J,$$

and if

$$\left|\left({}^{THT}_{a^+} \triangle_{\Xi}^{p,q,\lambda} \mathscr{A}\right)(t) - \mathbb{Q}\left(t, z(t), z\left(\frac{t}{\eta}\right)\right)\right| \leq \nu(t), \quad t \in J, \tag{6.14}$$

then there exists a solution $\mathscr{A} \in AC^n(J, \mathbb{R})$ of Eq. (6.1) with

$$|z - \mathscr{A}|_{AC^n(J,\mathbb{R})} \leq \phi_{\mathbb{Q},\nu}\nu(t), \quad t \in J.$$

Theorem 6.4. *Let $p > 0, \lambda \in \mathbb{R}$, and $\mathbb{Q} \in C([a, b] \times \mathbb{R} \times \mathbb{R}, \mathbb{R})$. If $\nu \in C(J, \mathbb{R}_+)$ is a nondecreasing function and $K > 0$ is a constant such that*

$${}^{T\mathbb{T}}_{a^+} \mathscr{I}_{\Xi}^{p,\lambda} \nu(t) \leq K\nu(t), \quad t \in J, \tag{6.15}$$

then the differential equation (6.1) *is UHR stable with respect to ν.*

Proof. Fix $\epsilon > 0$ and let $z \in AC^n(J, \mathbb{R})$ be any solution of the inequality

$$\left|\left({}^{THT}_{a^+} \triangle_{\Xi}^{p,q,\lambda} z\right)(t) - \mathbb{Q}\left(t, z(t), z\left(\frac{t}{\eta}\right)\right)\right| \leq \epsilon\nu(t), \quad t \in J. \tag{6.16}$$

Then, we can define $w \in AC^n(J, \mathbb{R})$ such that

$$\left({}^{THT}_{a^+} \triangle_{\Xi}^{p,q,\lambda}\right)(t)z(t) = \mathbb{Q}\left(t, z(t), z\left(\frac{t}{\eta}\right)\right) + w(t), \quad t \in J, \tag{6.17}$$

and $|w(t)| \leq \epsilon\nu(t)$, for all $t \in (a, b]$. Now let $\mathscr{A} \in AC^n(J, \mathbb{R})$ be a solution of the following boundary value problem:

$$\begin{cases} \left({}^{THT}_{a^+}\triangle^{p,q,\lambda}_{\Xi}\mathscr{A}\right)(t) = \mathbb{Q}\left(t, \mathscr{A}(t), \mathscr{A}\left(\frac{t}{\eta}\right)\right), & t \in J, \\ \mathscr{A}(a) = 0, \\ \mathscr{A}(b) = \sum_{j=1}^{m} \alpha_j \mathscr{A}(\zeta_j) + \sum_{i=1}^{n} \beta_i {}^{THT}_{a^+}\mathscr{I}^{\phi_i,\lambda}_{\Xi}\mathscr{A}(\theta_i) + \sum_{k=1}^{r} \varrho_k {}^{THT}_{a^+}\triangle^{\omega_k,\lambda}_{\Xi}\mathscr{A}(\mu_k). \end{cases} \tag{6.18}$$

The solution of Eq. (6.18) is

$$\begin{aligned} \mathscr{A}(t) = {}& {}^{T\mathbb{T}}_{a^+}\mathscr{I}^{p,q,\lambda}_{\Xi}\mathbb{Q}\left(t, \mathscr{A}(t), \mathscr{A}\left(\frac{t}{\eta}\right)\right) + \frac{e^{-\lambda(\Xi(t)-\Xi(a))}(\Xi(t)-\Xi(a))^{\vartheta-1}}{\wedge} \\ & \times \left\{ \sum_{j=1}^{m} \alpha_j {}^{T\mathbb{T}}_{a^+}\mathscr{I}^{p,q,\lambda}_{\Xi}\mathbb{Q}\left(\zeta_j, \mathscr{A}(\zeta_j), \mathscr{A}\left(\frac{\zeta_j}{\eta}\right)\right) \right. \\ & - \sum_{i=1}^{n} \beta_i {}^{T\mathbb{T}}_{a^+}\mathscr{I}^{p,q,\lambda}_{\Xi}\mathbb{Q}\left(\theta_i, \mathscr{A}(\theta_i), \mathscr{A}\left(\frac{\theta_i}{\eta}\right)\right) \\ & - \sum_{k=1}^{r} \varrho_k {}^{T\mathbb{T}}_{a^+}\mathscr{I}^{p-\omega_k,\lambda}_{\Xi}\mathbb{Q}\left(\mu_k, \mathscr{A}(\mu_k), \mathscr{A}\left(\frac{\mu_k}{\eta}\right)\right) \\ & \left. - {}^{T\mathbb{T}}_{a^+}\mathscr{I}^{p,\lambda}_{\Xi}\mathbb{Q}\left(b, \mathscr{A}(b), \mathscr{A}\left(\frac{b}{\eta}\right)\right) \right\}, \end{aligned} \tag{6.19}$$

while the solution of Eq. (6.16) is given by

$$\begin{aligned} z(t) = {}& {}^{T\mathbb{T}}_{a^+}\mathscr{I}^{p,\lambda}_{\Xi}\mathbb{Q}\left(t, z(t), z\left(\frac{t}{\eta}\right)\right) + \frac{e^{-\lambda(\Xi(t)-\Xi(a))}(\Xi(t)-\Xi(a))^{\vartheta-1}}{\wedge} \\ & \times \left\{ \sum_{j=1}^{m} \alpha_j {}^{T\mathbb{T}}_{a^+}\mathscr{I}^{p,\lambda}_{\Xi}\mathbb{Q}\left(\zeta_j, z(\zeta_j), z\left(\frac{\zeta_j}{\eta}\right)\right) \right. \\ & - \sum_{i=1}^{n} \beta_i {}^{T\mathbb{T}}_{a^+}\mathscr{I}^{p,q,\lambda}_{\Xi}\mathbb{Q}\left(\theta_i, z(\theta_i), z\left(\frac{\theta_i}{\eta}\right)\right) \\ & - \sum_{k=1}^{r} \varrho_k {}^{T\mathbb{T}}_{a^+}\mathscr{I}^{p-\omega_k,q,\lambda}_{\Xi}\mathbb{Q}\left(\mu_k, z(\mu_k), z\left(\frac{\mu_k}{\eta}\right)\right) \\ & \left. - {}^{T\mathbb{T}}_{a^+}\mathscr{I}^{p,q,\lambda}_{\Xi}\mathbb{Q}\left(b, z(b), z\left(\frac{b}{\eta}\right)\right) \right\}, \end{aligned} \tag{6.20}$$

where $h_1(\cdot) = \mathbb{Q}(\cdot, z(\cdot), z(\frac{\cdot}{\eta}))$.

From Eq. (6.20) and inequality (6.15), we have

$$\begin{aligned} & \left| z(t) - \left[{}^{T\mathbb{T}}_{a^+}\mathscr{I}^{p,\lambda}_{\Xi}\mathbb{Q}\left(t, z(t), z\left(\frac{t}{\eta}\right)\right) + \frac{e^{-\lambda(\Xi(t)-\Xi(a))}(\Xi(t)-\Xi(a))^{\vartheta-1}}{\wedge} \right. \right. \\ & \times \left\{ \sum_{j=1}^{m} \alpha_j {}^{T\mathbb{T}}_{a^+}\mathscr{I}^{p,\lambda}_{\Xi}\mathbb{Q}\left(\zeta_j, z(\zeta_j), z\left(\frac{\zeta_j}{\eta}\right)\right) - \sum_{i=1}^{n} \beta_i {}^{T\mathbb{T}}_{a^+}\mathscr{I}^{p,\lambda}_{\Xi}\mathbb{Q}\left(\theta_i, z(\theta_i), z\left(\frac{\theta_i}{\eta}\right)\right) \right. \end{aligned}$$

$$
\begin{aligned}
&- \sum_{k=1}^{r} \varrho_k {}^{TT}_{a^+}\mathscr{I}_{\Xi}^{p-\omega_k,\lambda} \mathbb{Q}\left(\mu_k, z(\mu_k), z\left(\frac{\mu_k}{\eta}\right)\right) - {}^{TT}_{a^+}\mathscr{I}_{\Xi}^{p,\lambda} \mathbb{Q}\left(b, z(b), z\left(\frac{b}{\eta}\right)\right)\Big\}\Big]\Big| \\
&\le \left| {}^{TT}_{a^+}\mathscr{I}_{\Xi}^{p,\lambda} w(t)\right| \\
&\le \epsilon {}^{TT}_{a^+}\mathscr{I}_{\Xi}^{p,\lambda} v(t) \\
&\le \epsilon K v(t).
\end{aligned}
$$

Using this together with Eq. (6.19) and Lipschitz condition for each $t \in J$, we have

$$
\begin{aligned}
|z(t) - \mathscr{A}(t)| = &\Bigg| z(t) - {}^{TT}_{a^+}\mathscr{I}_{\Xi}^{p,\lambda} \mathbb{Q}\left(t, \mathscr{A}(t), \mathscr{A}\left(\frac{t}{\eta}\right)\right) + \frac{e^{-\lambda(\Xi(t)-\Xi(a))}(\Xi(t)-\Xi(a))^{\vartheta-1}}{\wedge} \\
&\times \Bigg\{ \sum_{j=1}^{m} \alpha_j {}^{TT}_{a^+}\mathscr{I}_{\Xi}^{p,\lambda} \mathbb{Q}\left(\zeta_j, \mathscr{A}(\zeta_j), \mathscr{A}\left(\frac{\zeta_j}{\eta}\right)\right) \\
&- \sum_{i=1}^{n} \beta_i {}^{TT}_{a^+}\mathscr{I}_{\Xi}^{p,\lambda} \mathbb{Q}\left(\theta_i, \mathscr{A}(\theta_i), \mathscr{A}\left(\frac{\theta_i}{\eta}\right)\right) \\
&- \sum_{k=1}^{r} \varrho_k {}^{TT}_{a^+}\mathscr{I}_{\Xi}^{p-\omega_k,\lambda} \mathbb{Q}\left(\mu_k, \mathscr{A}(\mu_k), \mathscr{A}\left(\frac{\mu_k}{\eta}\right)\right) \\
&- {}^{TT}_{a^+}\mathscr{I}_{\Xi}^{p,\lambda} \mathbb{Q}\left(b, \mathscr{A}(b), \mathscr{A}\left(\frac{b}{\eta}\right)\right)\Bigg\}\Bigg| \\
&\le \epsilon K v(t) + \frac{1}{\Gamma(p)} \int_a^t \Xi'(s)(\Xi(t)-\Xi(s))^{p-1} e^{-\lambda(\Xi(t)-\Xi(s))} \\
&\times \left| \mathbb{Q}\left(s, z(s), z\left(\frac{s}{\eta}\right)\right) - \mathbb{Q}\left(s, \mathscr{A}(s), \mathscr{A}\left(\frac{s}{\eta}\right)\right)\right| \triangle s \\
&\le \epsilon K v(t) + \frac{L}{\Gamma(p)} \int_a^t \Xi'(s)(\Xi(t)-\Xi(s))^{p-1} e^{-\lambda(\Xi(t)-\Xi(s))} \\
&\times \left\{ |z(s) - \mathscr{A}(s)| + \left| z\left(\frac{s}{\eta}\right) - \mathscr{A}\left(\frac{s}{\eta}\right)\right| \right\} \triangle s.
\end{aligned}
$$

Applying Gronwall's inequality, with $u(t) = |z(t) - \mathscr{A}(t)|$, $v(t) = \epsilon K v(t)$, and $w(t) = \frac{L}{\Gamma(p)}$, we obtain

$$
\begin{aligned}
|z(t) - \mathscr{A}(t)| &\le \epsilon K v(t) E_p(L(\Xi(b) - \Xi(a))^p), \\
&\le \epsilon c_{\mathbb{Q},v} v(t), \quad t \in [a, b],
\end{aligned}
\tag{6.21}
$$

where $c_{\mathbb{Q},v} := K E_p(L(\Xi(b) - \Xi(a))^p)$. Thus our proposed problem (6.1) is UHR stable with respect to v. □

6.5 An example

Consider the following tempered Ξ-Hilfer type Ambartsumian boundary value problem on time scales:

$$\begin{cases} \left({}^{THT}_{a^+}\Delta_{\Xi}^{p,q,\lambda}\mathscr{A}\right)(t) = \mathbb{Q}\left(t, \mathscr{A}(t), \mathscr{A}\left(\frac{t}{\eta}\right)\right), & t \in [0,5], \\ \mathscr{A}(a) = 0, \\ \mathscr{A}(b) = \sum_{j=1}^{m} \alpha_j \mathscr{A}(\zeta_j) + \sum_{i=1}^{n} \beta_i {}^{THT}_{a^+}\mathscr{I}_{\Xi}^{\phi_i,\lambda}\mathscr{A}(\theta_i) + \sum_{k=1}^{r} \varrho_k {}^{THT}_{a^+}\Delta_{\Xi}^{\omega_k,\lambda}\mathscr{A}(\mu_k), \end{cases} \tag{6.22}$$

with $\Xi(t) = 1 + t^2, p = \frac{1}{3}, q = \frac{1}{4}, m = 4, n = 3, r = 2, \alpha_1 = \frac{1}{15}, \alpha_2 = \frac{1}{10}, \alpha_3 = \frac{2}{15}, \alpha_4 = \frac{2}{7}, \zeta_1 = \frac{1}{4}, \zeta_2 = \frac{1}{2}, \zeta_3 = \frac{2}{5}, \zeta_4 = 2, \beta_1 = \frac{1}{18}, \beta_2 = \frac{2}{11}, \beta_3 = \frac{2}{7}, \phi_1 = \frac{1}{2}, \phi_2 = \frac{1}{3}, \phi_3 = \frac{2}{5}, \theta_1 = \frac{1}{2}, \theta_2 = \frac{3}{4}, \theta_3 = 1, \mu_1 = 3, \mu_2 = \frac{7}{2}, \omega_1 = \frac{1}{4}, \omega_2 = \frac{1}{3}, \varrho_1 = \frac{1}{28}, \varrho_2 = \frac{1}{14}, \lambda = \frac{1}{11}, \vartheta = \frac{1}{2}, \eta = 1.175$.

From the above values we get $\wedge \approx -1.10200512$ and $\Omega \approx 0.3631064$. Also

$$\mathbb{Q}\left(t, \mathscr{A}(t), \mathscr{A}\left(\frac{t}{\eta}\right)\right) = \frac{1}{1.175}\mathscr{A}\left(\frac{t}{1.175}\right) - \mathscr{A}(t), \quad t \in [0,5], \tag{6.23}$$

$$\mathbb{Q}(t, u, v) = \frac{1}{1.175}(u - v), \quad u, v \in \mathbb{R}.$$

Clearly, $\mathbb{Q}$ is continuous.

For any $u, v, \overline{u}, \overline{v} \in \mathbb{R}$ and $t \in [0,5]$,

$$\begin{aligned} \left|\mathbb{Q}(t,u,v) - \mathbb{Q}(t,\overline{u},\overline{v})\right| &= \frac{1}{1.175}|u - v - \overline{u} + \overline{v}| \\ &\le \frac{1}{1.175}\{|u - \overline{u}| + |v - \overline{v}|\}. \end{aligned} \tag{6.24}$$

Hence, assumption (H_1) holds for the nonlinear function $\mathbb{Q}(t, \mathscr{A}(t), \mathscr{A}(\frac{t}{\eta}))$ with $l = \frac{1}{1.175}$ and also the condition $l\Omega \approx 0.3090267234 < 1$ holds. Hence our proposed problem (6.22) with (6.23) has a unique solution on $[0,5]$. Theorem 6.4 implies that (6.22) is UHR stable.

Bibliography

[1] I. Ahmed, P. Kumam, F. Jarad, P. Borisut and W. Jirakitpuwapat, On Hilfer generalized proportional fractional derivative, *Adv. Differ. Equ.*, 1 (2020), 1–18.

[2] S. Ahmad, A. Ullah, A. Akgul and M. De la Sen, A study of fractional order Ambartsumian equation involving exponential decay kernel, *AIMS Math.*, 6(9) (2021), 9981–9997.

[3] G. A. Anastassiou, Principles of delta fractional calculus on time scales and inequalities, *Math. Comput. Model.*, 52 (2010), 556–566.

[4] K. Deimling, Nonlinear Functional Analysis, Springer, New York, 1985.

[5] H. M. Fahad, A. Fernandez, M. ur Rehman and M. Siddiqi, Tempered and Hadamard-type fractional calculus with respect to functions, *Mediterr. J. Math.*, 18(4) (2021), 143.
[6] S. Harikrishnan, R. Ibrahim and K. Kanagarajan, An arbitrary order differential equations on time scales, *Univers. J. Math. Appl.*, 1 (2018), 262–266.
[7] F. Jarad, M. A. Alquadah and T. Abdeljawad, On more general forms of proportional fractional operators, *Open Math.*, 18 (2020), 167–176.
[8] F. Jarad, T. Abedljawad and D. Baleanu, On the generalized fractional derivatives and their Caputo modification, *J. Nonlinear Sci. Appl.*, 10 (2017), 2607–2619.
[9] F. Jarad, T. Abedljawad, S. Rashid and Z. Hammouch, More properties of the proportional fractional integrals and derivatives of a function with respect to another function, *Adv. Differ. Equ.*, 2020 (2020), 1–16.
[10] U. N. Katugampola, New approach to a generalized fractional integral, *Appl. Math. Comput.*, 218 (2011), 860–865.
[11] U. N. Katugampola, A new approach to generalized fractional derivatives, *Appl. Math. Comput.*, 6 (2014), 1–15.
[12] R. Khalil, M. A. Horani, A. Yousef and M. Sababheh, A new definition of fractional derivative, *J. Comput. Appl. Math.*, 264 (2014), 65–70.
[13] M. A. Krasnoselskii, Two remarks on the method of successive approximations, *Usp. Mat. Nauk*, 10 (1955), 123–127.
[14] K. D. Kucche, A. D. Mali, A. Fernandez and H. M. Fahad, On tempered Hilfer fractional derivative with respect to the functions and the associated fractional differential equations, *Chaos Solitons Fractals*, 163 (2022), 112547.
[15] D. Kumar, J. Singh, D. Baleanu and S. Rathore, Analysis of a fractional model of the Ambartsumian equation, *Eur. Phys. J. Plus*, 133 (2018), 259.
[16] A. D. Mali, K. D. Kucche, A. Fernandez and H. M. Fahad, On tempered fractional derivative with respect to the functions and the associated fractional differential equations, *Math. Methods Appl. Sci.*, 45(17) (2022), 11134–11157.
[17] I. Mallah, S. Alha, I. Ahmed, A. Akgul and F. Jarad, On ψ-Hilfer generalized proportional fractional operators, *AIMS Math.*, 7 (2021), 82–103.
[18] S. Manikandan, S. Sivasundaram, D. Vivek and K. Kanagarajan, Controllability and qualitative property results for Ambartsumian equation via Ξ-Hilfer generalized proportional fractional derivative on time scales, *Nonlinear Stud.*, 29(4) (2022), 1–23.
[19] M. Medved and E. Brestovanska, Differential equations with tempered ψ-Caputo fractional derivative, *Math. Model. Anal.*, 26(4) (2021), 631–650.
[20] S. K. Ntouyas and D. Vivek, Existence and uniqueness results for sequential ψ-Hilfer fractional differential equations with multi-point boundary conditions, *Acta Math. Univ. Comen.*, 2 (2021), 171–185.
[21] C. Nuchbong, S. K. Ntouyas and J. Tariboon, Boundary value problems for Hilfer fractional integro-differential equations and inclusions with nonlocal integro-multipoint boundary conditions, *Open Math.*, 18 (2020), 1879–1894.
[22] J. Patade and S. Bhalekar, On analytical solution of Ambartsumian equation, *Nat. Acad. Sci. Lett.*, 40 (2017), 291–293.
[23] S. Rashid, F. Jarad, M. A. Noor, H. Kalsoom and Y. M. Chu, Inequalities by means of generalized proportional fractional integral operators with respect to another function, *Math.*, 7 (2019), 1–16.
[24] S. Sitho, S. K. Ntouyas, A. Samadi and J. Tariboon, Boundary value problems for ψ-Hilfer type sequential fractional differential equations and inclusions with integral multi-point boundary conditions, *Math.*, 9 (2021), 1001.
[25] J. V. da, C. Sousa and E. C. de Oliveira, On the ψ-Hilfer fractional derivative, *Commun. Nonlinear Sci.*, 60 (2018), 72–91.
[26] J. V. da, C. Sousa, D. S. Oliveira, G. S. F. Frederico and D. F. M. Torres, Existence, uniqueness and controllability for g-Hilfer differential equations on time scales, *hal-03726777*, 2022.

[27] D. Vivek, K. Kanagarajan and E. Elasyed, Some existence and stability results for Hilfer fractional implicit differential equations with nonlocal conditions, *Mediterr. J. Math.*, 15 (2018), 15.
[28] A. Wongcharoen, B. Ahmad, S. K. Ntouyas and J. Tariboon, Three-point boundary value problems for Langevin equation with Hilfer fractional derivative, *Adv. Math. Phys.*, 2020 (2020), 9606428.
[29] A. Wongcharoen, S. K. Ntouyas and J. Tariboon, Nonlocal boundary value problems for Hilfer type pantograph fractional differential equations and inclusions, *Adv. Differ. Equ.*, 2020 (2020), 279.

Svetlin G. Georgiev and İnci M. Erhan

7 Series solution method on time scales and its applications

Abstract: In this chapter, a series solution method on general time scales is introduced. The derivation of the method and its application to dynamic and integral equations is discussed in details. Several examples illustrating the method are presented.

7.1 Introduction

In this chapter, we introduce the series solution method on a general time scale. The series solution method on time scales has been discussed for some particular cases in recent studies [4, 7–9]. The series solution method for certain integral equations has been used in recent books [3, 6]. Also, the series solution method has been applied to initial value problems associated with first-order dynamic equations and fractional dynamic equations in [5].

We present the series solution method on an arbitrary time scale and its application to linear dynamic equations and to nonlinear Volterra integral equations in which the nonlinear term is at most a quadratic function. We consider Volterra integral equations of both the first and second kind. Moreover, the method is applicable in cases when the graininess function is constant or nonconstant.

The chapter is organized as follows. In the next section, we recall some preliminary concepts such as the Taylor series and Cauchy product of power series on time scales, and derive other preliminary results. The series solution method for dynamic equations and its application to particular examples is discussed in Section 7.3. Finally, we introduce an application of the series solution method to a certain type nonlinear Volterra integral equations of the first and second kind and give specific examples in Sections 7.4 and 7.5.

7.2 Preliminary results

We first recall the Taylor series on general time scales. Let $\mathbb{T}$ be a time scale with forward jump operator and delta differentiation operator, σ and Δ, respectively. Let $s, t \in \mathbb{T}$. The monomials $h_k(t, s)$ are defined recursively as follows [2]:

Svetlin G. Georgiev, Sorbonne University, Paris, France, e-mail: svetlingeorgiev1@gmail.com
İnci M. Erhan, Department of Computer Engineering, Aydın Adnan Menderes University, Aydın, Turkey, e-mail: incierhan@gmail.com

https://doi.org/10.1515/9783111182971-007

$$h_0(t,s) = 1,$$
$$h_{k+1}(t,s) = \int_s^t h_k(\tau,s)\Delta\tau, \quad k = 0,1,2,\dots$$

Then it is obvious that

$$h_1(t,s) = \int_s^t h_0(\tau,s)\Delta\tau = \int_s^t \Delta\tau = t-s,$$
$$h_2(t,s) = \int_s^t h_1(\tau,s)\Delta\tau = \int_s^t (\tau - s)\Delta\tau,$$

and so on. Note that

$$h_k^\Delta(t,s) = h_{k-1}(t,s), \quad k \in \mathbb{N}.$$

Theorem 7.1 (Taylor formula, [1, 3]). *Let $n \in \mathbb{N}$. Suppose that f is n times differentiable on $\mathbb{T}^{\kappa^n}$. Let also $a \in \mathbb{T}^{\kappa^{n-1}}$, $t \in \mathbb{T}$. Then*

$$f(t) = \sum_{k=0}^{n-1} h_k(t,a) f^{\Delta^k}(a) + \int_a^{\rho^{n-1}(t)} h_{n-1}(t,\sigma(\tau)) f^{\Delta^n}(\tau)\Delta\tau.$$

Theorem 7.2 (Leibnitz formula, [1, 3]). *Let $S_k^{(n)}$ be the set consisting of all possible strings of length n, containing exactly k times σ and $n-k$ times Δ. If*

$$f^\Lambda \quad \text{exists for all } \Lambda \in S_k^{(n)},$$

then

$$(fg)^{\Delta^n} = \sum_{k=0}^{n}\Big(\sum_{\Lambda \in S_k^{(n)}} f^\Lambda\Big) g^{\Delta^k}.$$

The next theorem is an important result which is crucial in the application of the series solution method.

Theorem 7.3 ([3]). *For every $m, n \in \mathbb{N}_0$, we have*

$$h_n(t,a)h_m(t,a) = \sum_{l=m}^{m+n}\Big(\sum_{\Lambda_{l,m} \in S_m^{(l)}} h_n^{\Lambda_{l,m}}(a,a)\Big) h_l(t,a)$$

for every $t, a \in \mathbb{T}$.

Proof. If $m = 0$ or $n = 0$, the assertion is evident. Suppose that $m \neq 0$ and $n \neq 0$. By the Taylor formula, we have

$$h_n(t,\alpha)h_m(t,\alpha) = \sum_{l=0}^{\infty}(h_n(t,\alpha)h_m(t,\alpha))^{\Delta^l}|_{t=\alpha}h_l(t,\alpha), \quad t,\alpha \in \mathbb{T}.$$

By the Leibnitz formula, we have

$$(h_n(t,\alpha)h_m(t,\alpha))^{\Delta^l} = \sum_{k=0}^{l}\Big(\sum_{\Lambda_{l,k}\in S_k^{(l)}} h_n^{\Lambda_{l,k}}(t,\alpha)\Big)h_m^{\Delta^k}(t,\alpha), \quad t,\alpha \in \mathbb{T}.$$

If $l < m$, then

$$(h_n(t,\alpha)h_m(t,\alpha))^{\Delta^l} = \sum_{k=0}^{l}\Big(\sum_{\Lambda_{l,k}\in S_k^{(l)}} h_n^{\Lambda_{l,k}}(t,\alpha)\Big)h_{m-k}(t,\alpha), \quad t,\alpha \in \mathbb{T}.$$

From here, for $l < m$, we have $h_{m-k}(\alpha,\alpha) = 0$ and

$$(h_n(t,\alpha)h_m(t,\alpha))^{\Delta^l}|_{t=\alpha} = 0, \quad t,\alpha \in \mathbb{T}.$$

For $l \geq m$, using that $h_0(t,\alpha) = 1, \quad t,\alpha \in \mathbb{T}$, we get

$$\begin{aligned}(h_n(t,\alpha)h_m(t,\alpha))^{\Delta^l}|_{t=\alpha} &= \sum_{k=0}^{m-1}\Big(\sum_{\Lambda_{l,k}\in S_k^{(l)}} h_n^{\Lambda_{l,k}}(t,\alpha)\Big)h_{m-k}(t,\alpha)|_{t=\alpha} \\ &\quad + \sum_{\Lambda_{l,m}\in S_m^{(l)}} h_n^{\Lambda_{l,m}}(t,\alpha)|_{t=\alpha} \\ &= \sum_{\Lambda_{l,m}\in S_m^{(l)}} h_n^{\Lambda_{l,m}}(\alpha,\alpha), \quad t,\alpha \in \mathbb{T}.\end{aligned}$$

Hence, using that $\Lambda_{l,m}$ contains m times σ and $l-m$ times Δ, and

$$\begin{aligned} f^{\sigma} &= f \quad \text{or} \quad f^{\sigma} = f + \mu f^{\Delta}, \\ f^{\sigma\sigma} &= f \quad \text{or} \quad f^{\sigma\sigma} = f + \mu f^{\Delta} + \mu^{\sigma}(f^{\Delta} + \mu f^{\Delta^2}), \end{aligned}$$

and so on, we obtain

$$\begin{aligned} h_n(t,\alpha)h_m(t,\alpha) &= \sum_{l=m}^{\infty}(h_n(t,\alpha)h_m(t,\alpha))^{\Delta^l}|_{t=\alpha}h_l(t,\alpha) \\ &= \sum_{l=m}^{\infty}\Big(\sum_{\Lambda_{l,m}\in S_m^{(l)}} h_n^{\Lambda_{l,m}}(\alpha,\alpha)\Big)h_l(t,\alpha) \\ &= \sum_{l=m}^{m+n}\Big(\sum_{\Lambda_{l,m}\in S_m^{(l)}} h_n^{\Lambda_{l,m}}(\alpha,\alpha)\Big)h_l(t,\alpha), \quad t,\alpha \in \mathbb{T}, \end{aligned}$$

which completes the proof. □

Theorem 7.4. *We have*

$$\left(\sum_{i=0}^{\infty} A_i h_i(t,\alpha)\right)\left(\sum_{j=0}^{\infty} B_j h_j(t,\alpha)\right) = \sum_{r=0}^{\infty} C_r h_r(t,\alpha), \quad t,\alpha \in \mathbb{T},$$

where

$$C_r = \sum_{k=r}^{\infty}\left(\sum_{l=k-r}^{k} A_l B_{k-l}\left(\sum_{\Lambda_{r,k-l}\in S_{k-l}^{(r)}} h_l^{\Lambda_{r,k-l}}(\alpha,\alpha)\right)\right), \tag{7.1}$$

with A_i, B_i, $i \in \mathbb{N}_0$, being constants.

Proof. Using the Cauchy product of two infinite series and Theorem 7.3, we get

$$\begin{aligned}
&\left(\sum_{i=0}^{\infty} A_i h_i(t,\alpha)\right)\left(\sum_{j=0}^{\infty} B_j h_j(t,\alpha)\right)\\
&= \sum_{k=0}^{\infty}\left(\sum_{l=0}^{k} A_l h_l(t,\alpha) B_{k-l} h_{k-l}(t,\alpha)\right)\\
&= \sum_{k=0}^{\infty}\left(\sum_{l=0}^{k} A_l B_{k-l}\left(\sum_{r=k-l}^{k}\left(\sum_{\Lambda_{r,k-l}\in S_{k-l}^{(r)}} h_l^{\Lambda_{r,k-l}}(\alpha,\alpha)\right)h_r(t,\alpha)\right)\right), \quad t,\alpha \in \mathbb{T}.
\end{aligned} \tag{7.2}$$

Then the double sum in (7.2) becomes

$$\begin{aligned}
&\left(\sum_{i=0}^{\infty} A_i h_i(t,\alpha)\right)\left(\sum_{j=0}^{\infty} B_j h_j(t,\alpha)\right)\\
&= \sum_{k=0}^{\infty}\left(\sum_{l=0}^{k} A_l B_{k-l}\left(\sum_{r=k-l}^{k}\left(\sum_{\Lambda_{r,k-l}\in S_{k-l}^{(r)}} h_l^{\Lambda_{r,k-l}}(\alpha,\alpha)\right)h_r(t,\alpha)\right)\right), \quad t,\alpha \in \mathbb{T}.
\end{aligned} \tag{7.3}$$

If we reorder the triple sum $\sum_{k=0}^{\infty}\sum_{l=0}^{k}\sum_{r=k-l}^{k}$ in (7.3) as $\sum_{r=0}^{\infty}\sum_{k=r}^{\infty}\sum_{l=k-r}^{k}$ and denote

$$C_r = \sum_{k=r}^{\infty}\left(\sum_{l=k-r}^{k} A_l B_{k-l}\left(\sum_{\Lambda_{r,k-l}\in S_{k-l}^{(r)}} h_l^{\Lambda_{r,k-l}}(\alpha,\alpha)\right)\right), \quad r \in \mathbb{N}_0, \quad \alpha \in \mathbb{T}$$

for the sake of brevity, we conclude

$$\left(\sum_{i=0}^{\infty} A_i h_i(t,\alpha)\right)\left(\sum_{j=0}^{\infty} B_j h_j(t,\alpha)\right) = \sum_{r=0}^{\infty} C_r h_r(t,\alpha), \quad t,\alpha \in \mathbb{T}, \tag{7.4}$$

which completes the proof. □

In order to avoid the complicated structure of the constants C_r involved in the Cauchy product of infinite series, we define the constants $D_{n,m,l}$ as

$$D_{n,m,l} = \sum_{\Lambda_{l,m} \in S_m^{(l)}} h_n^{\Lambda_{l,m}}(\alpha, \alpha) \tag{7.5}$$

for $\alpha \in \mathbb{T}$, $n, m \in \mathbb{N}_0$ and $l \in \{m-n, \dots, m\}$. Then we have

$$h_n(t,\alpha)h_m(t,\alpha) = \sum_{l=m}^{m+n} D_{n,m,l} h_l(t,\alpha), \quad t,\alpha \in \mathbb{T},$$

and also

$$\left(\sum_{i=0}^{\infty} A_i h_i(t,\alpha)\right)\left(\sum_{j=0}^{\infty} B_j h_j(t,\alpha)\right) = \sum_{k=0}^{\infty}\sum_{l=0}^{k} A_l B_{k-l}\left(\sum_{r=k-l}^{k} D_{l,k-l,r}\right) h_r(t,\alpha), \quad t,\alpha \in \mathbb{T}.$$

The constants $D_{n,m,l}$ defined in (7.5) have some properties which we discuss below.

1. For all $m \in \mathbb{N}_0$ and $l = m, \dots, m+n$, we have

$$D_{0,m,l} = D_{m,0,l} = \begin{cases} 0 & \text{if } l \neq m, \\ 1 & \text{if } l = m. \end{cases} \tag{7.6}$$

Proof. This statement easily follows from the fact that

$$h_0(t,\alpha)h_m(t,\alpha) = h_m(t,\alpha)h_0(t,\alpha) = h_m(t,\alpha), \quad t,\alpha \in \mathbb{T}.$$

Indeed,

$$h_m(t,\alpha) = h_0(t,\alpha)h_m(t,\alpha) = \sum_{l=m}^{m} D_{0,m,l} h_l(t,\alpha) = D_{0,m,m} h_m(t,\alpha), \quad t,\alpha \in \mathbb{T}$$

implies $D_{0,m,m} = 1$, and

$$h_m(t,\alpha) = h_m(t,\alpha)h_0(t,\alpha) = \sum_{l=0}^{m} D_{m,0,l} h_l(t,\alpha), \quad t,\alpha \in \mathbb{T}$$

implies $D_{m,0,m} = 1$ and $D_{m,0,l} = 0$ for $l = 0, \dots, m-1$. □

2. For all $n, m \in \mathbb{N}_0$ and $l = m, \dots, n+m$ we have

$$\begin{aligned} D_{n,m,l} &= D_{m,n,l} \quad \text{for } n \geq m \text{ and } l \geq n, \\ D_{n,m,l} &= D_{m,n,l} = 0 \quad \text{for } n \geq m \text{ and } m < l < n. \end{aligned} \tag{7.7}$$

Proof. Since

$$h_n(t,\alpha)h_m(t,\alpha) = h_m(t,\alpha)h_n(t,\alpha), \quad t,\alpha \in \mathbb{T},$$

we have

$$\sum_{l=m}^{m+n} D_{n,m,l} h_l(t,\alpha) = \sum_{l=n}^{m+n} D_{m,n,l} h_l(t,\alpha), \quad t,\alpha \in \mathbb{T},$$

from which it follows that if $n \geq m$ we should have

$$D_{n,m,m} = D_{n,m,m+1} = \cdots = D_{n,m,n-1} = 0$$

and also

$$D_{n,m,l} = D_{m,n,l} \quad \text{for } l = n, n+1, \ldots, n+m. \qquad \square$$

We next give the computation of the constants $D_{n,m,l}$ for small values of the subscripts on arbitrary time scales. For larger values, the computation becomes long and complicated. Let $\mathbb{T}$ be a time scale with the forward jump operator σ, delta differentiation operator Δ, and the graininess function μ. Let $h_l(t,\alpha), t,\alpha \in \mathbb{T}, l = 0,1,2,3$ be the monomials on $\mathbb{T}$. We are using the following relations in the computations of the constants $D_{n,m,l}$:

$$\begin{aligned}
h_k^{\Delta}(t,\alpha) &= h_{k-1}(t,\alpha),\\
h_k^{\sigma}(t,\alpha) &= h_k(t,\alpha) + \mu(t)h_k^{\Delta}(t,\alpha) = h_k(t,\alpha) + \mu(t)h_{k-1}(t,\alpha),\\
h_k^{\Delta\sigma}(t,\alpha) &= h_{k-1}^{\sigma}(t,\alpha) = h_{k-1}(t,\alpha) + \mu(t)h_{k-2}(t,\alpha),\\
h_k^{\sigma\Delta}(t,\alpha) &= (1+\mu^{\Delta}(t))h_k^{\Delta\sigma}(t,\alpha),\\
h_k^{\sigma\sigma}(t,\alpha) &= h_k(t,\alpha) + \mu(t)h_k^{\Delta}(t,\alpha) + \mu^{\sigma}(t)[h_k^{\Delta}(t,\alpha) + \mu(t)h_k^{\Delta^2}(t,\alpha)]\\
&= h_k(t,\alpha) + (\mu(t)+\mu^{\sigma}(t))h_{k-1}(t,\alpha) + \mu^{\sigma}(t)\mu(t)h_{k-2}(t,\alpha),\\
h_k^{\Delta\Delta}(t,\alpha) &= h_{k-2}(t,\alpha),\\
h_k^{\sigma\sigma\sigma}(t,\alpha) &= (h_k(t,\alpha) + \mu(t)h_{k-1}(t,\alpha))^{\sigma\sigma}\\
&= h_k(t,\alpha)\\
&\quad + [\mu(t) + (\mu(t)+\mu^{\sigma}(t))(1+\mu^{\Delta}(t)) + \mu(t)\mu^{\sigma}(t)\mu^{\Delta\Delta}(t)]h_{k-1}(t,\alpha)\\
&\quad + [\mu^{\sigma}(t)(\mu(t)+\mu^{\sigma}(t)) + \mu(t)\mu^{\sigma}(t)(1+\mu^{\Delta\sigma}(t)+\mu^{\sigma\Delta}(t))]h_{k-2}(t,\alpha)\\
&\quad + \mu^{\sigma\sigma}(t)\mu^{\sigma}(t)\mu(t)h_{k-3}(t,\alpha), \quad t,\alpha \in \mathbb{T}.
\end{aligned}$$

Then we compute the following:

1. Let $l = 0$. Then from the property (7.6) we have

$$D_{0,0,0} = 1.$$

2. Let $l = 1$. Then from the property (7.6) we have

$$D_{1,0,1} = D_{0,1,1} = 1,$$

$$D_{1,1,1} = h_1^{\sigma}(\alpha,\alpha) = h_1(\alpha,\alpha) + \mu(\alpha)h_0(\alpha,\alpha) = \mu(\alpha), \quad \alpha \in \mathbb{T}.$$

3. Let $l = 2$. Then we have

$$\begin{aligned}
D_{2,0,2} &= D_{0,2,2} = 1,\\
D_{1,1,2} &= h_1^{\sigma\Delta}(\alpha,\alpha) + h_1^{\Delta\sigma}(\alpha,\alpha) = (2+\mu^{\Delta}(\alpha))h_1^{\Delta\sigma}(\alpha,\alpha)\\
&= (2+\mu^{\Delta}(\alpha))h_0^{\sigma}(\alpha,\alpha) = 2+\mu^{\Delta}(\alpha),\\
D_{2,1,2} &= D_{1,2,2} = h_1^{\sigma\sigma}(\alpha,\alpha)\\
&= h_1(\alpha,\alpha) + (\mu(\alpha)+\mu^{\sigma}(\alpha))h_0(\alpha,\alpha)\\
&= \mu(\alpha)+\mu^{\sigma}(\alpha), \quad \alpha \in \mathbb{T},
\end{aligned}$$

and

$$\begin{aligned}
D_{2,2,2} &= h_2^{\sigma\sigma}(\alpha,\alpha)\\
&= h_2(\alpha,\alpha) + (\mu(\alpha)+\mu^{\sigma}(\alpha))h_1(\alpha\alpha) + \mu(\alpha)\mu^{\sigma}(\alpha)h_0(\alpha,\alpha)\\
&= \mu(\alpha)\mu^{\sigma}(\alpha), \quad \alpha \in \mathbb{T}.
\end{aligned}$$

4. Let $l = 3$. Then we compute

$$\begin{aligned}
D_{3,0,3} &= D_{0,3,3} = 1,\\
D_{1,1,3} &= h_1^{\Delta\Delta\sigma}(\alpha,\alpha) + h_1^{\Delta\sigma\Delta}(\alpha,\alpha) + h_1^{\sigma\Delta\Delta}(\alpha,\alpha) = \mu^{\Delta\Delta}(\alpha),\\
D_{1,2,3} &= D_{2,1,3}(\alpha) = h_2^{\Delta\Delta\sigma}(\alpha,\alpha) + h_2^{\Delta\sigma\Delta}(\alpha,\alpha) + h_2^{\sigma\Delta\Delta}(\alpha,\alpha)\\
&= h_0^{\sigma}(\alpha,\alpha) + h_1^{\sigma\Delta}(\alpha,\alpha) + (h_2(\alpha,\alpha)+\mu(\alpha)h_1(\alpha,\alpha))^{\Delta\Delta}\\
&= 3+\mu^{\Delta}(\alpha)+\mu^{\Delta\sigma}(\alpha)+\mu^{\sigma\Delta}(\alpha),\\
D_{2,2,3} &= h_2^{\Delta\sigma\sigma}(\alpha,\alpha) + h_2^{\sigma\Delta\sigma}(\alpha,\alpha) + h_2^{\sigma\sigma\Delta}(\alpha,\alpha)\\
&= 2(\mu(\alpha)+\mu^{\sigma}(\alpha)) + (\mu(\alpha)+\mu^{\sigma}(\alpha))^{\sigma}\\
&\quad + (\mu(\alpha)\mu^{\sigma}(\alpha))^{\Delta} + \mu^{\Delta\sigma}(\alpha)(2+\mu^{\Delta}(\alpha)),\\
D_{3,1,3} &= D_{1,3,3} = h_1^{\sigma\sigma\sigma}(\alpha,\alpha)\\
&= h_1(\alpha,\alpha)\\
&\quad + [\mu(\alpha)+(\mu(\alpha)+\mu^{\sigma}(\alpha))(1+\mu^{\Delta}(\alpha))+\mu(\alpha)\mu^{\sigma}(\alpha)\mu^{\Delta\Delta}(\alpha)]h_0(\alpha,\alpha)\\
&= \mu(\alpha)+(\mu(\alpha)+\mu^{\sigma}(\alpha))(1+\mu^{\Delta}(\alpha))+\mu(\alpha)\mu^{\sigma}(\alpha)\mu^{\Delta\Delta}(\alpha),\\
D_{3,2,3} &= D_{2,3,3} = h_2^{\sigma\sigma\sigma}(\alpha,\alpha)\\
&= h_2(\alpha,\alpha)\\
&\quad + [\mu(\alpha)+(\mu(\alpha)+\mu^{\sigma}(\alpha))(1+\mu^{\Delta}(\alpha))+\mu(\alpha)\mu^{\sigma}(\alpha)\mu^{\Delta\Delta}(\alpha)]h_1(\alpha,\alpha)\\
&\quad + [\mu^{\sigma}(\alpha)(\mu(\alpha)+\mu^{\sigma}(\alpha))+\mu(\alpha)\mu^{\sigma}(\alpha)(1+\mu^{\Delta\sigma}(\alpha)+\mu^{\sigma\Delta}(\alpha))]h_0(\alpha,\alpha)\\
&= \mu^{\sigma}(\alpha)(\mu(\alpha)+\mu^{\sigma}(\alpha))+\mu(\alpha)\mu^{\sigma}(\alpha)(1+\mu^{\Delta\sigma}(\alpha)+\mu^{\sigma\Delta}(\alpha)),
\end{aligned}$$

$$\begin{aligned}D_{3,3,3} &= h_3^{\sigma\sigma\sigma}(\alpha,\alpha)\\ &= h_3(\alpha,\alpha)\\ &\quad + [\mu(\alpha) + (\mu(\alpha)+\mu^\sigma(\alpha))(1+\mu^\Delta(\alpha)) + \mu(\alpha)\mu^\sigma(\alpha)\mu^{\Delta\Delta}(\alpha)]h_2(\alpha,\alpha)\\ &\quad + [\mu^\sigma(\alpha)(\mu(\alpha)+\mu^\sigma(\alpha)) + \mu(\alpha)\mu^\sigma(\alpha)(1+\mu^{\Delta\sigma}(\alpha)+\mu^{\sigma\Delta}(\alpha))]h_1(\alpha,\alpha)\\ &\quad + \mu(\alpha)\mu^\sigma(\alpha)\mu^{\sigma\sigma}(\alpha)h_0(\alpha,\alpha)\\ &= \mu(\alpha)\mu^\sigma(\alpha)\mu^{\sigma\sigma}(\alpha), \quad \alpha\in\mathbb{T}.\end{aligned}$$

As seen above, the computation of the constants $D_{n,m,l}$ becomes hard for large values of the subscripts and depends solely on the forward jump operator and the graininess function of the time scale under consideration.

7.3 Series solution method for dynamic equations

In this section, we use the series solution method to solve linear dynamic equations. Let $\mathbb{T}$ be a time scale with forward jump operator and delta differentiation operator, σ and Δ, respectively. Suppose that the graininess function μ is differentiable.

Consider the dynamic equation

$$y^{\Delta^n}(t) + a_1(t)y^{\Delta^{n-1}}(t) + \cdots + a_n(t)y(t) = b(t), \quad t\in\mathbb{T}, \tag{7.8}$$

where

$$a_m(t) = \sum_{i=0}^{\infty} a_i^m h_i(t,\alpha), \quad b(t) = \sum_{i=0}^{\infty} b_i h_i(t,\alpha), \quad t,\alpha\in\mathbb{T},$$

and $a_i^m, b_i, i\in\mathbb{N}_0, m\in\{1,\dots,n\}$, are given constants.

We will search for a solution y of equation (7.8) in the form

$$y(t) = \sum_{i=0}^{\infty} c_i h_i(t,\alpha), \quad t,\alpha\in\mathbb{T}, \tag{7.9}$$

where c_i, $i\in\mathbb{N}_0$, are constants that will be determined below. We have

$$y^{\Delta^r}(t) = \sum_{i=0}^{\infty} c_i h_i^{\Delta^r}(t,\alpha) = \sum_{i=r}^{\infty} c_i h_{i-r}(t,\alpha) = \sum_{i=0}^{\infty} c_{i+r} h_i(t,\alpha), \quad t,\alpha\in\mathbb{T}, \quad r\in\mathbb{N}_0.$$

By Theorem 7.4, we obtain

$$\begin{aligned}
a_m(t)y^{\Delta^{n-m}}(t) &= \left(\sum_{i=0}^{\infty} a_i^m h_i(t,\alpha)\right)\left(\sum_{j=0}^{\infty} c_{j+n-m} h_j(t,\alpha)\right) \\
&= \sum_{k=0}^{\infty}\left(\sum_{l=0}^{k} a_l^m c_{k-l+n-m} h_l(t,\alpha) h_{k-l}(t,\alpha)\right) \\
&= \sum_{k=0}^{\infty}\left(\sum_{l=0}^{k} a_l^m c_{k-l+n-m}\left(\sum_{r=k-l}^{k}\left(\sum_{\Lambda_{r,k-l}\in S_{k-l}^{(r)}} h_l^{\Lambda_{r,k-l}}(\alpha,\alpha)\right) h_r(t,\alpha)\right)\right) \\
&= \sum_{k=0}^{\infty}\left(\sum_{l=0}^{k} a_l^m c_{k-l+n-m}\left(\sum_{r=k-l}^{k} D_{l,k-l,r} h_r(t,\alpha)\right)\right), \quad t,\alpha\in\mathbb{T}, \qquad (7.10)
\end{aligned}$$

where $D_{l,k-l,r}$ can be computed using (7.5) for $k = 0,1,\dots, l = 0,1,\dots,k$, and $r = k-l,\dots,k$.

Note that the double sum $\sum_{l=0}^{k}\sum_{r=k-l}^{k}$ can be reordered as $\sum_{r=0}^{k}\sum_{l=k-r}^{k}$. In addition, the sum $\sum_{k=0}^{\infty}\sum_{r=0}^{k}$ can be reordered as $\sum_{r=0}^{\infty}\sum_{k=r}^{\infty}$ using the Fubini theorem. Therefore, we rewrite the triple sum $\sum_{k=0}^{\infty}\sum_{l=0}^{k}\sum_{r=k-l}^{k}$ in (7.10) as $\sum_{r=0}^{\infty}\sum_{k=r}^{\infty}\sum_{l=k-r}^{k}$, which yields

$$a_m(t)y^{\Delta^{n-m}}(t) = \sum_{r=0}^{\infty}\left(\sum_{k=r}^{\infty}\sum_{l=k-r}^{k} a_l^m c_{k-l+n-m} D_{l,k-l,r}\right) h_r(t,\alpha), \quad t,\alpha\in\mathbb{T}. \qquad (7.11)$$

Then, equation (7.8) implies

$$\begin{aligned}
& y^{\Delta^n}(t) + \sum_{m=1}^{n} a_m(t) y^{\Delta^{n-m}}(t) = b(t) \\
\Longrightarrow \quad & \sum_{r=0}^{\infty} c_{r+n} h_r(t,\alpha) + \sum_{r=0}^{\infty}\left(\sum_{m=1}^{n}\sum_{k=r}^{\infty}\sum_{l=k-r}^{r} a_l^m c_{k-l+n-m} D_{l,k-l,r}\right) h_r(t,\alpha) = \sum_{r=0}^{\infty} b_r h_r(t,\alpha),
\end{aligned}$$

for $t,\alpha\in\mathbb{T}$. Hence, we deduce the following recurrence relation for the computation of the coefficients $c_j, j\geq 0$:

$$c_{r+n} = -\sum_{m=1}^{n}\sum_{k=r}^{\infty}\sum_{l=k-r}^{r} a_l^m c_{k-l+n-m} D_{l,k-l,r} + b_r. \qquad (7.12)$$

Solving this recurrence relation will lead to a complete determination of the coefficients $c_j, j\geq 0$. Having determined the coefficients $c_j, j\geq 0$, the series solution follows immediately upon substituting the derived coefficients into (7.9). The exact solution may be obtained if such an exact solution exists. If an exact solution does not exist, then the infinite series can be used for numerical purposes. In this case, the more terms we determine, the higher accuracy level we achieve.

Our first example is a dynamic equation with constant coefficients. It provides an opportunity for testing the series solution method presented above.

Example. Consider the second-order constant-coefficient dynamic equation

$$y^{\Delta\Delta} - 3y^{\Delta} + 2y = 0, \quad t \in \mathbb{T} = 2^{\mathbb{N}_0}. \tag{7.13}$$

Its characteristic equation is

$$\lambda^2 - 3\lambda + 2 = 0,$$

with the solutions $\lambda = 1, \lambda = 2$. Then the linearly independent solutions of the dynamic equation for $\alpha = 1$ are $y_1(t) = e_1(t,1), y_2(t) = e_2(t,1), t \in \mathbb{T}$, [1].

We will apply the series solution method to the equation (7.13). Take $\alpha = 1$ and propose a solution in the form

$$y(t) = \sum_{n=0}^{\infty} c_n h_n(t,1), \quad t \in \mathbb{T}.$$

Then we have

$$y^{\Delta}(t) = \sum_{n=0}^{\infty} c_{n+1} h_n(t,1), \quad y^{\Delta\Delta}(t) = \sum_{n=0}^{\infty} c_{n+2} h_n(t,1), \quad t \in \mathbb{T}.$$

Put $y(t), y^{\Delta}(t)$, and $y^{\Delta\Delta}(t)$ into the equation which gives

$$\sum_{n=0}^{\infty} c_{n+2} h_n(t,1) - 3\sum_{n=0}^{\infty} c_{n+1} h_n(t,1) + 2\sum_{n=0}^{\infty} c_n h_n(t,1) = 0, \quad t \in \mathbb{T},$$

and upon combining the series we get

$$\sum_{n=0}^{\infty} (c_{n+2} - 3c_{n+1} + 2c_n) h_n(t,1) = 0, \quad t \in \mathbb{T}.$$

Then we have the following recurrence relation:

$$c_{n+2} = 3c_{n+1} - 2c_n, \quad \text{for } n \geq 0.$$

We compute first few terms, for arbitrary c_0 and c_1, as follows:

$$\begin{aligned}
n &= 0, \quad c_2 = 3c_1 - 2c_0, \\
n &= 1, \quad c_3 = 3c_2 - 2c_1 = 7c_1 - 6c_0, \\
n &= 2, \quad c_4 = 3c_3 - 2c_2 = 15c_1 - 14c_0, \\
n &= 3, \quad c_5 = 3c_4 - 2c_3 = 31c_1 - 30c_0, \\
n &= 4, \quad c_6 = 3c_5 - 2c_4 = 63c_1 - 62c_0,
\end{aligned}$$

so that the recurrence relation can be generalized as

$$c_n = (2^n - 1)c_1 - (2^n - 2)c_0,$$

for arbitrary c_0, c_1 and $n \geq 2$. Then the solution $y(t)$ becomes

$$\begin{aligned} y(t) &= \sum_{n=0}^{\infty} c_n h_n(t,1) \\ &= c_0 h_0(t,1) + c_1 h_1(t,1) + \sum_{n=2}^{\infty} c_n h_n(t,1) \\ &= c_0 h_0(t,1) + c_1 h_1(t,1) + \sum_{n=2}^{\infty} [(2^n - 1)c_1 - (2^n - 2)c_0] h_n(t,1), \quad t \in \mathbb{T} \end{aligned}$$

and can be arranged as

$$y(t) = (c_1 - c_0) \sum_{n=0}^{\infty} 2^n h_n(t,1) + (2c_0 - c_1) \sum_{n=0}^{\infty} h_n(t,1), \quad t \in \mathbb{T}.$$

Recalling that $e_\lambda(t,1) = \sum_{n=0}^{\infty} \lambda^n h_n(t,1)$, $t \in \mathbb{T}$, we conclude

$$y(t) = d_1 e_2(t,1) + d_2 e_1(t,1), \quad t \in \mathbb{T}, \quad d_1 = c_1 - c_0, \quad d_2 = 2c_0 - c_1,$$

which is the general solution of the dynamic equation.

We now apply the series solution method to a dynamic equation with nonconstant coefficients.

Example. Consider the second-order dynamic equation

$$[1 - (\sigma(t))^2] y^{\Delta\Delta}(t) - (\sigma(t) + t) y^{\Delta}(t) + k(k+1) y(t) = 0, \quad t \in \mathbb{T}, \tag{7.14}$$

on some time scale $\mathbb{T}$ with differentiable graininess function, where k is a real constant.

We assume that

$$\begin{aligned} 1 - (\sigma(t))^2 &= \sum_{m=0}^{\infty} a_m h_m(t,\alpha), \\ -(\sigma(t) + t) &= \sum_{m=0}^{\infty} b_m h_m(t,\alpha), \end{aligned} \tag{7.15}$$

for some $t, \alpha \in \mathbb{T}$ and propose the following series expansion for the solution y:

$$y(t) = \sum_{n=0}^{\infty} c_n h_n(t,\alpha). \tag{7.16}$$

As we did in the previous example, we substitute the series (7.15) and (7.16) into the dynamic equation (7.14) and, using the Cauchy product for infinite series, we get

$$\sum_{n=0}^{\infty}\sum_{m=0}^{n}(a_m c_{n-m+2}+b_m c_{n-m+1})h_m(t,a)h_{n-m}(t,a)+k(k+1)\sum_{n=0}^{\infty}c_n h_n(t,a)=0,\quad t,a\in\mathbb{T}.$$

Note that the product $h_m(t,a)h_{n-m}(t,a)$, $t,a\in\mathbb{T}$ involved in the last relation can be written as

$$h_m(t,a)h_{n-m}(t,a)=\sum_{l=n-m}^{n}D_{m,n-m,l}h_l(t,a),\quad t,a\in\mathbb{T},\tag{7.17}$$

for all $n=0,1,\ldots,$ and $m=0,1,\ldots,n$ where the constants $D_{m,n-m,l}$ are defined in (7.5). Employing (7.17) and (7.5), we obtain

$$\begin{aligned}\sum_{n=0}^{\infty}\sum_{m=0}^{n}\sum_{l=n-m}^{n}&[(a_m c_{n-m+2}+b_m c_{n-m+1})D_{m,n-m,l}]h_l(t,a)\\&+k(k+1)\sum_{n=0}^{\infty}c_n h_n(t,a)=0,\quad t,a\in\mathbb{T}.\end{aligned}$$

Upon reordering the triple sum as $\sum_{l=0}^{\infty}\sum_{n=l}^{\infty}\sum_{m=n-l}^{n}$, we rewrite the above equation as

$$\sum_{l=0}^{\infty}\left[\left(\sum_{n=l}^{\infty}\sum_{m=n-l}^{n}(a_m c_{n-m+2}+b_m c_{n-m+1})D_{m,n-m,l}\right)+k(k+1)c_l\right]h_l(t,a)=0,\quad t,a\in\mathbb{T},$$

which implies the following recurrence relation for the computation of the coefficients c_l, $l=0,1,2,\ldots$:

$$\sum_{n=l}^{\infty}\sum_{m=n-l}^{n}((a_m c_{n-m+2}+b_m c_{n-m+1})D_{m,n-m,l})+k(k+1)c_l=0,\quad l=0,1,2,\ldots\tag{7.18}$$

For computational purposes, we consider the particular case of $\mathbb{T}=s\mathbb{N}_0$, $s>0$ and choose $a=0$. Then we have

$$\sigma(t)=t+s,\quad \mu(t)=s,\quad t\in\mathbb{T},$$

and

$$h_0(t,0)=1,\quad h_1(t,0)=t,\quad h_2(t,0)=\frac{t(t-s)}{2},\quad t\in\mathbb{T}.$$

The dynamic equation (7.14) can be written as

$$(1-s^2+2st-t^2)y^{\Delta\Delta}-(2t+s)y^{\Delta}+k(k+1)y=0,\quad t\in\mathbb{T},\tag{7.19}$$

where

$$\begin{aligned}1-(\sigma(t))^2&=1-s^2+2st-t^2=(1-s^2)h_0(t,0)+sh_1(t,0)-2h_2(t,0),\\-t-\sigma(t)&=-2t-s=-sh_0(t,0)-2h_1(t,0),\quad t\in\mathbb{T},\end{aligned}\tag{7.20}$$

that is, $a_0 = 1 - s^2$, $a_1 = s$, $a_2 = -2$, $b_0 = -s$, and $b_1 = -2$. For arbitrary c_0 and c_1, the recurrence relation (7.18) becomes

$$\sum_{n=l}^{l+2} \sum_{m=n-l}^{2} [(a_m c_{n-m+2} + b_m c_{n-m+1}) D_{m,n-m,l}] + k(k+1)c_l = 0, \quad l = 0, 1, 2, \ldots, \tag{7.21}$$

which yields

$$\begin{aligned}
&(a_0 c_2 + b_0 c_1) D_{0,0,0} + k(k+1)c_0 = 0, \quad \text{for } l = 0,\\
&(a_0 c_3 + b_0 c_2) D_{0,1,1} + (a_1 c_2 + b_1 c_1) D_{1,0,1}\\
&\quad + (a_1 c_3 + b_1 c_2) D_{1,1,1} + k(k+1)c_1 = 0, \quad \text{for } l = 1,
\end{aligned}$$

and

$$\begin{aligned}
&(a_0 c_{l+2} + b_0 c_{l+1}) D_{0,l,l} + (a_1 c_{l+1} + b_1 c_l) D_{1,l-1,l} + a_2 c_l D_{2,l-2,l}\\
&\quad + (a_1 c_{l+2} + b_1 c_{l+1}) D_{1,l,l} + a_2 c_{l+1} D_{2,l-1,l} + a_2 c_{l+2} D_{2,l,l}\\
&\quad + k(k+1)c_l = 0,
\end{aligned}$$

for $l \geq 2$. Therefore, for $s \neq 1$ we compute

$$\begin{aligned}
c_2 &= \frac{s}{1-s^2} c_1 - \frac{k(k+1)}{1-s^2} c_0,\\
c_3 &= 2sc_2 - (k(k+1) - 2)c_1,
\end{aligned}$$

and

$$\begin{aligned}
c_{l+2} = &- \frac{sD_{1,l-1,l} - 2D_{2,l-1,l} - sD_{0,l,l} - 2D_{1,l,l}}{(1-s^2)D_{0,l,l} + sD_{1,l,l} - 2D_{2,l,l}} c_{l+1}\\
&- \frac{k(k+1) - 2D_{2,l-2,l} - 2D_{1,l-1,l}}{(1-s^2)D_{0,l,l} + sD_{1,l,l} - 2D_{2,l,l}} c_l
\end{aligned}$$

for $l \geq 2$.

On the other hand, if $s = 1$, for arbitrary c_0 and c_2 we have

$$\begin{aligned}
c_1 &= -k(k+1)c_0,\\
c_3 &= 2c_2 - (k(k+1) - 2)c_1,
\end{aligned}$$

and for $l = 2, 3, 4, \ldots$ we obtain

$$\begin{aligned}
c_{l+2} = &- \frac{D_{1,l-1,l} - 2D_{2,l-1,l} - D_{0,l,l} - 2D_{1,l,l}}{D_{1,l,l} - 2D_{2,l,l}} c_{l+1}\\
&- \frac{k(k+1) - 2D_{2,l-2,l} - 2D_{1,l-1,l}}{D_{1,l,l} - 2D_{2,l,l}} c_l.
\end{aligned}$$

Clearly, a generalization of the recurrence relation is not possible for this example. On the other hand, one can get a good approximate solution by computing many terms of the sequence of coefficients.

7.4 Volterra integral equations of the first kind

In this section we develop the series solution method for the nonlinear Volterra integral equation of the first kind

$$u(x) = \int_a^x K(x, t, \sigma(x), \sigma(t)) F(\varphi(t)) \Delta t, \quad x \in \mathbb{T}, \tag{7.22}$$

where $K : \mathbb{T} \times \mathbb{T} \times \mathbb{T} \times \mathbb{T} \to \mathbb{R}$, $u : \mathbb{T} \to \mathbb{R}$, and $F : \mathbb{R} \to \mathbb{R}$ are given functions and $\varphi : \mathbb{T} \to \mathbb{R}$ is unknown.

We suppose that the nonlinear function F is a quadratic function in φ, that is,

$$F(\varphi(x)) = \alpha(\varphi(x))^2 + \beta\varphi(x) + \gamma, \quad x \in \mathbb{T}.$$

Assume that the unknown function $\varphi(x)$ has a Taylor series expansion about $x = a$ of the form

$$\varphi(x) = \sum_{n=0}^{\infty} f_n h_n(x, a), \quad x \in \mathbb{T}, \tag{7.23}$$

where the coefficients f_n will be determined from the equation. Then we have

$$(\varphi(x))^2 = \sum_{n=0}^{\infty} \sum_{m=0}^{\infty} f_n f_m h_n(x, a) h_m(x, a), \quad x \in \mathbb{T}. \tag{7.24}$$

By Theorem 7.4, the double series in (7.24) can be written as

$$(\varphi(x))^2 = \sum_{r=0}^{\infty} g_r h_r(x, a), \quad x \in \mathbb{T},$$

where

$$g_r = \sum_{k=r}^{\infty} \sum_{l=k-r}^{k} f_l f_{k-l} D_{r,k,l},$$

and

$$D_{r,k,l} = \sum_{\Lambda_{r,k-l} \in S_{k-l}^{(r)}} h_l^{\Lambda_{r,k-l}}(a, a), \quad a \in \mathbb{T}.$$

Note that in the above relation, the definition of the constants $D_{r,k,l}$ is slightly different from the definition given in (7.5), however, this is only a notational difference which does not affect the properties and computation of the constants. Then we get

$$\begin{aligned} F(\varphi(x)) &= \alpha(\varphi(x))^2 + \beta\varphi(x) + \gamma h_0(x,a) \\ &= \sum_{r=0}^{\infty} (\alpha g_r + \beta f_r + \gamma\delta_{r0}) h_r(x,a) \\ &= \sum_{r=0}^{\infty} Q_r h_r(x,a), \quad x \in \mathbb{T}, \end{aligned} \tag{7.25}$$

where $Q_r = \alpha g_r + \beta f_r + \gamma\delta_{r0}$ for $r \in \mathbb{N}_0$, and δ_{r0} is the Kronecker delta.

We also assume that the kernel $K(x,t,\sigma(x),\sigma(t))$, $x,t \in \mathbb{T}$, has a series representation of the form

$$K(x,t,\sigma(x),\sigma(t)) = \sum_{m=0}^{\infty} a_m h_m(x,a) \sum_{i=0}^{\infty} b_i h_i(t,a), \quad x,t \in \mathbb{T}, \tag{7.26}$$

where a_m, $m \in \mathbb{N}_0$, and b_i, $i \in \mathbb{N}_0$, are given constants.

Finally, we suppose that the Taylor series of the function $u(x)$ is also known to be

$$u(x) = \sum_{n=0}^{\infty} u_n h_n(x,a), \quad x \in \mathbb{T}. \tag{7.27}$$

If we substitute the series in (7.25), (7.26), and (7.27) into the nonlinear integral equation (7.22), we obtain

$$\sum_{n=0}^{\infty} u_n h_n(x,a) = \int_a^x \sum_{m=0}^{\infty} a_m h_m(x,a) \sum_{i=0}^{\infty} b_i h_i(t,a) \sum_{r=0}^{\infty} Q_r h_r(t,a)\Delta t, \quad x \in \mathbb{T}. \tag{7.28}$$

Now, we take the x-dependent series out of the integral sign and use Theorem 7.4 to write the product of the two series inside the integral as a single one, which gives

$$\sum_{n=0}^{\infty} u_n h_n(x,a) = \sum_{m=0}^{\infty} a_m h_m(x,a) \int_a^x \sum_{p=0}^{\infty} y_p h_p(t,a)\Delta t, \quad x \in \mathbb{T},$$

and upon integrating,

$$\sum_{n=0}^{\infty} u_n h_n(x,a) = \sum_{m=0}^{\infty} a_m h_m(x,a) \sum_{p=0}^{\infty} y_p h_{p+1}(x,a), \quad x \in \mathbb{T}, \tag{7.29}$$

where

$$y_p = \sum_{k=p}^{\infty} \sum_{l=k-p}^{k} b_l Q_{k-l} D_{p,k,l}.$$

Finally, we employ again Theorem 7.4 to write the product on the right-hand side as a single series. This yields

$$\sum_{n=0}^{\infty} u_n h_n(x,a) = \sum_{n=1}^{\infty} w_n h_n(x,a), \quad x \in \mathbb{T}, \tag{7.30}$$

where

$$w_n = \sum_{k=n-1}^{\infty} \sum_{r=k-n+1}^{k+1} a_r y_{k-r} D_{n,k+1,r-1} \quad \text{and we set} \quad y_{-1} \equiv 0.$$

We equate the coefficients of $h_n(x,a)$, $x \in \mathbb{T}$, of both sides. This will result in a nonlinear recurrence relation for the computation of the coefficients f_n of the Taylor series of the unknown function φ.

We present a specific example to illustrate the method.

Example. Consider the equation

$$u(x) = \int_0^x h_2(x,0)(\phi(t))^2 \Delta t, \quad x \in \mathbb{T}, \tag{7.31}$$

on any time scale $\mathbb{T}$ containing 0. We will search for a solution of equation (7.31) in the form

$$\phi(x) = \sum_{i=0}^{\infty} f_i h_i(x,0), \quad x \in \mathbb{T},$$

where f_i are constants to be determined from the equation. We assume that the function $u(x)$ has the form

$$u(x) = \sum_{n=1}^{\infty} u_n h_n(x,0), \quad x \in \mathbb{T}.$$

Observe that

$$(\phi(x))^2 = \sum_{r=0}^{\infty} g_r h_r(x,0), \quad x \in \mathbb{T},$$

where

$$g_r = \sum_{k=r}^{\infty} \sum_{l=k-r}^{k} f_l f_{k-l} D_{r,k,l}.$$

Then, equation (7.31) can be written as

$$\sum_{n=1}^{\infty} u_n h_n(x,0) = h_2(x,0) \int_0^x \sum_{r=0}^{\infty} g_r h_r(t,0)\Delta t, \quad x \in \mathbb{T},$$

and upon integration,

$$\sum_{n=1}^{\infty} u_n h_n(x,0) = \sum_{r=0}^{\infty} g_r h_2(x,0) h_{r+1}(x,0), \quad x \in \mathbb{T}. \tag{7.32}$$

We notice that according to Theorems 7.3 and 7.4,

$$h_2(x,0)h_{r+1}(x,0) = \sum_{n=r+1}^{r+3} D_{n,r+3,2} h_n(x,0), \quad x \in \mathbb{T}.$$

Hence, we have

$$\begin{aligned}
\sum_{r=0}^{\infty} g_r h_2(x,0)h_{r+1}(x,0) &= \sum_{r=0}^{\infty} \sum_{n=r+1}^{r+3} g_r D_{n,r+3,2} h_n(x,0) \\
&= \sum_{r=0}^{\infty} g_r D_{r+1,r+3,2} h_{r+1}(x,0) + \sum_{r=0}^{\infty} g_r D_{r+2,r+3,2} h_{r+2}(x,0) \\
&\quad + \sum_{r=0}^{\infty} g_r D_{r+3,r+3,2} h_{r+3}(x,0) \\
&= \sum_{r=1}^{\infty} g_{r-1} D_{r,r+2,2} h_r(x,0) + \sum_{r=2}^{\infty} g_{r-2} D_{r,r+1,2} h_r(x,0) \\
&\quad + \sum_{r=3}^{\infty} g_{r-3} D_{r,r,2} h_r(x,0) \\
&= g_0 D_{1,3,2} h_1(x,0) + (g_1 D_{2,4,2} + g_0 D_{2,3,2}) h_2(x,0) \\
&\quad + \sum_{r=3}^{\infty} (g_{r-1} D_{r,r+2,2} + g_{r-2} D_{r,r+1,2} + g_{r-3} D_{r,r,2}) h_r(x,0), \quad x \in \mathbb{T}.
\end{aligned}$$

Thus, equation (7.32) yields

$$\begin{aligned}
\sum_{n=1}^{\infty} u_n h_n(x,0) &= g_0 D_{1,3,2} h_1(x,0) + (g_1 D_{2,4,2} + g_0 D_{2,3,2}) h_2(x,0) \\
&\quad + \sum_{r=3}^{\infty} (g_{r-1} D_{r,r+2,2} + g_{r-2} D_{r,r+1,2} + g_{r-3} D_{r,r,2}) h_r(x,0), \quad x \in \mathbb{T},
\end{aligned} \tag{7.33}$$

and we obtain

$$u_1 = g_0 D_{1,3,2},$$

$$u_2 = g_0 D_{2,3,2} + g_1 D_{2,4,2},$$
$$u_r = (g_{r-1} D_{r,r+2,2} + g_{r-2} D_{r,r+1,2} + g_{r-3} D_{r,r,2}), \quad r \geq 3.$$

7.5 Volterra integral equations of the second kind

In this section we consider the Volterra integral equation of the second kind

$$\varphi(x) = u(x) + \int_a^x K(x,t,\sigma(x),\sigma(t))F(\varphi(t))\Delta t, \quad x \in \mathbb{T}, \tag{7.34}$$

on an arbitrary time scale, where $K : \mathbb{T} \times \mathbb{T} \times \mathbb{T} \times \mathbb{T} \to \mathbb{R}$, $u : \mathbb{T} \to \mathbb{R}$, and $F : \mathbb{R} \to \mathbb{R}$ are given functions and $\varphi : \mathbb{T} \to \mathbb{R}$ is unknown.

We make the same assumption about the nonlinear function F inside the integral, that is, let

$$F(\varphi(x)) = \alpha\varphi^2(x) + \beta\varphi(x) + \gamma, \quad x \in \mathbb{T}.$$

As in the case of Volterra integral equations of the first kind, we assume that the unknown function φ has a Taylor series expansion about $x = a$ of the form

$$\varphi(x) = \sum_{n=0}^{\infty} f_n h_n(x,a), \quad x \in \mathbb{T}, \tag{7.35}$$

where the coefficients f_n will be determined from the equation.

We assume that the kernel $K(x,t,\sigma(x),\sigma(t))$, $x \in \mathbb{T}$, and the given function u have the Taylor series representations given in (7.26) and (7.27). We substitute these series and (7.25) into the nonlinear integral equation (7.34), which results in

$$\sum_{n=0}^{\infty} f_n h_n(x,a) = \sum_{n=0}^{\infty} u_n h_n(x,a) + \int_a^x \sum_{m=0}^{\infty} a_m h_m(x,a) \sum_{i=0}^{\infty} b_i h_i(t,a) \sum_{r=0}^{\infty} Q_r h_r(t,a)\Delta t, \quad x,a \in \mathbb{T}. \tag{7.36}$$

Now, we arrange the equation as

$$\sum_{n=0}^{\infty} f_n h_n(x,a) = \sum_{n=0}^{\infty} u_n h_n(x,a) + \sum_{m=0}^{\infty} a_m h_m(x,a) \int_a^x \sum_{p=0}^{\infty} y_p h_p(t,a)\Delta t, \quad x,a \in \mathbb{T},$$

and upon integrating,

$$\sum_{n=0}^{\infty} f_n h_n(x,a) = \sum_{n=0}^{\infty} u_n h_n(x,a) + \sum_{m=0}^{\infty} a_m h_m(x,a) \sum_{p=0}^{\infty} y_p h_{p+1}(x,a), \quad x,a \in \mathbb{T}, \tag{7.37}$$

where

$$y_p = \sum_{k=p}^{\infty} \sum_{l=k-p}^{k} b_l Q_{k-l} D_{p,k,l}.$$

Finally, we combine the series on the right-hand side as a single one to get

$$\sum_{n=0}^{\infty} f_n h_n(x,a) = \sum_{n=0}^{\infty} [u_n + w_n] h_n(x,a), \quad x, a \in \mathbb{T}, \tag{7.38}$$

where

$$w_n = \sum_{k=n-1}^{\infty} \sum_{r=k-n+1}^{k} a_r y_{k-r} D_{n,k,r-1}.$$

We equate the coefficients of $h_n(x,a)$, $x, a \in \mathbb{T}$ of both sides. This will result in a nonlinear recurrence relation for the computation of the coefficients f_n of the Taylor series of the unknown function φ.

Remark 7.1. Notice that the series solution method simplifies considerably when applied to linear integral equations. In this case, the recurrence relation and the constants involved in this relation have a simpler form and their computation is easier.

Bibliography

[1] M. Bohner and A. Peterson, Dynamic Equations on Time Scales: An Introduction with Applications, Birkhäuser, Boston, 2003.

[2] M. Bohner and R. P. Agarwal, Basic Calculus on Time Scales and some of its Applications, *Results Math.*, 35 (1999), 3–22.

[3] S. Georgiev, Integral Equations on Time Scales, Atlantis Press, 2016.

[4] S. Georgiev, Fractional Dynamic Calculus and Fractional Dynamic Equations on Time Scales, Springer, 2017.

[5] S. Georgiev and İ. M. Erhan, Series Solution Method for Cauchy Problems with Fractional Δ-derivative on Time Scales, *Fract. Differ. Calc.*, 9 (2019), 243–261.

[6] S. Georgiev and İ. M. Erhan, Nonlinear Integral Equations on Time Scales, Nova Science Publ., 2019.

[7] D. Mozyrska and E. Pawluszewicz, Hermite's Equations on Time Scales, *Appl. Math. Lett.*, 22 (2009), 1217–1219.

[8] H.-K. Liu, Developing a Series Solution Method of q-difference Equation, *J. Appl. Math.*, 2013 (2013), 743973.

[9] B. D. Haile and L. M. Hall, Polynomial and Series Solutions of Dynamic Equations on Time Scales, *Dyn. Syst. Appl.*, 12 (2003), 149–157. https://web.mst.edu/~bohner/bimadsa/hall.pdf

Zeynep Kayar, Billur Kaymakçalan, and Neslihan Nesliye Pelen

8 Generalized diamond alpha Bennett–Leindler-type dynamic inequalities

Abstract: The dual results; delta and nabla inequalities and their special cases; continuous and discrete inequalities are unified into diamond alpha case and new forms of such results as well as new diamond alpha Bennett–Leindler-type dynamic inequalities are established by developing a novel method, which does not require the Integration by Parts Formula and the Fundamental Theorem of Calculus. These theorems are standard arguments in the proofs of Bennett–Leindler-type dynamic inequalities in the delta and nabla approaches but do not follow naturally in the diamond alpha calculus.

8.1 Introduction

Since Hardy's inequality is one of those inequalities which turns information about derivatives of functions into information about the size of the function, it is essential part of all areas of mathematics and useful in various applications.

In 1920, when Hardy tried to find a simple and elementary proof of Hilbert's inequality [39],

$$\sum_{n=1}^{\infty}\sum_{m=1}^{\infty}\frac{a_m c_n}{m+n}\leq \pi\left(\sum_{m=1}^{\infty}a_m^2\right)^{1/2}\left(\sum_{n=1}^{\infty}c_n^2\right)^{1/2},$$

where $a_m, c_n \geq 0$ and $\sum_{m=1}^{\infty} a_m^2$ and $\sum_{n=1}^{\infty} c_n^2$ are convergent, he showed the following pioneering discrete inequality [24]:

$$\sum_{j=1}^{\infty}\left(\frac{1}{j}\sum_{i=1}^{m}c(i)\right)^{\zeta}\leq\left(\frac{\zeta}{\zeta-1}\right)^{\zeta}\sum_{j=1}^{\infty}c^{\zeta}(j),\quad c(j)\geq 0,\zeta>1, \tag{8.1}$$

and a pioneering continuous inequality [24] for a nonnegative function ϕ and for a real constant $\zeta > 1$ stated as follows:

$$\int_0^{\infty}\left(\frac{1}{t}\int_0^{t}\phi(s)ds\right)^{\zeta}dt\leq\left(\frac{\zeta}{\zeta-1}\right)^{\zeta}\int_0^{\infty}\phi^{\zeta}(t)dt, \tag{8.2}$$

Zeynep Kayar, Department of Mathematics, Van Yüzüncü Yıl University, Van, Turkey, e-mail: zeynepkayar@yyu.edu.tr
Billur Kaymakçalan, Department of Mathematics, Çankaya University, 06810 Ankara, Turkey, e-mail: billurkaymakcalan@gmail.com
Neslihan Nesliye Pelen, Department of Mathematics, Ondokuz Mayıs University, Samsun, Turkey, e-mail: nesliyeaykir@gmail.com

https://doi.org/10.1515/9783111182971-008

where $\int_0^\infty \phi^\zeta(t)dt < \infty$. In fact, Hardy only stated inequality (8.2) in [24] but did not prove it. In 1925, the proof of inequality (8.2), which depends on the calculus of variations, was shown by Hardy in [25].

The constant $(\frac{\zeta}{\zeta-1})^\zeta$ that is seen in the above inequalities also has been found as the best possible constant, because if we replace it by a smaller constant, then inequalities (8.1) and (8.2) are no longer satisfied for the involved sequences and functions, respectively.

Later in [27, Theorem 330], Hardy et al. generalized inequality (8.2) to

$$\int_0^\infty \frac{\Phi^\zeta(t)}{t^\theta}dt \le \left|\frac{\zeta}{\theta-1}\right|^\zeta \int_0^\infty \frac{\phi^\zeta(t)}{t^{\theta-\zeta}}dt, \quad \zeta > 1, \tag{8.3}$$

where ϕ is a nonnegative function and $\Phi(t) = \int_0^t \phi(s)ds$, if $\theta > 1$ and $\Phi(t) = \int_t^\infty \phi(s)ds$, if $\theta < 1$.

The discrete and continuous Hardy inequalities have been improved and generalized in many directions and used in many applications; see the books [9, 27, 39, 40, 44] and the references therein.

The investigation of the reverse Hardy–Copson inequalities, which are called Bennett–Leindler inequalities, started almost at the same time with the original inequalities.

The first reverse discrete Hardy–Copson inequalities were obtained by Hardy and Littlewood [26] in 1927 for $0 < \zeta < 1$ without finding the best possible constants. Then Copson [16], Bennett [10], and Leindler [41] established discrete Bennett–Leindler inequalities by means of the following: Assume that the sequences z and h are nonnegative. If $0 < \zeta < 1$, then

$$\sum_{m=1}^\infty \frac{z(m)}{[\overline{G}(m)]^\theta}\left(\sum_{j=m}^\infty h(j)z(j)\right)^\zeta \ge \zeta^\zeta \sum_{m=1}^\infty \frac{z(m)h^\zeta(m)}{[\overline{G}(m)]^{\theta-\zeta}}, \quad 0 \le \theta < 1,$$

where $\overline{G}(m) = \sum_{j=1}^m z(j)$ and

$$\sum_{m=1}^\infty \frac{z(m)}{[\overline{G}(m)]^\theta}\left(\sum_{j=m}^\infty h(j)z(j)\right)^\zeta \ge \left(\frac{\zeta}{1-\theta}\right)^\zeta \sum_{m=1}^\infty \frac{z(m)h^\zeta(m)}{[\overline{G}(m)]^{\theta-\zeta}}, \quad \theta < 0, \tag{8.4}$$

and for $0 < L \le \frac{z(m)}{z(m+1)}$,

$$\sum_{m=1}^\infty \frac{z(m)}{[\overline{G}(m)]^\theta}\left(\sum_{j=1}^m h(j)z(j)\right)^\zeta \ge \left(\frac{L\zeta}{\theta-1}\right)^\zeta \sum_{m=1}^\infty \frac{z(m)h^\zeta(m)}{[\overline{G}(m)]^{\theta-\zeta}}, \quad \theta > 1. \tag{8.5}$$

There are some results in [42] about the reverse discrete Hardy–Copson inequalities different from those above and in [20] about finding conditions on the sequence $z(m)$ for $0 < \zeta < 1$ to obtain the best possible constant.

The first continuous Bennett–Leindler inequality, which is the reverse version of the continuous Hardy–Copson inequality (8.3), when $\theta = \zeta$, was established in [27, Theorem 337] for $0 < \zeta < 1$ and for $\overline{H}(t) = \int_t^\infty h(s)ds$ as

$$\int_0^\infty \frac{\overline{H}^\zeta(t)}{t^\zeta} dt \geq \left(\frac{\zeta}{1-\zeta}\right)^\zeta \int_0^\infty h^\zeta(t)dt, \quad h(t) \geq 0.$$

Later Copson derived continuous analogues of the discrete Bennett–Leindler inequalities (8.4) and (8.5), which are called continuous Bennett–Leindler inequalities, in [17, Theorems 4 and 2], respectively, for $z(t) \geq 0$ and $h(t) \geq 0$, and $\overline{G}(t) = \int_0^t z(s)ds, H(t) = \int_0^t z(s)h(s)ds, \overline{H}(t) = \int_t^\infty z(s)h(s)ds$ in the following manner: If $0 < \zeta \leq 1, \theta < 1$ then

$$\int_0^b \frac{z(t)}{[\overline{G}(t)]^\theta} [\overline{H}(t)]^\zeta dt \geq \left(\frac{\zeta}{1-\theta}\right)^\zeta \int_0^b z(t)[\overline{G}(t)]^{\zeta-\theta} h^\zeta(t)dt, \quad 0 < b \leq \infty.$$

If $0 < \zeta \leq 1 < \theta, a > 0$, then

$$\int_a^\infty \frac{z(t)}{[\overline{G}(t)]^\theta} [H(t)]^\zeta dt \geq \left(\frac{\zeta}{\theta-1}\right)^\zeta \int_a^\infty z(t)[\overline{G}(t)]^{\zeta-\theta} h^\zeta(t)dt.$$

Following the development of the time scale concept [8, 14, 15, 21, 22], the analysis of dynamic inequalities has become a popular research area and most classical inequalities have been extended to an arbitrary time scale. The surveys [1, 51] and the monograph [3] can be used to see these extended dynamic inequalities for the delta approach. Although the nabla dynamic inequalities are less attractive compared to the delta ones, some of the nabla dynamic inequalities can be found in [6, 12, 23, 47, 48].

The growing interest to Hardy–Copson type inequalities take place in the time scale calculus as well, and delta unifications of these inequalities are established in the book [4] and in the articles [2, 19, 49, 52, 54–58], whereas their nabla counterparts and extensions can be seen in [30–32].

In the delta time scale calculus, the reverse Hardy–Copson type inequalities, which are called delta Bennett–Leindler inequalities, can be found in [18, 53, 58, 59] for $0 < \zeta < 1$. These results are unifications of discrete and continuous Bennett–Leindler inequalities mentioned above. In addition to the delta calculus, the above discrete and continuous Bennett–Leindler inequalities can be unified in the nabla time scale calculus for $0 < \zeta < 1$, see [29, 37]. Some delta and nabla Bennett–Leindler-type inequalities can be given as below.

Let us define the following functions and the constant in the delta calculus:

$$\begin{aligned} &G(t) = \int_t^\infty z(s)\Delta s, \quad H(t) = \int_a^t z(s)h(s)\Delta s, \\ &\overline{G}(t) = \int_a^t z(s)\Delta s, \quad \overline{H}(t) = \int_t^\infty z(s)h(s)\Delta s, \\ &\exists M_1 > 0 \quad \text{such that } \frac{G(t)}{G^\sigma(t)} \leq M_1 \text{ for } t \in (a,\infty)_\mathbb{T}, \end{aligned} \tag{8.6}$$

where $z, h \geq 0$ are rd-continuous, Δ-differentiable, and locally delta integrable functions, and $0 < a \in \mathbb{T}$.

Theorem 8.1 ([34, 37, 59]). *For the functions $G(t)$ and $H(t)$ and the constant M_1 defined in* (8.6),

(i) ([34, 59]) *if $\theta \leq 0 < \zeta < 1$, then we have*

$$\int_a^\infty \frac{z(t)[H^\sigma(t)]^\zeta}{[G(t)]^\theta}\Delta t \geq \frac{\zeta}{1-\theta}\int_a^\infty \frac{z(t)h(t)[H^\sigma(t)]^{\zeta-1}}{[G(t)]^{\theta-1}}\Delta t, \tag{8.7}$$

$$\int_a^\infty \frac{z(t)[H^\sigma(t)]^\zeta}{[G(t)]^\theta}\Delta t \geq \left[\frac{\zeta}{1-\theta}\right]^\zeta \int_a^\infty \frac{z(t)h^\zeta(t)}{[G(t)]^{\theta-\zeta}}\Delta t. \tag{8.8}$$

(ii) ([34]) *if $0 < \zeta < 1 < \theta$, then we have*

$$\int_a^\infty \frac{z(t)[H^\sigma(t)]^\zeta}{[G(t)]^\theta}\Delta t \geq \frac{\zeta}{\theta-1}\int_a^\infty \frac{z(t)h(t)[H^\sigma(t)]^{\zeta-1}}{[G(t)]^{\theta-1}}\Delta t, \tag{8.9}$$

$$\int_a^\infty \frac{z(t)[H^\sigma(t)]^\zeta}{[G(t)]^\theta}\Delta t \geq \left[\frac{\zeta}{\theta-1}\right]^\zeta \int_a^\infty \frac{z(t)h^\zeta(t)}{[G(t)]^{\theta-\zeta}}\Delta t. \tag{8.10}$$

(iii) ([37]) *if $\zeta > 1$, $\eta \geq 0$, $\eta + \theta \leq 0$, then we have*

$$\int_a^\infty \frac{z(t)[H(t)]^{\eta+\zeta}}{[G(t)]^{\eta+\theta}}\Delta t \geq \frac{\eta+\zeta}{1-\eta-\theta}\int_a^\infty \frac{z(t)h(t)[H(t)]^{\eta+\zeta-1}}{[G(t)]^{\eta+\theta-1}}\Delta t, \tag{8.11}$$

$$\int_a^\infty \frac{z(t)[H(t)]^{\eta+\zeta}}{[G(t)]^{\eta+\theta}}\Delta t \geq \left[\frac{M_1^{\eta+\theta-1}(\eta+\zeta)}{1-\eta-\theta}\right]^{1/\zeta} \int_a^\infty \frac{z(t)h^{1/\zeta}(t)[H(t)]^{\eta+\zeta-\frac{1}{\zeta}}}{[G(t)]^{\eta+\theta-\frac{1}{\zeta}}}\Delta t. \tag{8.12}$$

(iv) ([37]) *if $\zeta > 1$, $\eta \geq 0$, $0 \leq \eta + \theta < 1$, then we have*

$$\int_a^\infty \frac{z(t)[H(t)]^{\eta+\zeta}}{[G^\sigma(t)]^{\eta+\theta}}\Delta t \geq \frac{\eta+\zeta}{1-\eta-\theta}\int_a^\infty \frac{z(t)h(t)[H(t)]^{\eta+\zeta-1}}{[G^\sigma(t)]^{\eta+\theta-1}}\Delta t, \tag{8.13}$$

$$\int_a^\infty \frac{z(t)[H(t)]^{\eta+\zeta}}{[G^\sigma(t)]^{\eta+\theta}}\Delta t \geq \left[\frac{\eta+\zeta}{1-\eta-\theta}\right]^{1/\zeta} \int_a^\infty \frac{z(t)h^{1/\zeta}(t)[H(t)]^{\eta+\zeta-\frac{1}{\zeta}}}{[G^\sigma(t)]^{\eta+\theta-\frac{1}{\zeta}}}\Delta t. \tag{8.14}$$

Similar to the delta case, let us define the following functions and the constant in the nabla calculus:

$$\begin{gathered} G(t)=\int_t^\infty z(s)\nabla s, \quad H(t)=\int_a^t z(s)h(s)\nabla s, \\ \overline{G}(t)=\int_a^t z(s)\nabla s, \quad \overline{H}(t)=\int_t^\infty z(s)h(s)\nabla s, \\ \exists L_1>0 \quad \text{such that } \frac{\overline{G}(t)}{\overline{G}^\rho(t)} \le L_1 \text{ for } t\in(a,\infty)_{\mathbb{T}}, \end{gathered} \tag{8.15}$$

where $z, h \geq 0$ are ld-continuous, ∇-differentiable, and locally nabla integrable functions, and $0 < a \in \mathbb{T}$.

Theorem 8.2 ([29, 34, 37]). *For the functions* $\overline{G}(t)$ *and* $\overline{H}(t)$ *and the constant* L_1 *defined in* (8.15),

(i) ([29]) *if* $\theta \leq 0 < \zeta < 1$, *then we have*

$$\int_a^\infty \frac{z(t)[\overline{H}^\rho(t)]^\zeta}{[\overline{G}(t)]^\theta}\nabla t \ge \frac{\zeta}{1-\theta}\int_a^\infty \frac{z(t)h(t)[\overline{H}^\rho(t)]^{\zeta-1}}{[\overline{G}(t)]^{\theta-1}}\nabla t, \tag{8.16}$$

$$\int_a^\infty \frac{z(t)[\overline{H}^\rho(t)]^\zeta}{[\overline{G}(t)]^\theta}\nabla t \ge \left[\frac{\zeta}{1-\theta}\right]^\zeta \int_a^\infty \frac{z(t)h^\zeta(t)}{[\overline{G}(t)]^{\theta-\zeta}}\nabla t. \tag{8.17}$$

(ii) ([34]) *if* $0 < \zeta < 1 < \theta$, *then we have*

$$\int_a^\infty \frac{z(t)[\overline{H}^\rho(t)]^\zeta}{[\overline{G}(t)]^\theta}\nabla t \ge \frac{\zeta}{\theta-1}\int_a^\infty \frac{z(t)h(t)[\overline{H}^\rho(t)]^{\zeta-1}}{[\overline{G}(t)]^{\theta-1}}\nabla t, \tag{8.18}$$

$$\int_a^\infty \frac{z(t)[\overline{H}^\rho(t)]^\zeta}{[\overline{G}(t)]^\theta}\nabla t \ge \left[\frac{\zeta}{\theta-1}\right]^\zeta \int_a^\infty \frac{z(t)h^\zeta(t)}{[\overline{G}(t)]^{\theta-\zeta}}\nabla t. \tag{8.19}$$

(iii) ([37]) *if* $\zeta > 1$, $\eta \geq 0$, $\eta + \theta \leq 0$, *then we have*

$$\int_a^\infty \frac{z(t)[\overline{H}(t)]^{\eta+\zeta}}{[\overline{G}(t)]^{\eta+\theta}}\nabla t \ge \frac{\eta+\zeta}{1-\eta-\theta}\int_a^\infty \frac{z(t)h(t)[\overline{H}(t)]^{\eta+\zeta-1}}{[\overline{G}^\rho(t)]^{\eta+\theta-1}}\nabla t, \tag{8.20}$$

$$\int_a^\infty \frac{z(t)[\overline{H}(t)]^{\eta+\zeta}}{[\overline{G}(t)]^{\eta+\theta}}\nabla t \ge \left[\frac{L_1^{\eta+\theta-1}(\eta+\zeta)}{1-\eta-\theta}\right]^{1/\zeta} \int_a^\infty \frac{z(t)h^{1/\zeta}(t)[\overline{H}(t)]^{\eta+\zeta-\frac{1}{\zeta}}}{[\overline{G}(t)]^{\eta+\theta-\frac{1}{\zeta}}}\nabla t. \tag{8.21}$$

(iv) ([37]) *if* $\zeta > 1$, $\eta \geq 0$, $0 \leq \eta + \theta < 1$, *then we have*

$$\int_a^\infty \frac{z(t)[\overline{H}(t)]^{\eta+\zeta}}{[\overline{G}^\rho(t)]^{\eta+\theta}}\nabla t \ge \frac{\eta+\zeta}{1-\eta-\theta}\int_a^\infty \frac{z(t)h(t)[\overline{H}(t)]^{\eta+\zeta-1}}{[\overline{G}^\rho(t)]^{\eta+\theta-1}}\nabla t, \tag{8.22}$$

$$\int_a^\infty \frac{z(t)[\overline{H}(t)]^{\eta+\zeta}}{[\overline{G}^\rho(t)]^{\eta+\theta}}\nabla t \ge \left[\frac{\eta+\zeta}{1-\eta-\theta}\right]^{1/\zeta} \int_a^\infty \frac{z(t)h^{1/\zeta}(t)[\overline{H}(t)]^{\eta+\zeta-\frac{1}{\zeta}}}{[\overline{G}^\rho(t)]^{\eta+\theta-\frac{1}{\zeta}}}\nabla t. \tag{8.23}$$

The diamond alpha time scale unifications of the classical Bennett–Leindler-type inequalities ($0 < \zeta < 1$) on an arbitrary time scale are given in [34] while their complements are obtained in [38] for $\zeta > 1$, some of which are given in the next two theorems.

Let us define the following functions in the diamond alpha calculus:

$$G(t) = \int_t^\infty z(s)\diamond_\alpha s, \quad H(t) = \int_a^t z(s)h(s)\diamond_\alpha s,$$
$$\overline{G}(t) = \int_a^t z(s)\diamond_\alpha s, \quad \overline{H}(t) = \int_t^\infty z(s)h(s)\diamond_\alpha s, \tag{8.24}$$

where $z, h \geq 0$ are diamond alpha differentiable and locally diamond alpha integrable functions, and $0 < a \in \mathbb{T}$.

Theorem 8.3 ([34]). *For the functions $G(t), \overline{G}(t)$ and $H(t), \overline{H}(t)$ defined in* (8.24),

(i) *if $\theta \leq 0 < \zeta < 1$, then we have*

$$\int_a^\infty \frac{z(t)[H^\sigma(t)]^\zeta}{[G(t)]^\theta}\diamond_\alpha t \geq \left[\frac{\alpha\zeta}{1-\theta}\right]^\zeta \int_a^\infty \frac{z(t)h^\zeta(t)}{[G(t)]^{\theta-\zeta}}\diamond_\alpha t, \tag{8.25}$$

where z is a nonincreasing function on $[a,\infty)_{\mathbb{T}}$ and

$$\int_a^\infty \frac{z(t)[\overline{H}^\rho(t)]^\zeta}{[\overline{G}(t)]^\theta}\diamond_\alpha t \geq \left[\frac{(1-\alpha)\zeta}{1-\theta}\right]^\zeta \int_a^\infty \frac{z(t)h^\zeta(t)}{[\overline{G}(t)]^{\theta-\zeta}}\diamond_\alpha t, \tag{8.26}$$

where z is a nondecreasing function on $[a,\infty)_{\mathbb{T}}$.

(ii) *if $0 < \zeta < 1 < \theta$, then we have*

$$\int_a^\infty \frac{z(t)[H^\sigma(t)]^\zeta}{[G(t)]^\theta}\diamond_\alpha t \geq \left[\frac{\alpha\zeta}{\theta-1}\right]^\zeta \int_a^\infty \frac{z(t)h^\zeta(t)}{[G(t)]^{\theta-\zeta}}\diamond_\alpha t, \tag{8.27}$$

where z is a nondecreasing function on $[a,\infty)_{\mathbb{T}}$ and

$$\int_a^\infty \frac{z(t)[\overline{H}^\rho(t)]^\zeta}{[\overline{G}(t)]^\theta}\diamond_\alpha t \geq \left[\frac{(1-\alpha)\zeta}{\theta-1}\right]^\zeta \int_a^\infty \frac{z(t)h^\zeta(t)}{[\overline{G}(t)]^{\theta-\zeta}}\diamond_\alpha t, \tag{8.28}$$

where z is a nonincreasing function on $[a,\infty)_{\mathbb{T}}$.

Theorem 8.4 ([38]). *For the functions $G(t), \overline{G}(t)$ and $H(t), \overline{H}(t)$ defined in* (8.24), *assume that there exist $M_1, L_1 > 0$ such that $\frac{G(t)}{G^\sigma(t)} \leq M_1$ and $\frac{\overline{G}(t)}{\overline{G}^\rho(t)} \leq L_1$ for $t \in (a,\infty)_{\mathbb{T}}$. Let $\zeta > 1$, $\eta \geq 0$ be real constants.*

(i) *If $\eta + \theta \leq 0$, then we have*

$$\int_a^\infty \frac{z(t)[H(t)]^{\eta+\zeta}}{[G(t)]^{\eta+\theta}}\diamond_\alpha t \geq \left[\frac{\alpha(\eta+\zeta)M_1^{\eta+\theta-1}}{1-\eta-\theta}\right]^{1/\zeta} \int_a^\infty \frac{z(t)h^{1/\zeta}(t)[H(t)]^{\eta+\zeta-\frac{1}{\zeta}}}{[G(t)]^{\eta+\theta-\frac{1}{\zeta}}}\diamond_\alpha t, \tag{8.29}$$

where z is a nonincreasing function on $[a,\infty)_{\mathbb{T}}$ *and*

$$\int_a^\infty \frac{z(t)[\overline{H}(t)]^{\eta+\zeta}}{[\overline{G}(t)]^{\eta+\theta}}\diamond_\alpha t \geq \left[\frac{L_1^{\eta+\theta-1}(1-\alpha)(\eta+\zeta)}{1-\eta-\theta}\right]^{1/\zeta}\int_a^\infty \frac{z(t)h^{1/\zeta}(t)[\overline{H}(t)]^{\eta+\zeta-\frac{1}{\zeta}}}{[\overline{G}(t)]^{\eta+\theta-\frac{1}{\zeta}}}\diamond_\alpha t, \tag{8.30}$$

where z is a nondecreasing function on $[a,\infty)_{\mathbb{T}}$.

(ii) *If* $0 \leq \eta+\theta < 1$, *then we have*

$$\int_a^\infty \frac{z(t)[H(t)]^{\eta+\zeta}}{[G^\sigma(t)]^{\eta+\theta}}\diamond_\alpha t \geq \left[\frac{\alpha(\eta+\zeta)}{1-\eta-\theta}\right]^{1/\zeta}\int_a^\infty \frac{z(t)h^{1/\zeta}(t)[H(t)]^{\eta+\zeta-\frac{1}{\zeta}}}{[G^\sigma(t)]^{\eta+\theta-\frac{1}{\zeta}}}\diamond_\alpha t, \tag{8.31}$$

where z is a nonincreasing function on $[a,\infty)_{\mathbb{T}}$ *and*

$$\int_a^\infty \frac{z(t)[\overline{H}(t)]^{\eta+\zeta}}{[\overline{G}^\rho(t)]^{\eta+\theta}}\diamond_\alpha t \geq \left[\frac{(1-\alpha)(\eta+\zeta)}{1-\eta-\theta}\right]^{1/\zeta}\int_a^\infty \frac{z(t)h^{1/\zeta}(t)[\overline{H}(t)]^{\eta+\zeta-\frac{1}{\zeta}}}{[\overline{G}^\rho(t)]^{\eta+\theta-\frac{1}{\zeta}}}\diamond_\alpha t, \tag{8.32}$$

where z is a nondecreasing function on $[a,\infty)_{\mathbb{T}}$.

(iii) *If* $\eta+\theta > 1$, *then we have*

$$\int_a^\infty \frac{z(t)[H(t)]^{\eta+\zeta}}{[G(t)]^{\eta+\theta}}\diamond_\alpha t \geq \left[\frac{\alpha(\eta+\zeta)M_1^{\eta+\theta-1}}{\eta+\theta-1}\right]^{1/\zeta}\int_a^\infty \frac{z(t)h^{1/\zeta}(t)[H(t)]^{\eta+\zeta-\frac{1}{\zeta}}}{[G(t)]^{\eta+\theta-\frac{1}{\zeta}}}\diamond_\alpha t, \tag{8.33}$$

where z is a nondecreasing function on $[a,\infty)_{\mathbb{T}}$ *and*

$$\int_a^\infty \frac{z(t)[\overline{H}(t)]^{\eta+\zeta}}{[\overline{G}(t)]^{\eta+\theta}}\diamond_\alpha t \geq \left[\frac{L_1^{\eta+\theta-1}(1-\alpha)(\eta+\zeta)}{\eta+\theta-1}\right]^{1/\zeta}\int_a^\infty \frac{z(t)h^{1/\zeta}(t)[\overline{H}(t)]^{+\zeta+\frac{1}{\zeta}}}{[\overline{G}(t)]^{\eta+\theta-\frac{1}{\zeta}}}\diamond_\alpha t, \tag{8.34}$$

where z is a nonincreasing function on $[a,\infty)_{\mathbb{T}}$.

Since there is a gap in the literature for diamond alpha Bennett–Leindler-type inequalities for $0 < \zeta < 1$, $\eta \geq 0$, our aim is to fill this gap by obtaining new diamond alpha Bennett–Leindler-type inequalities to generalize the results in [34] and to complement the current literature given in [38]. Another significant contribution of this chapter is to employ a new method, which depends on using the diamond alpha calculus rather than algebra in contrast to the literature, for establishing new diamond alpha Bennett–Leindler-type inequalities. By this new method, we do not need Integration by Parts Formula and the Fundamental Theorem of Calculus, which are standard arguments in the proofs of Bennett–Leindler-type dynamic inequalities in the delta and nabla approaches but do not follow through in the diamond alpha calculus. Moreover, since both sides of the diamond alpha Bennett–Leindler-type inequalities include only single diamond alpha integrals, these inequalities have more compact forms. In addition, we notice that special cases of $\eta+\theta \leq 0$ and $\eta+\theta > 1$ corresponding to $\eta = 0$ were considered in [34], while the case $0 \leq \eta+\theta < 1$ was not investigated therein. By taking account a constant $\eta \geq 0$, we not only generalize the diamond alpha Bennett–Leindler-type inequalities presented in [34] for $\eta \geq 0$, but also obtain complement inequalities of those in [38] established for $\zeta > 1$.

As a result, our novel technique allows us to generalize the results in [34], to unify the foregoing delta and nabla Bennett–Leindler-type inequalities and derive new ones, and to obtain complementary inequalities for the existing diamond alpha Bennett–Leindler-type inequalities in [38].

8.2 Preliminaries

This section is devoted to the main definitions and theorems of delta, nabla, and diamond alpha calculi. We refer the reader to [8, 14, 15, 21, 22] for the concept of time scale calculus in the delta and nabla sense.

A nonempty closed subset of $\mathbb{R}$ is called a time scale which is denoted by $\mathbb{T}$. Since the delta and nabla time scale calculi are very well known [8, 14, 15], we skip the details of them and consider only main properties which will be used in the sequel.

Theorem 8.5 ([14]). *Suppose that $\Lambda, \Gamma : \mathbb{T} \to \mathbb{R}$ and $s \in \mathbb{T}^{\kappa}$. For $\mu(s) = \sigma(s) - s$, we have the following statements:*

1. *If Λ is delta differentiable at s, then $\Lambda(\sigma(s)) = \Lambda^{\sigma}(s) = \Lambda(s) + \mu(s)\Lambda^{\Delta}(s)$.*
2. *The product $\Lambda\Gamma : \mathbb{T} \to \mathbb{R}$ is differentiable at s with*

$$(\Lambda\Gamma)^{\Delta}(s) = \Lambda^{\Delta}(s)\Gamma(s) + \Lambda(\sigma(s))\Gamma^{\Delta}(s) = \Lambda(s)\Gamma^{\Delta}(s) + \Lambda^{\Delta}(s)\Gamma(\sigma(s)). \tag{8.35}$$

Lemma 8.1 (Chain rule for the delta derivative, [14]). *If $\Gamma \in C^1(\mathbb{R}, \mathbb{R})$ and $\Lambda \in C(\mathbb{T}, \mathbb{R})$ is delta differentiable on $\mathbb{T}^{\kappa}$, then $\Gamma \circ \Lambda$ is delta differentiable and*

(i) *one can find $c \in [s, \sigma(s)]$ with*

$$(\Gamma \circ \Lambda)^{\Delta}(s) = \Gamma'(\Lambda(c))\Lambda^{\Delta}(s); \tag{8.36}$$

(ii) *the following relation holds:*

$$(\Gamma \circ \Lambda)^{\Delta}(s) = \Lambda^{\Delta}(s) \int_0^1 \Gamma'(\Lambda(s) + w\mu(s)\Lambda^{\Delta}(s))dw. \tag{8.37}$$

Theorem 8.6 ([14]). *Suppose that $\Gamma : \mathbb{T} \to \mathbb{R}$ and $s \in \mathbb{T}_{\kappa}$. For $\nu(s) = s - \rho(s)$, we have the following statements:*

1. *If Γ is nabla differentiable at s, then $\Gamma(\rho(s)) = \Gamma^{\rho}(s) = \Gamma(s) - \nu(s)\Gamma^{\nabla}(s)$.*
2. *The product $\Lambda\Gamma : \mathbb{T} \to \mathbb{R}$ is differentiable at s with*

$$(\Lambda\Gamma)^{\nabla}(s) = \Lambda^{\nabla}(s)\Gamma(s) + \Lambda(\rho(s))\Gamma^{\nabla}(s) = \Lambda(s)\Gamma^{\nabla}(s) + \Lambda^{\nabla}(s)\Gamma(\rho(s)). \tag{8.38}$$

Lemma 8.2 (Chain rule for the nabla derivative, [23]). *If $\Gamma \in C^1(\mathbb{R}, \mathbb{R})$ and $\Lambda \in C(\mathbb{T}, \mathbb{R})$ is nabla differentiable on $\mathbb{T}_{\kappa}$, then $\Gamma \circ \Lambda$ is nabla differentiable and the following holds:*

$$(\Gamma \circ \Lambda)^{\nabla}(s) = \Lambda^{\nabla}(s) \int_0^1 \Gamma'(\Lambda(\rho(s)) + w\nu(s)\Lambda^{\nabla}(s))dw. \tag{8.39}$$

The next lemmas play crucial roles in the proofs of the main theorems.

Lemma 8.3 ([8, 14]). *If Γ is continuous, then*

(i) $\left(\int_{t_1}^{t} \Gamma(s)\Delta s\right)^{\Delta} = \Gamma(t)$ *and* $\left(\int_{t_1}^{t} \Gamma(s)\nabla s\right)^{\Delta} = \Gamma(\sigma(t))$,

(ii) $\left(\int_{t_1}^{t} \Gamma(s)\nabla s\right)^{\nabla} = \Gamma(t)$ *and* $\left(\int_{t_1}^{t} \Gamma(s)\Delta s\right)^{\nabla} = \Gamma(\rho(t))$.

Lemma 8.4 ([22]). *If Γ is continuous for all $t_1, t_2 \in \mathbb{T}$ with $t_1 < t_2$, then*

$$\int_{t_1}^{t_2} \Gamma(t)\Delta t = \int_{t_1}^{t_2} \Gamma(\rho(t))\nabla t \quad \text{and} \quad \int_{t_1}^{t_2} \Gamma(t)\nabla t = \int_{t_1}^{t_2} \Gamma(\sigma(t))\Delta t.$$

The diamond alpha time scale calculus has been introduced by Sheng et al. in [60]. This calculus deals with diamond alpha, denoted by $\diamond_\alpha$, differentiable and diamond alpha integrable functions which are convex linear combinations of delta and nabla differentiable and integrable functions, respectively. For some developments of this calculus and for some dynamic inequalities in this calculus, we refer to [5, 7, 11–13, 28, 33–36, 38, 43, 45–47, 50] and the references therein.

Definition 8.1 ([60]). The diamond alpha derivative and integral have been introduced as follows.

(i) Let $\rho(s) - \tau = a_{s\tau}$ and $\sigma(s) - \tau = b_{s\tau}$. Then the $\diamond_\alpha$-derivative of $\Lambda : \mathbb{T} \to \mathbb{R}$ at the point $s \in \mathbb{T}_\kappa^\kappa$ denoted by $\Lambda^{\diamond_\alpha}(s)$ is the number enjoying the property that for all $\epsilon > 0$, there exists a neighborhood $V \subset \mathbb{T}$ of $s \in \mathbb{T}_\kappa^\kappa$ such that for any $\tau \in V$,

$$\big|\alpha|\Lambda(\sigma(s)) - \Lambda(\tau)||a_{s\tau}| + (1-\alpha)|\Lambda(\rho(s)) - \Lambda(\tau)||b_{s\tau}| - \Lambda^{\diamond_\alpha}(s)|a_{s\tau}||b_{s\tau}|\big| \le \epsilon |a_{s\tau}||b_{s\tau}|.$$

(ii) $\Lambda : \mathbb{T} \to \mathbb{R}$ is $\diamond_\alpha$-differentiable at $s \in \mathbb{T}_\kappa^\kappa$ provided it is both delta and nabla differentiable at s. Moreover, for $0 \le \alpha \le 1$, such a function satisfies

$$\Lambda^{\diamond_\alpha}(s) = \alpha\Lambda^{\Delta}(s) + (1-\alpha)\Lambda^{\nabla}(s).$$

(iii) $\Lambda : \mathbb{T} \to \mathbb{R}$ is diamond alpha integrable provided it is continuous. Moreover, for $0 \le \alpha \le 1$, we have

$$\int_{s_1}^{s_2} \Lambda(s)\diamond_\alpha s = \alpha \int_{s_1}^{s_2} \Lambda(s)\Delta s + (1-\alpha)\int_{s_1}^{s_2} \Lambda(s)\nabla s.$$

Lemma 8.5 ([22, 46]). *For all $s \in \mathbb{T}$, a time scale $\mathbb{T}$ is said to be regular provided $\sigma(\rho(s)) = \rho(\sigma(s)) = s$. A regular time scale $\mathbb{T}$ satisfies $\mathbb{T}^\kappa_\kappa = \mathbb{T}_\kappa = \mathbb{T}^\kappa = \mathbb{T}$. Moreover, $\sigma(\mathbb{T}) = \rho(\mathbb{T}) = \mathbb{T}$ in such a time scale.*

Lemma 8.6 (Diamond alpha Hölder's inequality, [5, 47]). *Let $s_1, s_2 \in \mathbb{T}$ and $\lambda_1, \lambda_2 > 1$ be given satisfying $\frac{1}{\lambda_1} + \frac{1}{\lambda_2} = 1$. For $\phi, \psi \in C([a,b]_\mathbb{T}, [0,\infty))$ with $\int_{s_1}^{s_2} \psi^{\lambda_2}(s) \diamond_\alpha s > 0$, the diamond alpha Hölder's inequality*

$$\int_{s_1}^{s_2} \phi(s)\psi(s)\diamond_\alpha s \le \left(\int_{s_1}^{s_2} \phi^{\lambda_1}(s)\diamond_\alpha s\right)^{1/\lambda_1} \left(\int_{s_1}^{s_2} \psi^{\lambda_2}(s)\diamond_\alpha s\right)^{1/\lambda_2}$$

holds.

If $\lambda_1 < 0$ or $0 < \lambda_1 < 1$ with $\frac{1}{\lambda_1} + \frac{1}{\lambda_2} = 1$, then the diamond alpha reverse Hölder's inequality

$$\int_{s_1}^{s_2} \phi(s)\psi(s)\diamond_\alpha s \ge \left(\int_{s_1}^{s_2} \phi^{\lambda_1}(s)\diamond_\alpha s\right)^{1/\lambda_1} \left(\int_{s_1}^{s_2} \psi^{\lambda_2}(s)\diamond_\alpha s\right)^{1/\lambda_2} \tag{8.40}$$

is satisfied.

8.3 Diamond alpha Bennett–Leindler type dynamic inequalities

This section is devoted to deriving new diamond alpha Bennett–Leindler-type integral inequalities, which are established using the properties of the delta, nabla and diamond alpha derivatives and integrals.

Let $\mathbb{T}$ be a regular time scale and $a \in [0,\infty)_\mathbb{T}$.

The following theorem asserts not only novel diamond alpha and delta Bennett–Leindler-type inequalities when $0 < \zeta < 1$, $\eta \ge 0$, and $\eta + \theta \le 0$, but also complements the diamond alpha Bennett–Leindler type inequalities given in [38, Theorem 9] established for $\zeta > 1$, $\eta \ge 0$, and $\eta + \theta \le 0$ and generalizations of the diamond alpha Bennett–Leindler-type inequalities given in [34, Theorem 15] proven for $0 < \zeta < 1$, $\eta = 0$, and $\theta \le 0$. This novelty is caused by the condition $\eta + \zeta \ge 1$, which has not been considered until this current contribution to the literature.

Theorem 8.7. *Suppose that z is a nonincreasing function on $[a,\infty)_\mathbb{T}$. For the functions $G(t)$ and $H(t)$ defined in (8.24), assume that there exists $M_1 > 0$ such that $\frac{G(t)}{G^\sigma(t)} \le M_1$ for $t \in (a,\infty)_\mathbb{T}$. Let $0 < \zeta < 1$, $\eta \ge 0$, and $\eta + \theta \le 0$ be real constants.*

(i) *If $0 < \eta + \zeta \le 1$, then*

$$\int_a^\infty \frac{z(t)[H^\sigma(t)]^{\eta+\zeta}}{[G(t)]^{\eta+\theta}}\diamond_\alpha t \ge \frac{\alpha(\eta+\zeta)}{1-\eta-\theta}\int_a^\infty \frac{z(t)h(t)[H^\sigma(t)]^{\eta+\zeta-1}}{[G(t)]^{\eta+\theta-1}}\diamond_\alpha t \tag{8.41}$$

and

$$\int_a^\infty \frac{z(t)[H^\sigma(t)]^{\eta+\zeta}}{[G(t)]^{\eta+\theta}} \diamond_\alpha t \geq \left[\frac{\alpha(\eta+\zeta)}{1-\eta-\theta}\right]^\zeta \int_a^\infty \frac{z(t)h^\zeta(t)[H^\sigma(t)]^\eta}{[G(t)]^{\eta+\theta-\zeta}} \diamond_\alpha t. \tag{8.42}$$

(ii) *If* $\eta + \zeta \geq 1$, *then*

$$\int_a^\infty \frac{z(t)[H(t)]^{\eta+\zeta}}{[G(t)]^{\eta+\theta}} \diamond_\alpha t \geq \frac{\alpha(\eta+\zeta)}{1-\eta-\theta} \int_a^\infty \frac{z(t)h(t)[H(t)]^{\eta+\zeta-1}}{[G^\sigma(t)]^{\eta+\theta-1}} \diamond_\alpha t \tag{8.43}$$

and

$$\int_a^\infty \frac{z(t)[H(t)]^{\eta+\zeta}}{[G(t)]^{\eta+\theta}} \diamond_\alpha t \geq \left[\frac{\alpha(\eta+\zeta)M_1^{\eta+\theta-1}}{1-\eta-\theta}\right]^\zeta \int_a^\infty \frac{z(t)h^\zeta(t)[H(t)]^\eta}{[G(t)]^{\eta+\theta-\zeta}} \diamond_\alpha t. \tag{8.44}$$

Proof. First we obtain some inequalities which will be used in the sequel.

By utilizing Lemma 8.3, one can obtain

$$\begin{aligned} H^\Delta(t) &= \left[\int_a^t z(s)h(s)\diamond_\alpha s\right]^\Delta \\ &= \alpha\left[\int_a^t z(s)h(s)\Delta s\right]^\Delta + (1-\alpha)\left[\int_a^t z(s)h(s)\nabla s\right]^\Delta \\ &= \alpha z(t)h(t) + (1-\alpha)z^\sigma(t)h^\sigma(t) \geq 0. \end{aligned} \tag{8.45}$$

By taking into account equation (8.45) and by employing formula (8.36), one can observe that for $t \leq c \leq \sigma(t)$,

$$\begin{aligned} [H^{\eta+\zeta}(t)]^\Delta &= (\eta+\zeta)H^\Delta(t)H^{\eta+\zeta-1}(c) \\ &= (\eta+\zeta)[\alpha z(t)h(t) + (1-\alpha)z^\sigma(t)h^\sigma(t)]H^{\eta+\zeta-1}(c) \end{aligned}$$

holds. Then two estimates can be obtained depending on whether $0 < \eta+\zeta \leq 1$ or $\eta+\zeta \geq 1$ for the function $[H^{\eta+\zeta}(t)]^\Delta$ as follows:

(i) If $0 < \eta+\zeta \leq 1$, then

$$[H^{\eta+\zeta}(t)]^\Delta \geq (\eta+\zeta)\alpha z(t)h(t)[H^\sigma(t)]^{\eta+\zeta-1}; \tag{8.46}$$

(ii) If $\eta+\zeta \geq 1$, then

$$[H^{\eta+\zeta}(t)]^\Delta \geq (\eta+\zeta)\alpha z(t)h(t)[H(t)]^{\eta+\zeta-1}, \tag{8.47}$$

where $H(t) \leq H(c) \leq H^\sigma(t)$ has been used for $t \leq c \leq \sigma(t)$.

By Lemma 8.3, note that

$$\begin{aligned} G^{\Delta}(t) &= \left[\int_t^{\infty} z(s)\diamond_{\alpha}s\right]^{\Delta} = \alpha\left[\int_t^{\infty} z(s)\Delta s\right]^{\Delta} + (1-\alpha)\left[\int_t^{\infty} z(s)\nabla s\right]^{\Delta} \\ &= -\alpha z(t) - (1-\alpha)z^{\sigma}(t) \le 0. \end{aligned} \tag{8.48}$$

It follows from (8.48) and formula (8.36) that for $\eta + \theta \le 0$ we have

$$\begin{aligned} \left[G^{1-\eta-\theta}(t)\right]^{\Delta} &= (1-\eta-\theta)G^{\Delta}(t)\int_0^1 \frac{dw}{[(1-w)G(t) + wG^{\sigma}(t)]^{\eta+\theta}} \\ &= \int_0^1 \frac{(1-\eta-\theta)[-\alpha z(t) - (1-\alpha)z^{\sigma}(t)]}{[(1-w)G(t) + wG^{\sigma}(t)]^{\eta+\theta}}dw \\ &\ge -(1-\eta-\theta)\frac{z(t)}{[G(t)]^{\eta+\theta}}, \end{aligned} \tag{8.49}$$

where $G^{\sigma}(t) \le G(t)$ and the nonincreasingness property of z have been used.

(i) Let us define $u(t) = [H(t)]^{\eta+\zeta}[G(t)]^{1-\eta-\theta}$ for $t \in [a, \infty)$. If we take the delta derivative of the function u using formula (8.35), we get

$$u^{\Delta}(t) = \left[H^{\eta+\zeta}(t)\right]^{\Delta}\left[G^{1-\eta-\theta}(t)\right] + \left[H^{\sigma}(t)\right]^{\eta+\zeta}\left[G^{1-\eta-\theta}(t)\right]^{\Delta}. \tag{8.50}$$

By making use of inequalities (8.46) and (8.49) in equation (8.50), we obtain

$$u^{\Delta}(t) \ge -(1-\eta-\theta)\frac{z(t)}{[G(t)]^{\eta+\theta}}\left[H^{\sigma}(t)\right]^{\eta+\zeta} + \frac{\alpha(\eta+\zeta)z(t)h(t)[H^{\sigma}(t)]^{\eta+\zeta-1}}{[G(t)]^{\eta+\theta-1}},$$

or

$$\int_a^{\infty} \frac{z(t)[H^{\sigma}(t)]^{\eta+\zeta}}{[G(t)]^{\eta+\theta}}\diamond_{\alpha}t \ge \frac{\alpha(\eta+\zeta)}{1-\eta-\theta}\int_a^{\infty} \frac{z(t)h(t)[H^{\sigma}(t)]^{\eta+\zeta-1}}{[G(t)]^{\eta+\theta-1}}\diamond_{\alpha}t - \frac{1}{1-\eta-\theta}\int_a^{\infty} u^{\Delta}(t)\diamond_{\alpha}t. \tag{8.51}$$

The definition of u implies $u(\infty) = u(a) = 0$ and, by employing Lemma 8.4, we obtain

$$\begin{aligned} \int_a^{\infty} u^{\Delta}(t)\diamond_{\alpha}t &= \alpha\int_a^{\infty} u^{\Delta}(t)\Delta t + (1-\alpha)\int_a^{\infty} u^{\Delta}(t)\nabla t \\ &= \alpha\left[u(\infty) - u(a)\right] + (1-\alpha)\left[u(\sigma(\infty)) - u(\sigma(a))\right] \le 0, \end{aligned} \tag{8.52}$$

where we have imposed that $(-1)^{\eta+\theta} = 1$. Therefore we can infer that inequality (8.51) becomes

$$\int_a^{\infty} \frac{z(t)[H^{\sigma}(t)]^{\eta+\zeta}}{[G(t)]^{\eta+\theta}}\diamond_{\alpha}t \ge \frac{\alpha(\eta+\zeta)}{1-\eta-\theta}\int_a^{\infty} \frac{z(t)h(t)[H^{\sigma}(t)]^{\eta+\zeta-1}}{[G(t)]^{\eta+\theta-1}}\diamond_{\alpha}t,$$

which is the desired inequality (8.41). Since

$$\int_a^\infty \frac{z(t)h(t)[H^\sigma(t)]^{\eta+\zeta-1}}{[G(t)]^{\eta+\theta-1}}\diamond_a t = \int_a^\infty \frac{z^{\frac{1}{\zeta}}(t)h(t)[H^\sigma(t)]^{\frac{\eta}{\zeta}}}{[G(t)]^{\frac{\eta+\theta-\zeta}{\zeta}}}\left[\frac{z(t)[H^\sigma(t)]^{\eta+\zeta}}{[G^\sigma(t)]^{\eta+\theta}}\right]^{\frac{\zeta-1}{\zeta}}\diamond_a t,$$

applying the reverse Hölder's inequality (8.40) with the constants $0 < \zeta < 1$ and $\zeta/(\zeta-1) < 0$ to the right-hand side of the above equation leads to inequality (8.42).

(ii) Let us define $u(t) = [H(t)]^{\eta+\zeta}[G(t)]^{1-\eta-\theta}$ for $t \in [a,\infty)$. If we take the delta derivative of the function u using formula (8.35), we get

$$u^\Delta(t) = [H^{\eta+\zeta}(t)][G^{1-\eta-\theta}(t)]^\Delta + [H^{\eta+\zeta}(t)]^\Delta[G^\sigma(t)]^{1-\eta-\theta}. \tag{8.53}$$

Using inequalities (8.47) and (8.49) in equation (8.53) yields

$$u^\Delta(t) \geq -(1-\eta-\theta)\frac{z(t)}{[G(t)]^{\eta+\theta}}[H(t)]^{\eta+\zeta} + \frac{\alpha(\eta+\zeta)z(t)h(t)[H(t)]^{\eta+\zeta-1}}{[G^\sigma(t)]^{\eta+\theta-1}},$$

or

$$\int_a^\infty \frac{z(t)[H(t)]^{\eta+\zeta}}{[G(t)]^{\eta+\theta}}\diamond_a t \geq \frac{\alpha(\eta+\zeta)}{1-\eta-\theta}\int_a^\infty \frac{z(t)h(t)[H(t)]^{\eta+\zeta-1}}{[G^\sigma(t)]^{\eta+\theta-1}}\diamond_a t - \frac{1}{1-\eta-\theta}\int_a^\infty u^\Delta(t)\diamond_a t.$$

If we employ inequality (8.52), we obtain from the above inequality that

$$\int_a^\infty \frac{z(t)[H(t)]^{\eta+\zeta}}{[G(t)]^{\eta+\theta}}\diamond_a t \geq \frac{\alpha(\eta+\zeta)}{1-\eta-\theta}\int_a^\infty \frac{z(t)h(t)[H(t)]^{\eta+\zeta-1}}{[G^\sigma(t)]^{\eta+\theta-1}}\diamond_a t,$$

which is the desired inequality (8.43).
After using $\frac{G(t)}{G^\sigma(t)} \leq M_1$ on the right-hand side of inequality (8.43) and applying the reverse Hölder's inequality (8.40) with the constants $0 < \zeta < 1$ and $\zeta/(\zeta-1) < 0$ to the resulting integral, one can obtain inequality (8.44). □

Remark 8.1. The diamond alpha Bennett–Leindler-type inequalities (8.41)–(8.42) obtained for $0 < \zeta < 1$, $\eta \geq 0$, and $\eta + \theta \leq 0$ are generalizations of the diamond alpha Bennett–Leindler-type inequality (8.25) established for $0 < \zeta < 1$, $\eta = 0$, and $\theta \leq 0$ given in [34, Theorem 15]. The diamond alpha Bennett–Leindler-type inequalities (8.43)–(8.44) obtained for $0 < \zeta < 1$, $\eta \geq 0$, and $\eta + \theta \leq 0$ are complements of the diamond alpha Bennett–Leindler-type inequality (8.29) established for $\zeta > 1$, $\eta \geq 0$, and $\eta + \theta \leq 0$ given in [38, Theorem 9].

Remark 8.2. Although the special case of the condition $\eta + \zeta \leq 1$ is automatically satisfied in [34, Theorem 15], the other case, $\eta + \zeta \geq 1$ with $0 < \zeta < 1$, has not appeared in the literature previously even for the special cases. This is one of the gaps in the literature that this theorem aims to fill. By the novel diamond alpha Bennett–Leindler inequalities (8.43)–(8.44) obtained for $0 < \zeta < 1$, $\eta \geq 0$, $\eta + \theta \leq 0$, and $\eta + \zeta \geq 1$ for the first time, this aim is achieved.

Remark 8.3. Special cases of the diamond alpha Bennett–Leindler-type inequalities (8.41)–(8.44) can be seen below.

(i) Expressing inequalities (8.41)–(8.42) in terms of delta and nabla integrals for $0 < \zeta < 1$, $\eta \geq 0$, $\eta + \theta \leq 0$, and $\eta + \zeta \leq 1$ and then choosing $\alpha = 1$ in those inequalities yields

$$\int_a^\infty \frac{z(t)[H^\sigma(t)]^{\eta+\zeta}}{[G(t)]^{\eta+\theta}} \Delta t \geq \frac{\eta+\zeta}{1-\eta-\theta} \int_a^\infty \frac{z(t)h(t)[H^\sigma(t)]^{\eta+\zeta-1}}{[G(t)]^{\eta+\theta-1}} \Delta t \tag{8.54}$$

and

$$\int_a^\infty \frac{z(t)[H^\sigma(t)]^{\eta+\zeta}}{[G(t)]^{\eta+\theta}} \Delta t \geq \left[\frac{\eta+\zeta}{1-\eta-\theta}\right]^\zeta \int_a^\infty \frac{z(t)h^\zeta(t)[H^\sigma(t)]^{\eta}}{[G(t)]^{\eta+\theta-\zeta}} \Delta t, \tag{8.55}$$

respectively, where $H(t)$ and $G(t)$ are defined in (8.6). Inequalities (8.54)–(8.55) generalize delta Bennett–Leindler-type inequalities (8.7)–(8.8) established for $0 < \zeta < 1$, $\eta = 0$, and $\theta \leq 0$ given in [59, Theorem 2.1] and [34, Remark 5] and complement delta Bennett–Leindler-type inequalities (8.11)–(8.12) in [37, Corollary 1] and [38, Remark 3] established for $\zeta > 1$, $\eta \geq 0$, and $\eta + \theta \leq 0$. We can conclude that (i) of Theorem 8.7 is a diamond alpha unification of Theorem 8.1 which is given in [59, Theorem 2.1] and [34, Remark 5] and is a generalization of the diamond alpha Bennett–Leindler-type inequality (8.25) established for $0 < \zeta < 1$, $\eta = 0$, and $\theta \leq 0$ given in [34, Theorem 15] and is a completion of the diamond alpha Bennett–Leindler-type inequality (8.29) established for $\zeta > 1$, $\eta \geq 0$, and $\eta + \theta \leq 0$ given in [38, Theorem 9].

(ii) Expressing inequalities (8.43)–(8.44) in terms of delta and nabla integrals for $0 < \zeta < 1$, $\eta \geq 0, \eta + \theta \leq 0$, and $\eta + \zeta \geq 1$, and then choosing $\alpha = 1$ in those inequalities yields

$$\int_a^\infty \frac{z(t)[H(t)]^{\eta+\zeta}}{[G(t)]^{\eta+\theta}} \Delta t \geq \frac{\eta+\zeta}{1-\eta-\theta} \int_a^\infty \frac{z(t)h(t)[H(t)]^{\eta+\zeta-1}}{[G^\sigma(t)]^{\eta+\theta-1}} \Delta t \tag{8.56}$$

and

$$\int_a^\infty \frac{z(t)[H(t)]^{\eta+\zeta}}{[G(t)]^{\eta+\theta}} \Delta t \geq \left[\frac{(\eta+\zeta)M_1^{\eta+\theta-1}}{1-\eta-\theta}\right]^\zeta \int_a^\infty \frac{z(t)h^\zeta(t)[H(t)]^{\eta}}{[G(t)]^{\eta+\theta-\zeta}} \Delta t, \tag{8.57}$$

respectively, where $H(t), G(t)$ and M_1 are defined in (8.6). Inequalities (8.56)–(8.57) are novel even in the delta calculus. These novel inequalities generalize delta Bennett–Leindler-type inequalities (8.7)–(8.8) given in [59, Theorem 2.1] and [34, Remark 5] established for $0 < \zeta < 1$, $\eta = 0$, and $\theta \leq 0$ and complement delta Bennett–Leindler-type inequalities (8.11)–(8.12) given in [37, Corollary 1] and [38, Remark 3] established for $\zeta > 1$, $\eta \geq 0$ and $\eta + \theta \leq 0$. We can conclude that (ii) of Theorem 8.7 is a diamond alpha unification of Theorem 8.1 which is given in [59, Theorem 2.1] and

[34, Remark 5] and is a generalization of the diamond alpha Bennett–Leindler-type inequality (8.25) established for $0 < \zeta < 1$, $\eta = 0$, and $\theta \le 0$ given in [34, Theorem 15] and is a completion of the diamond alpha Bennett–Leindler-type inequality (8.29) established for $\zeta > 1$, $\eta \ge 0$, and $\eta + \theta \le 0$ given in [38, Theorem 9].

The following theorem asserts not only novel diamond alpha and delta Bennett–Leindler-type inequalities when $0 < \zeta < 1$, $\eta \ge 0$, and $0 \le \eta + \theta < 1$, but also complements the diamond alpha Bennett–Leindler-type inequalities given in [38, Theorem 9] established for $\zeta > 1$, $\eta \ge 0$, and $0 \le \eta + \theta < 1$, and generalizations of the diamond alpha Bennett–Leindler-type inequalities given in [34, Theorem 15] established for $0 < \zeta < 1$, $\eta = 0$, and $\theta \le 0$. This novelty is caused by the condition $\eta + \zeta \ge 1$, which has not been considered so far.

Theorem 8.8. *Suppose that z is a nonincreasing function on $[a, \infty)_{\mathbb{T}}$. For the functions $G(t)$ and $H(t)$ defined in (8.24), let $0 < \zeta < 1$, $\eta \ge 0$, and $0 \le \eta + \theta < 1$ be real constants.*

(i) *If $0 < \eta + \zeta \le 1$, then*

$$\int_a^\infty \frac{z(t)[H^\sigma(t)]^{\eta+\zeta}}{[G^\sigma(t)]^{\eta+\theta}} \diamond_\alpha t \ge \frac{\alpha(\eta+\zeta)}{1-\eta-\theta} \int_a^\infty \frac{z(t)h(t)[H^\sigma(t)]^{\eta+\zeta-1}}{[G(t)]^{\eta+\theta-1}} \diamond_\alpha t \tag{8.58}$$

and

$$\int_a^\infty \frac{z(t)[H^\sigma(t)]^{\eta+\zeta}}{[G^\sigma(t)]^{\eta+\theta}} \diamond_\alpha t \ge \left[\frac{\alpha(\eta+\zeta)}{1-\eta-\theta}\right]^\zeta \int_a^\infty \frac{z(t)h^\zeta(t)[H^\sigma(t)]^{\eta}}{[G^\sigma(t)]^{\eta+\theta-\zeta}} \diamond_\alpha t. \tag{8.59}$$

(ii) *If $\eta + \zeta \ge 1$, then*

$$\int_a^\infty \frac{z(t)[H(t)]^{\eta+\zeta}}{[G^\sigma(t)]^{\eta+\theta}} \diamond_\alpha t \ge \frac{\alpha(\eta+\zeta)}{1-\eta-\theta} \int_a^\infty \frac{z(t)h(t)[H(t)]^{\eta+\zeta-1}}{[G^\sigma(t)]^{\eta+\theta-1}} \diamond_\alpha t \tag{8.60}$$

and

$$\int_a^\infty \frac{z(t)[H(t)]^{\eta+\zeta}}{[G^\sigma(t)]^{\eta+\theta}} \diamond_\alpha t \ge \left[\frac{\alpha(\eta+\zeta)}{1-\eta-\theta}\right]^\zeta \int_a^\infty \frac{z(t)h^\zeta(t)[H(t)]^{\eta}}{[G^\sigma(t)]^{\eta+\theta-\zeta}} \diamond_\alpha t. \tag{8.61}$$

Proof. It follows from (8.48) and formula (8.37) for $0 \le \eta + \theta < 1$ that we have

$$\begin{aligned}
[G^{1-\eta-\theta}(t)]^\Delta &= (1-\eta-\theta)G^\Delta(t) \int_0^1 \frac{dw}{[(1-w)G(t)+wG^\sigma(t)]^{\eta+\theta}} \\
&= \int_0^1 \frac{(1-\eta-\theta)[-\alpha z(t)-(1-\alpha)z^\sigma(t)]}{[(1-w)G(t)+wG^\sigma(t)]^{\eta+\theta}} dw \\
&\ge -(1-\eta-\theta)\frac{z(t)}{[G^\sigma(t)]^{\eta+\theta}},
\end{aligned} \tag{8.62}$$

where $G^\sigma(t) \le G(t)$ and the nonincreasingness property of z have been used.

(i) Using inequalities (8.46) and (8.62) in equation (8.50) yields

$$u^{\Delta}(t) \geq -(1-\eta-\theta)\frac{z(t)}{[G^{\sigma}(t)]^{\eta+\theta}}[H^{\sigma}(t)]^{\eta+\zeta} + \frac{\alpha(\eta+\zeta)z(t)h(t)[H^{\sigma}(t)]^{\eta+\zeta-1}}{[G(t)]^{\eta+\theta-1}},$$

or

$$\int_a^{\infty} \frac{z(t)[H^{\sigma}(t)]^{\eta+\zeta}}{[G^{\sigma}(t)]^{\eta+\theta}} \diamond_a t \geq \frac{\alpha(\eta+\zeta)}{1-\eta-\theta}\int_a^{\infty} \frac{z(t)h(t)[H^{\sigma}(t)]^{\eta+\zeta-1}}{[G(t)]^{\eta+\theta-1}} \diamond_a t - \frac{1}{1-\eta-\theta}\int_a^{\infty} u^{\Delta}(t)\diamond_a t.$$

If we employ inequality (8.52), we obtain from the above inequality that

$$\int_a^{\infty} \frac{z(t)[H^{\sigma}(t)]^{\eta+\zeta}}{[G^{\sigma}(t)]^{\eta+\theta}} \diamond_a t \geq \frac{\alpha(\eta+\zeta)}{1-\eta-\theta}\int_a^{\infty} \frac{z(t)h(t)[H^{\sigma}(t)]^{\eta+\zeta-1}}{[G(t)]^{\eta+\theta-1}} \diamond_a t,$$

which is the desired inequality (8.58).

Since

$$\begin{aligned}\int_a^{\infty} \frac{z(t)[H^{\sigma}(t)]^{\eta+\zeta}}{[G^{\sigma}(t)]^{\eta+\theta}} \diamond_a t &\geq \frac{\alpha(\eta+\zeta)}{1-\eta-\theta}\int_a^{\infty} \frac{z(t)h(t)[H^{\sigma}(t)]^{\eta+\zeta-1}}{[G(t)]^{\eta+\theta-1}} \diamond_a t \\ &\geq \frac{\alpha(\eta+\zeta)}{1-\eta-\theta}\int_a^{\infty} \frac{z(t)h(t)[H^{\sigma}(t)]^{\eta+\zeta-1}}{[G^{\sigma}(t)]^{\eta+\theta-1}} \diamond_a t,\end{aligned}$$

applying the reverse Hölder's inequality (8.40) with the constants $0 < \zeta < 1$ and $\zeta/(\zeta-1) < 0$ to the right-hand side of the above inequality leads to inequality (8.59).

(ii) Using inequalities (8.47) and (8.62) in equation (8.53) yields

$$u^{\Delta}(t) \geq -(1-\eta-\theta)\frac{z(t)}{[G^{\sigma}(t)]^{\eta+\theta}}[H(t)]^{\eta+\zeta} + \frac{\alpha(\eta+\zeta)z(t)h(t)[H(t)]^{\eta+\zeta-1}}{[G^{\sigma}(t)]^{\eta+\theta-1}},$$

or

$$\int_a^{\infty} \frac{z(t)[H(t)]^{\eta+\zeta}}{[G^{\sigma}(t)]^{\eta+\theta}} \diamond_a t \geq \frac{\alpha(\eta+\zeta)}{1-\eta-\theta}\int_a^{\infty} \frac{z(t)h(t)[H(t)]^{\eta+\zeta-1}}{[G^{\sigma}(t)]^{\eta+\theta-1}} \diamond_a t - \frac{1}{1-\eta-\theta}\int_a^{\infty} u^{\Delta}(t)\diamond_a t.$$

If we employ inequality (8.52), we obtain from the above inequality that

$$\int_a^{\infty} \frac{z(t)[H(t)]^{\eta+\zeta}}{[G^{\sigma}(t)]^{\eta+\theta}} \diamond_a t \geq \frac{\alpha(\eta+\zeta)}{1-\eta-\theta}\int_a^{\infty} \frac{z(t)h(t)[H(t)]^{\eta+\zeta-1}}{[G^{\sigma}(t)]^{\eta+\theta-1}} \diamond_a t,$$

which is the desired inequality (8.60).

Applying the reverse Hölder's inequality (8.40) with the constants $0 < \zeta < 1$ and $\zeta/(\zeta-1) < 0$ to the right-hand side of inequality (8.60), one can obtain inequality (8.61). □

Remark 8.4. The diamond alpha Bennett–Leindler-type inequalities (8.58)–(8.61) obtained for $0 < \zeta < 1$, $\eta \geq 0$, and $0 \leq \eta + \theta < 1$ are derived for the first time due to the condition $0 \leq \eta + \theta < 1$. Moreover, these inequalities generalize the diamond alpha Bennett–Leindler-type inequality (8.25) given in [34, Theorem 15] established for $0 < \zeta < 1$, $\eta = 0$, and $\theta \leq 0$ and complement the diamond alpha Bennett–Leindler-type inequality (8.31) given in [38, Theorem 9] established for $\zeta > 1$, $\eta \geq 0$, and $0 \leq \eta + \theta < 1$.

Remark 8.5. The condition $0 \leq \eta+\theta < 1$ with $0 < \zeta < 1$ has not appeared in the literature before even in the special cases. This is one of the gaps in the literature that this theorem aims to fill. By the novel diamond alpha Bennett–Leindler inequalities (8.58)–(8.61) obtained for $0 < \zeta < 1$, $\eta \geq 0$, and $0 \leq \eta + \theta < 1$ for the first time, this aim is achieved.

Remark 8.6. Special cases of the diamond alpha Bennett–Leindler-type inequalities (8.58)–(8.61) can be seen below.

(i) Expressing inequalities (8.58)–(8.59) in terms of delta and nabla integrals for $0 < \zeta < 1$, $\eta \geq 0, 0 \leq \eta + \theta < 1$, and $\eta + \zeta \leq 1$, and then choosing $\alpha = 1$ in those inequalities yields

$$\int_a^\infty \frac{z(t)[H^\sigma(t)]^{\eta+\zeta}}{[G^\sigma(t)]^{\eta+\theta}}\Delta t \geq \frac{\eta+\zeta}{1-\eta-\theta}\int_a^\infty \frac{z(t)h(t)[H^\sigma(t)]^{\eta+\zeta-1}}{[G(t)]^{\eta+\theta-1}}\Delta t \tag{8.63}$$

and

$$\int_a^\infty \frac{z(t)[H^\sigma(t)]^{\eta+\zeta}}{[G^\sigma(t)]^{\eta+\theta}}\Delta t \geq \left[\frac{\eta+\zeta}{1-\eta-\theta}\right]^\zeta \int_a^\infty \frac{z(t)h^\zeta(t)[H^\sigma(t)]^{\eta}}{[G^\sigma(t)]^{\eta+\theta-\zeta}}\Delta t, \tag{8.64}$$

respectively, where $H(t)$ and $G(t)$ are defined in (8.6). Inequalities (8.63)–(8.64) are novel even in the delta calculus. These novel inequalities generalize delta Bennett–Leindler-type inequalities (8.7)–(8.8) given in [59, Theorem 2.1] and [34, Remark 5] established for $0 < \zeta < 1$, $\eta = 0$, and $\theta \leq 0$ and complement delta Bennett–Leindler-type inequalities (8.13)–(8.14) given in [37, Corollary 2] and [38, Remark 3] established for $\zeta > 1$, $\eta \geq 0$ and $0 \leq \eta + \theta < 1$. We can conclude that (i) of Theorem 8.8 is a diamond alpha unification of Theorem 8.1 which is given in [59, Theorem 2.1] and [34, Remark 5] and is a generalization of the diamond alpha Bennett–Leindler-type inequality (8.25) established for $0 < \zeta < 1$, $\eta = 0$, and $\theta \leq 0$ given in [34, Theorem 15] and is a completion of the diamond alpha Bennett–Leindler-type inequality (8.31) established for $\zeta > 1$, $\eta \geq 0$, and $0 \leq \eta + \theta < 1$ given in [38, Theorem 9].

(ii) Expressing inequalities (8.60)–(8.61) in terms of delta and nabla integrals for $0 < \zeta < 1, \eta \geq 0, 0 \leq \eta+\theta < 1$, and $\eta+\zeta \geq 1$, and then choosing $\alpha = 1$ in those inequalities yields

$$\int_a^\infty \frac{z(t)[H(t)]^{\eta+\zeta}}{[G^\sigma(t)]^{\eta+\theta}} \Delta t \geq \frac{\eta+\zeta}{1-\eta-\theta} \int_a^\infty \frac{z(t)h(t)[H(t)]^{\eta+\zeta-1}}{[G^\sigma(t)]^{\eta+\theta-1}} \Delta t \tag{8.65}$$

and

$$\int_a^\infty \frac{z(t)[H(t)]^{\eta+\zeta}}{[G^\sigma(t)]^{\eta+\theta}} \Delta t \geq \left[\frac{\eta+\zeta}{1-\eta-\theta}\right]^\zeta \int_a^\infty \frac{z(t)h^\zeta(t)[H(t)]^{\eta}}{[G^\sigma(t)]^{\eta+\theta-\zeta}} \Delta t, \tag{8.66}$$

respectively, where $H(t), G(t)$ are defined in (8.6). Inequalities (8.65)–(8.66) are novel even in the delta calculus. These novel inequalities generalize delta Bennett–Leindler-type inequalities (8.7)–(8.8) given in [59, Theorem 2.1] and [34, Remark 5] established for $0 < \zeta < 1$, $\eta = 0$, and $\theta \leq 0$ and complement delta Bennett–Leindler-type inequalities (8.13)–(8.14) given in [37, Corollary 2] and [38, Remark 3] established for $\zeta > 1$, $\eta \geq 0$ and $0 \leq \eta+\theta < 1$. We can conclude that (ii) of Theorem 8.8 is a diamond alpha unification of Theorem 8.1 which is given in [59, Theorem 2.1] and [34, Remark 5] and is a generalization of the diamond alpha Bennett–Leindler-type inequality (8.25) established for $0 < \zeta < 1$, $\eta = 0$, and $\theta \leq 0$ given in [34, Theorem 15] and is a completion of the diamond alpha Bennett–Leindler-type inequality (8.31) established for $\zeta > 1$, $\eta \geq 0$, and $0 \leq \eta+\theta < 1$ given in [38, Theorem 9].

The following theorem asserts not only novel diamond alpha and delta Bennett–Leindler-type inequalities when $0 < \zeta < 1$, $\eta \geq 0$, and $\eta+\theta > 1$, but also complements the diamond alpha Bennett–Leindler-type inequalities given in [38, Theorem 10] established for $\zeta > 1$, $\eta \geq 0$, and $\eta+\theta > 1$, and generalizations of the diamond alpha Bennett–Leindler-type inequalities given in [34, Theorem 17] established for $0 < \zeta < 1$, $\eta = 0$, and $\theta > 1$. This novelty is caused by the condition $\eta+\zeta \geq 1$, which has not been considered so far.

Theorem 8.9. *Suppose that z is a nondecreasing function on $[a,\infty)_\mathbb{T}$. For the functions $G(t)$ and $H(t)$ defined in (8.24), let $0 < \zeta < 1$, $\eta \geq 0$, and $\eta+\theta > 1$.*

(i) *If $0 \leq \eta+\zeta < 1$, then*

$$\int_a^\infty \frac{z(t)[H^\sigma(t)]^{\eta+\zeta}}{[G(t)]^{\eta+\theta}} \diamond_\alpha t \geq \frac{\alpha(\eta+\zeta)}{\eta+\theta-1} \int_a^\infty \frac{z(t)h(t)[H^\sigma(t)]^{\eta+\zeta-1}}{[G(t)]^{\eta+\theta-1}} \diamond_\alpha t \tag{8.67}$$

and

$$\int_a^\infty \frac{z(t)[H^\sigma(t)]^{\eta+\zeta}}{[G(t)]^{\eta+\theta}} \diamond_\alpha t \geq \left[\frac{\alpha(\eta+\zeta)}{\eta+\theta-1}\right]^\zeta \int_a^\infty \frac{z(t)h^\zeta(t)[H^\sigma(t)]^{\eta}}{[G(t)]^{\eta+\theta-\zeta}} \diamond_\alpha t. \tag{8.68}$$

(ii) *If* $\eta+\zeta>1$, *then*

$$\int_a^\infty \frac{z(t)[H(t)]^{\eta+\zeta}}{[G(t)]^{\eta+\theta}}\diamond_\alpha t \geq \frac{\alpha(\eta+\zeta)}{\eta+\theta-1}\int_a^\infty \frac{z(t)h(t)[H(t)]^{\eta+\zeta-1}}{[G^\sigma(t)]^{\eta+\theta-1}}\diamond_\alpha t \tag{8.69}$$

and

$$\int_a^\infty \frac{z(t)[H(t)]^{\eta+\zeta}}{[G(t)]^{\eta+\theta}}\diamond_\alpha t \geq \left[\frac{\alpha(\eta+\zeta)}{\eta+\theta-1}\right]^\zeta \int_a^\infty \frac{z(t)h^\zeta(t)[H(t)]^{\eta}}{[G(t)]^{\eta+\theta-\zeta}}\diamond_\alpha t. \tag{8.70}$$

Proof. It follows from (8.48) and the formula (8.36) for $\eta+\theta>1$ that

$$\begin{aligned} \left[G^{1-\eta-\theta}(t)\right]^\Delta &= (1-\eta-\theta)G^\Delta(t)\int_0^1 \frac{dw}{[(1-w)G(t)+wG^\sigma(t)]^{\eta+\theta}} \\ &= \int_0^1 \frac{(1-\eta-\theta)[-\alpha z(t)-(1-\alpha)z^\sigma(t)]}{[(1-w)G(t)+wG^\sigma(t)]^{\eta+\theta}}dw \\ &\geq (\eta+\theta-1)\frac{z(t)}{[G(t)]^{\eta+\theta}}, \end{aligned} \tag{8.71}$$

where $G^\sigma(t)\leq G(t)$ and the nondecreasingness property of z have been used.

(i) Using inequalities (8.46) and (8.71) in equation (8.50) yields

$$u^\Delta(t) \geq (\eta+\theta-1)\frac{z(t)}{[G(t)]^{\eta+\theta}}\left[H^\sigma(t)\right]^{\eta+\zeta} + \frac{\alpha(\eta+\zeta)z(t)h(t)[H^\sigma(t)]^{\eta+\zeta-1}}{[G(t)]^{\eta+\theta-1}}$$

or

$$\begin{aligned} \int_a^\infty \frac{z(t)[H^\sigma(t)]^{\eta+\zeta}}{[G(t)]^{\eta+\theta}}\diamond_\alpha t &\geq \frac{\alpha(\eta+\zeta)}{1-\eta-\theta}\int_a^\infty \frac{z(t)h(t)[H^\sigma(t)]^{\eta+\zeta-1}}{[G(t)]^{\eta+\theta-1}}\diamond_\alpha t \\ &\quad - \frac{1}{1-\eta-\theta}\int_a^\infty u^\Delta(t)\diamond_\alpha t. \end{aligned}$$

If we employ inequality (8.52), we obtain from the above inequality that

$$\int_a^\infty \frac{z(t)[H^\sigma(t)]^{\eta+\zeta}}{[G(t)]^{\eta+\theta}}\diamond_\alpha t \geq \frac{\alpha(\eta+\zeta)}{1-\eta-\theta}\int_a^\infty \frac{z(t)h(t)[H^\sigma(t)]^{\eta+\zeta-1}}{[G(t)]^{\eta+\theta-1}}\diamond_\alpha t$$

which is the desired inequality (8.67).

Applying the reverse Hölder inequality (8.40) with the constants $0<\zeta<1$ and $\zeta/(\zeta-1)<0$ to the right hand side of inequality (8.67), one can obtain inequality (8.68).

(ii) Using inequalities (8.47) and (8.71) in equation (8.53) yields

$$u^{\Delta}(t) \geq (\eta+\theta-1)\frac{z(t)}{[G(t)]^{\eta+\theta}}[H(t)]^{\eta+\zeta} + \frac{\alpha(\eta+\zeta)z(t)h(t)[H(t)]^{\eta+\zeta-1}}{[G^{\sigma}(t)]^{\eta+\theta-1}}$$
$$\geq -(\eta+\theta-1)\frac{z(t)}{[G(t)]^{\eta+\theta}}[H(t)]^{\eta+\zeta} + \frac{\alpha(\eta+\zeta)z(t)h(t)[H(t)]^{\eta+\zeta-1}}{[G^{\sigma}(t)]^{\eta+\theta-1}}$$

or

$$\int_a^{\infty} \frac{z(t)[H(t)]^{\eta+\zeta}}{[G(t)]^{\eta+\theta}}\diamond_a t \geq \frac{\alpha(\eta+\zeta)}{\eta+\theta-1}\int_a^{\infty} \frac{z(t)h(t)[H(t)]^{\eta+\zeta-1}}{[G^{\sigma}(t)]^{\eta+\theta-1}}\diamond_a t$$
$$-\frac{1}{1-\eta-\theta}\int_a^{\infty} u^{\Delta}(t)\diamond_a t.$$

If we employ inequality (8.52), we obtain from the above inequality that

$$\int_a^{\infty} \frac{z(t)[H(t)]^{\eta+\zeta}}{[G(t)]^{\eta+\theta}}\diamond_a t \geq \frac{\alpha(\eta+\zeta)}{\eta+\theta-1}\int_a^{\infty} \frac{z(t)h(t)[H(t)]^{\eta+\zeta-1}}{[G^{\sigma}(t)]^{\eta+\theta-1}}\diamond_a t,$$

which is the desired inequality (8.69).

Since

$$\int_a^{\infty} \frac{z(t)[H(t)]^{\eta+\zeta}}{[G(t)]^{\eta+\theta}}\diamond_a t \geq \frac{\alpha(\eta+\zeta)}{\eta+\theta-1}\int_a^{\infty} \frac{z(t)h(t)[H(t)]^{\eta+\zeta-1}}{[G^{\sigma}(t)]^{\eta+\theta-1}}\diamond_a t$$
$$\geq \frac{\alpha(\eta+\zeta)}{\eta+\theta-1}\int_a^{\infty} \frac{z(t)h(t)[H(t)]^{\eta+\zeta-1}}{[G(t)]^{\eta+\theta-1}}\diamond_a t,$$

applying the reverse Hölder inequality (8.40) with the constants $0 < \zeta < 1$ and $\zeta/(\zeta-1) < 0$ to the right hand side of the above inequality leads to inequality (8.70). □

Remark 8.7. The diamond alpha Bennett–Leindler-type inequalities (8.67)–(8.70) obtained for $0 < \zeta < 1$, $\eta \geq 0$, and $\eta + \theta > 1$ are generalizations of the diamond alpha Bennett–Leindler-type inequality (8.27) established for $0 < \zeta < 1$, $\eta = 0$, and $\theta > 1$ given in [34, Theorem 17] and complements of the diamond alpha Bennett–Leindler-type inequality (8.33) established for $\zeta > 1$, $\eta \geq 0$, and $\eta + \theta > 1$ given in [38, Theorem 10].

Remark 8.8. Although the special case of the condition $\eta + \zeta \leq 1$ is automatically satisfied in [34, Theorem 17], the other case, $\eta + \zeta \geq 1$ with $0 < \zeta < 1$, has not appeared in the literature before even in the special cases. This is one of the gaps in the literature that this theorem aims to fill. By the novel diamond alpha Bennett–Leindler inequalities (8.69)–(8.70) obtained for $0 < \zeta < 1$, $\eta \geq 0, \eta + \theta > 1$, and $\eta + \zeta \geq 1$ for the first time, this aim is achieved.

Remark 8.9. Special cases of the diamond alpha Bennett–Leindler-type inequalities (8.67)–(8.70) can be seen below.

(i) Expressing inequalities (8.67)–(8.68) in terms of delta and nabla integrals for $0 < \zeta < 1$, $\eta \geq 0$, $\eta + \theta > 1$, and $0 \leq \eta + \zeta \leq 1$, and then choosing $\alpha = 1$ in those inequalities yields

$$\int_a^\infty \frac{z(t)[H^\sigma(t)]^{\eta+\zeta}}{[G(t)]^{\eta+\theta}} \Delta t \geq \frac{\eta+\zeta}{\eta+\theta-1} \int_a^\infty \frac{z(t)h(t)[H^\sigma(t)]^{\eta+\zeta-1}}{[G(t)]^{\eta+\theta-1}} \Delta t \tag{8.72}$$

and

$$\int_a^\infty \frac{z(t)[H^\sigma(t)]^{\eta+\zeta}}{[G(t)]^{\eta+\theta}} \Delta t \geq \left[\frac{\eta+\zeta}{\eta+\theta-1}\right]^\zeta \int_a^\infty \frac{z(t)h^\zeta(t)[H^\sigma(t)]^{\eta}}{[G(t)]^{\eta+\theta-\zeta}} \Delta t, \tag{8.73}$$

respectively, where $H(t), G(t)$ are defined in (8.6). Inequalities (8.72)–(8.73) generalize delta Bennett–Leindler-type inequalities (8.9)–(8.10) established for $0 < \zeta < 1$, $\eta = 0$, and $\theta > 1$ given in [34, Remark 7] and complement delta Bennett–Leindler-type inequalities in [38, Remark 5] established for $\zeta > 1$, $\eta \geq 0$, and $\eta + \theta > 1$. We can conclude that (i) of Theorem 8.9 is a diamond alpha unification of Theorem 8.1 given in [34, Remark 7] and a generalization of the diamond alpha Bennett–Leindler-type inequality (8.27) established for $0 < \zeta < 1$, $\eta = 0$, and $\theta > 1$ given in [34, Theorem 17] and a completion of the diamond alpha Bennett–Leindler-type inequality (8.33) established for $\zeta > 1$, $\eta \geq 0$, and $\eta + \theta > 1$ given in [38, Theorem 10].

(ii) Expressing inequalities (8.69)–(8.70) in terms of delta and nabla integrals for $0 < \zeta < 1$, $\eta \geq 0$, $\eta + \theta > 1$, and $\eta + \zeta \geq 1$, and then choosing $\alpha = 1$ in those inequalities yields

$$\int_a^\infty \frac{z(t)[H(t)]^{\eta+\zeta}}{[G(t)]^{\eta+\theta}} \Delta t \geq \frac{\eta+\zeta}{\eta+\theta-1} \int_a^\infty \frac{z(t)h(t)[H(t)]^{\eta+\zeta-1}}{[G^\sigma(t)]^{\eta+\theta-1}} \Delta t \tag{8.74}$$

and

$$\int_a^\infty \frac{z(t)[H(t)]^{\eta+\zeta}}{[G(t)]^{\eta+\theta}} \Delta t \geq \left[\frac{\eta+\zeta}{\eta+\theta-1}\right]^\zeta \int_a^\infty \frac{z(t)h^\zeta(t)[H(t)]^{\eta}}{[G(t)]^{\eta+\theta-\zeta}} \Delta t, \tag{8.75}$$

respectively, where $H(t), G(t)$ are defined in (8.6). Inequalities (8.74)–(8.75) are novel even in the delta calculus. These novel inequalities generalize delta Bennett–Leindler-type inequalities (8.9)–(8.10) given in [34, Remark 7] established for $0 < \zeta < 1$, $\eta = 0$, and $\theta > 1$ and complement delta Bennett–Leindler-type inequalities [38, Remark 5] established for $\zeta > 1$, $\eta \geq 0$, and $\eta + \theta > 1$. We can conclude that (ii) of Theorem 8.9 is a diamond alpha unification of Theorem 8.1 which is given in [34, Remark 5] and is a generalization of the diamond alpha Bennett–Leindler-type inequality (8.27) established for $0 < \zeta < 1$, $\eta = 0$, and $\theta > 1$ given in [34, Theorem 17] and is a completion of the diamond alpha Bennett–Leindler-type inequality (8.33) established for $\zeta > 1$, $\eta \geq 0$, and $\eta + \theta > 1$ given in [38, Theorem 10].

The following theorem asserts not only novel diamond alpha and nabla Bennett–Leindler-type inequalities when $0 < \zeta < 1$, $\eta \geq 0$, and $\eta+\theta \leq 0$, but also complements the diamond alpha Bennett–Leindler-type inequalities given in [38, Theorem 11] established for $\zeta > 1$, $\eta \geq 0$, and $\eta + \theta \leq 0$, and generalizations of the diamond alpha Bennett–Leindler-type inequalities given in [34, Theorem 16] established for $0 < \zeta < 1$, $\eta = 0$, and $\theta \leq 0$. This novelty is caused by the condition $\eta + \zeta \geq 1$, which has not been considered so far.

Theorem 8.10. *Suppose that z is a nondecreasing function on $[a,\infty)_{\mathbb{T}}$. For the functions $\overline{G}(t)$ and $\overline{H}(t)$ defined in (8.24), assume that there exists $L_1 > 0$ such that $\frac{\overline{G}(t)}{\overline{G}^{\rho}(t)} \leq L_1$ for $t \in (a,\infty)_{\mathbb{T}}$. Let $0 < \zeta < 1$, $\eta \geq 0$, and $\eta + \theta \leq 0$ be real constants.*

(i) *If $0 \leq \eta + \zeta \leq 1$, then*

$$\int_a^{\infty} \frac{z(t)[\overline{H}^{\rho}(t)]^{\eta+\zeta}}{[\overline{G}(t)]^{\eta+\theta}} \diamond_\alpha t \geq \frac{(1-\alpha)(\eta+\zeta)}{1-\eta-\theta} \int_a^{\infty} \frac{z(t)h(t)[\overline{H}^{\rho}(t)]^{\eta+\zeta-1}}{[\overline{G}(t)]^{\eta+\theta-1}} \diamond_\alpha t \tag{8.76}$$

and

$$\int_a^{\infty} \frac{z(t)[\overline{H}^{\rho}(t)]^{\eta+\zeta}}{[\overline{G}(t)]^{\eta+\theta}} \diamond_\alpha t \geq \left[\frac{(1-\alpha)(\eta+\zeta)}{1-\eta-\theta}\right]^{\zeta} \int_a^{\infty} \frac{z(t)h^{\zeta}(t)[\overline{H}^{\rho}(t)]^{\eta}}{[\overline{G}(t)]^{\eta+\theta-\zeta}} \diamond_\alpha t. \tag{8.77}$$

(ii) *If $\eta + \zeta \geq 1$, then*

$$\int_a^{\infty} \frac{z(t)[\overline{H}(t)]^{\eta+\zeta}}{[\overline{G}(t)]^{\eta+\theta}} \diamond_\alpha t \geq \frac{(1-\alpha)(\eta+\zeta)}{1-\eta-\theta} \int_a^{\infty} \frac{z(t)h(t)[\overline{H}(t)]^{\eta+\zeta-1}}{[\overline{G}^{\rho}(t)]^{\eta+\theta-1}} \diamond_\alpha t \tag{8.78}$$

and

$$\int_a^{\infty} \frac{z(t)[\overline{H}(t)]^{\eta+\zeta}}{[\overline{G}(t)]^{\eta+\theta}} \diamond_\alpha t \geq \left[\frac{L_1^{\eta+\theta-1}(1-\alpha)(\eta+\zeta)}{1-\eta-\theta}\right]^{\zeta} \int_a^{\infty} \frac{z(t)h^{\zeta}(t)[\overline{H}(t)]^{\eta}}{[\overline{G}(t)]^{\eta+\theta-\zeta}} \diamond_\alpha t. \tag{8.79}$$

Proof. First we obtain some inequalities which will be used in the sequel. By utilizing Lemma 8.3, one can obtain

$$\begin{aligned} \overline{H}^{\nabla}(t) &= \left[\int_t^{\infty} z(s)h(s)\diamond_\alpha s\right]^{\nabla} \\ &= \alpha\left[\int_t^{\infty} z(s)h(s)\Delta s\right]^{\nabla} + (1-\alpha)\left[\int_t^{\infty} z(s)h(s)\nabla s\right]^{\nabla} \\ &= -\alpha z^{\rho}(t)h^{\rho}(t) - (1-\alpha)z(t)h(t) \leq 0. \end{aligned} \tag{8.80}$$

By taking account of (8.80) and employing formula (8.39), one can observe that

$$
\begin{aligned}
[\overline{H}^{\eta+\zeta}(t)]^{\nabla} &= (\eta+\zeta)\overline{H}^{\nabla}(t)\int_0^1 [w\overline{H}(t)+(1-w)\overline{H}^{\rho}(t)]^{\eta+\zeta-1}dw \\
&= (\eta+\zeta)[-\alpha z^{\rho}(t)h^{\rho}(t)-(1-\alpha)z(t)h(t)] \\
&\quad\times\int_0^1 [w\overline{H}(t)+(1-w)\overline{H}^{\rho}(t)]^{\eta+\zeta-1}dw.
\end{aligned}
$$

Then two estimates can be obtained depending on whether $\eta+\zeta\le 1$ or $\eta+\zeta\ge 1$ for the function $[\overline{H}^{\eta+\zeta}(t)]^{\nabla}$ as follows:

(i) If $\eta+\zeta\le 1$, then

$$
[\overline{H}^{\eta+\zeta}(t)]^{\nabla} \le -(1-\alpha)(\eta+\zeta)z(t)h(t)[\overline{H}^{\rho}(t)]^{\eta+\zeta-1}; \tag{8.81}
$$

(ii) If $\eta+\zeta\ge 1$, then

$$
[\overline{H}^{\eta+\zeta}(t)]^{\nabla} \le -(1-\alpha)(\eta+\zeta)z(t)h(t)[\overline{H}(t)]^{\eta+\zeta-1}, \tag{8.82}
$$

where $\overline{H}^{\rho}(t)\ge\overline{H}(t)$ has been used.

Similarly, by Lemma 8.3, we note that

$$
\begin{aligned}
\overline{G}^{\nabla}(t) &= \left[\int_a^t z(s)\diamond_{\alpha}s\right]^{\nabla} = \alpha\left[\int_a^t z(s)\Delta s\right]^{\nabla} + (1-\alpha)\left[\int_a^t z(s)\nabla s\right]^{\nabla} \\
&= \alpha z^{\rho}(t)+(1-\alpha)z(t)\ge 0.
\end{aligned} \tag{8.83}
$$

It follows from (8.83) and formula (8.39) for $\eta+\theta\le 0$ that we have

$$
\begin{aligned}
[\overline{G}^{1-\eta-\theta}(t)]^{\nabla} &= \int_0^1 \frac{(1-\eta-\theta)\overline{G}^{\nabla}(t)dw}{[w\overline{G}(t)+(1-w)\overline{G}^{\rho}(t)]^{\eta+\theta}} \\
&= \int_0^1 \frac{(1-\eta-\theta)[\alpha z^{\rho}(t)+(1-\alpha)z(t)]dw}{[w\overline{G}(t)+(1-w)\overline{G}^{\rho}(t)]^{\eta+\theta}} \\
&\le \frac{(1-\eta-\theta)z(t)}{[\overline{G}(t)]^{\eta+\theta}},
\end{aligned} \tag{8.84}
$$

where $\overline{G}(t)\ge\overline{G}(\rho(t))$ and the nondecreasingness property of z have been used.

(i) Let us define $u(t)=[\overline{H}(t)]^{\eta+\zeta}[\overline{G}(t)]^{1-\eta-\theta}$ for $t\in[a,\infty)$. If we take the nabla derivative of the function u using formula (8.38), we get

$$
u^{\nabla}(t) = [\overline{H}^{\eta+\zeta}(t)]^{\nabla}[\overline{G}^{1-\eta-\theta}(t)] + [\overline{H}^{\rho}(t)]^{\eta+\zeta}[\overline{G}^{1-\eta-\theta}(t)]^{\nabla}. \tag{8.85}
$$

Using inequalities (8.81) and (8.84) in equation (8.85) yields

$$u^\nabla(t) \le \frac{(1-\eta-\theta)z(t)}{[\overline{G}(t)]^{\eta+\theta}}\left[\overline{H}^\rho(t)\right]^{\eta+\zeta} - \frac{(1-\alpha)(\eta+\zeta)z(t)h(t)[\overline{H}^\rho(t)]^{\eta+\zeta-1}}{[\overline{G}(t)]^{\eta+\theta-1}},$$

or

$$\begin{aligned}\int_a^\infty \frac{z(t)[\overline{H}^\rho(t)]^{\eta+\zeta}}{[\overline{G}(t)]^{\eta+\theta}}\diamond_\alpha t &\ge \frac{(1-\alpha)(\eta+\zeta)}{1-\eta-\theta}\int_a^\infty \frac{z(t)h(t)[\overline{H}^\rho(t)]^{\eta+\zeta-1}}{[\overline{G}(t)]^{\eta+\theta-1}}\diamond_\alpha t \\ &\quad + \frac{1}{1-\eta-\theta}\int_a^\infty u^\nabla(t)\diamond_\alpha t.\end{aligned} \tag{8.86}$$

The definition of u implies $u(\infty) = u(a) = 0$ and, by using Lemma 8.4, we obtain

$$\begin{aligned}\int_a^\infty u^\nabla(t)\diamond_\alpha t &= \alpha\int_a^\infty u^\nabla(t)\Delta t + (1-\alpha)\int_a^\infty u^\nabla(t)\nabla t \\ &= \alpha\big[u(\rho(\infty)) - u(\rho(a))\big] + (1-\alpha)\big[u(\infty) - u(a)\big] \ge 0,\end{aligned} \tag{8.87}$$

where we have imposed that $(-1)^{\eta+\theta} = 1$.

Therefore we can infer that inequality (8.86) becomes

$$\int_a^\infty \frac{z(t)[\overline{H}^\rho(t)]^{\eta+\zeta}}{[\overline{G}(t)]^{\eta+\theta}}\diamond_\alpha t \ge \frac{(1-\alpha)(\eta+\zeta)}{1-\eta-\theta}\int_a^\infty \frac{z(t)h(t)[\overline{H}^\rho(t)]^{\eta+\zeta-1}}{[\overline{G}(t)]^{\eta+\theta-1}}\diamond_\alpha t,$$

which is the desired inequality (8.76).

Inequality (8.77) can be obtained by applying the reverse Hölder's inequality (8.40) with the constants $\zeta < 1$ and $\zeta/(\zeta-1) < 0$ to the right-hand side of inequality (8.76).

(ii) Let us define $u(t) = [\overline{H}^{\eta+\zeta}(t)][\overline{G}^{1-\eta-\theta}(t)]$ for $t \in [a,\infty)$. If we take the nabla derivative of the function u by using the formula (8.38), we get

$$u^\nabla(t) = [\overline{H}^{\eta+\zeta}(t)][\overline{G}^{1-\eta-\theta}(t)]^\nabla + [\overline{H}^{\eta+\zeta}(t)]^\nabla [\overline{G}^\rho(t)]^{1-\eta-\theta}. \tag{8.88}$$

Using inequalities (8.82) and (8.84) in equation (8.88) yields

$$u^\nabla(t) \le \frac{(1-\eta-\theta)z(t)[\overline{H}(t)]^{\eta+\zeta}}{[\overline{G}(t)]^{\eta+\theta}} - \frac{(1-\alpha)(\eta+\zeta)z(t)h(t)[\overline{H}(t)]^{\eta+\zeta-1}}{[\overline{G}^\rho(t)]^{\eta+\theta-1}},$$

or

$$\begin{aligned}\int_a^\infty \frac{z(t)[\overline{H}(t)]^{\eta+\zeta}}{[\overline{G}(t)]^{\eta+\theta}}\diamond_\alpha t &\ge \frac{(1-\alpha)(\eta+\zeta)}{1-\eta-\theta}\int_a^\infty \frac{z(t)h(t)[\overline{H}(t)]^{\eta+\zeta-1}}{[\overline{G}^\rho(t)]^{\eta+\theta-1}}\diamond_\alpha t \\ &\quad + \frac{1}{1-\eta-\theta}\int_a^\infty u^\nabla(t)\diamond_\alpha t.\end{aligned}$$

If we employ inequality (8.87), we obtain from the above inequality that

$$\int_a^\infty \frac{z(t)[\overline{H}(t)]^{\eta+\zeta}}{[\overline{G}(t)]^{\eta+\theta}} \diamond_\alpha t \geq \frac{(1-\alpha)(\eta+\zeta)}{1-\eta-\theta} \int_a^\infty \frac{z(t)h(t)[\overline{H}(t)]^{\eta+\zeta-1}}{[\overline{G}^\rho(t)]^{\eta+\theta-1}} \diamond_\alpha t,$$

which is the desired inequality (8.78).

Using $\frac{\overline{G}(t)}{\overline{G}^\rho(t)} \leq L_1$ on the right-hand side of inequality (8.78) and applying the reverse Hölder inequality (8.40) with the constants $\zeta < 1$ and $\zeta/(\zeta-1) < 0$ to the resulting integral, one can obtain inequality (8.79). □

Remark 8.10. The diamond alpha Bennett–Leindler-type inequalities (8.76)–(8.79) obtained for $0 < \zeta < 1$, $\eta \geq 0$, and $\eta + \theta \leq 0$ are generalizations of the diamond alpha Bennett–Leindler-type inequality (8.26) given in [34, Theorem 16] established for $0 < \zeta < 1$, $\eta = 0$, and $\theta \leq 0$, and complements of the diamond alpha Bennett–Leindler-type inequality (8.30) given in [38, Theorem 11] established for $\zeta > 1$, $\eta \geq 0$, and $\eta+\theta \leq 0$.

Remark 8.11. Although a special case of the condition $\eta+\zeta \leq 1$ is automatically satisfied in [34, Theorem 16], the other case, $\eta+\zeta \geq 1$ with $0 < \zeta < 1$, has not appeared in the literature before even for the special cases. This is one of the gaps in the literature that this theorem aims to fill. By the novel diamond alpha Bennett–Leindler inequalities (8.78)–(8.79) obtained for $0 < \zeta < 1$, $\eta \geq 0, \eta+\theta \leq 0$, and $\eta+\zeta \geq 1$ for the first time, this aim is achieved.

Remark 8.12. Special cases of the diamond alpha Bennett–Leindler-type inequalities (8.76)–(8.79) can be seen below.

(i) Expressing inequalities (8.76)–(8.77) in terms of delta and nabla integrals for $0 < \zeta < 1$, $\eta \geq 0, \eta + \theta \leq 0$, and $\eta + \zeta \leq 1$, and then choosing $\alpha = 0$ in those inequalities yields

$$\int_a^\infty \frac{z(t)[\overline{H}^\rho(t)]^{\eta+\zeta}}{[\overline{G}(t)]^{\eta+\theta}} \nabla t \geq \frac{\eta+\zeta}{1-\eta-\theta} \int_a^\infty \frac{z(t)h(t)[\overline{H}^\rho(t)]^{\eta+\zeta-1}}{[\overline{G}(t)]^{\eta+\theta-1}} \nabla t \tag{8.89}$$

and

$$\int_a^\infty \frac{z(t)[\overline{H}^\rho(t)]^{\eta+\zeta}}{[\overline{G}(t)]^{\eta+\theta}} \nabla t \geq \left[\frac{\eta+\zeta}{1-\eta-\theta}\right]^\zeta \int_a^\infty \frac{z(t)h^\zeta(t)[\overline{H}^\rho(t)]^{\eta}}{[\overline{G}(t)]^{\eta+\theta-\zeta}} \nabla t, \tag{8.90}$$

respectively, where $\overline{H}(t)$ and $\overline{G}(t)$ are defined in (8.24). Inequalities (8.89)–(8.90) generalize nabla Bennett–Leindler-type inequalities (8.16)–(8.17) established for $0 < \zeta < 1$, $\eta = 0$, and $\theta \leq 0$ given in [29, Theorem 3.9] and [34, Remark 6] and complement nabla Bennett–Leindler-type inequalities (8.20)–(8.21) established for $\zeta > 1$, $\eta \geq 0$, and $\eta + \theta \leq 0$ given in [37, Theorem 7] and [38, Remark 7]. We can conclude that (i) of Theorem 8.10 is a diamond alpha unification of Theorem 8.2 which is given in [29, Theorem 3.9] and [34, Remark 6] and is a generalization of the diamond alpha Bennett–Leindler-type inequality (8.26) established for $0 < \zeta < 1$, $\eta = 0$, and $\theta \leq 0$ given in [34, Theorem 16] and is a completion of the diamond alpha Bennett–

Leindler-type inequality (8.30) established for $\zeta > 1$, $\eta \geq 0$, and $\eta + \theta \leq 0$ given in [38, Theorem 11].

(ii) Expressing inequalities (8.78)–(8.79) in terms of delta and nabla integrals for $0 < \zeta < 1$, $\eta \geq 0, \eta+\theta \leq 0$, and $\eta+\zeta \geq 1$, and then choosing $\alpha = 0$ in those inequalities yields

$$\int_a^\infty \frac{z(t)[\overline{H}(t)]^{\eta+\zeta}}{[\overline{G}(t)]^{\eta+\theta}} \nabla t \geq \frac{\eta+\zeta}{1-\eta-\theta} \int_a^\infty \frac{z(t)h(t)[\overline{H}(t)]^{\eta+\zeta-1}}{[\overline{G}^\rho(t)]^{\eta+\theta-1}} \nabla t \tag{8.91}$$

and

$$\int_a^\infty \frac{z(t)[\overline{H}(t)]^{\eta+\zeta}}{[\overline{G}(t)]^{\eta+\theta}} \nabla t \geq \left[\frac{L_1^{\eta+\theta-1}\eta+\zeta}{1-\eta-\theta}\right]^\zeta \int_a^\infty \frac{z(t)h^\zeta(t)[\overline{H}(t)]^\eta}{[\overline{G}(t)]^{\eta+\theta-\zeta}} \nabla t, \tag{8.92}$$

respectively, where $\overline{H}(t)$ and $\overline{G}(t)$ and L_1 are defined in (8.24). Inequalities (8.91)–(8.92) are novel even in the nabla calculus. These novel inequalities generalize nabla Bennett–Leindler-type inequalities (8.16)–(8.17) in [29, Theorem 3.9] and [34, Remark 6] established for $0 < \zeta < 1$, $\eta = 0$ and $\theta \leq 0$ and complement nabla Bennett–Leindler-type inequalities (8.20)–(8.21) in [37, Theorem 7] and [38, Remark 7] established for $\zeta > 1$, $\eta \geq 0$, and $\eta + \theta \leq 0$. We can conclude that (ii) of Theorem 8.10 is a diamond alpha unification of Theorem 8.2 which is given in [29, Theorem 3.9] and [34, Remark 6] and is a generalization of the diamond alpha Bennett–Leindler-type inequality (8.26) established for $0 < \zeta < 1$, $\eta = 0$, and $\theta \leq 0$ given in [34, Theorem 16] and is a completion of the diamond alpha Bennett–Leindler-type inequality (8.30) established for $\zeta > 1$, $\eta \geq 0$, and $\eta + \theta \leq 0$ given in [38, Theorem 11].

The following theorem asserts not only novel diamond alpha and nabla Bennett–Leindler-type inequalities when $0 < \zeta < 1$, $\eta \geq 0$, and $0 \leq \eta + \theta < 1$, but also complements the diamond alpha Bennett–Leindler-type inequalities given in [38, Theorem 11] established for $\zeta > 1$, $\eta \geq 0$, and $0 \leq \eta + \theta < 1$, and generalization of the diamond alpha Bennett–Leindler-type inequalities given in [34, Theorem 16] established for $0 < \zeta < 1$, $\eta = 0$, and $\theta \leq 0$. This novelty is caused by the condition $\eta + \zeta \geq 1$, which has not been considered so far.

Theorem 8.11. *Suppose that z is a nondecreasing function on $[a,\infty)_\mathbb{T}$. For the functions $\overline{G}(t)$ and $\overline{H}(t)$ defined in (8.24), let $0 < \zeta < 1$, $\eta \geq 0$ and $0 \leq \eta + \theta < 1$ be real constants.*

(i) *If $0 \leq \eta + \zeta \leq 1$, then*

$$\int_a^\infty \frac{z(t)[\overline{H}^\rho(t)]^{\eta+\zeta}}{[\overline{G}^\rho(t)]^{\eta+\theta}} \diamond_\alpha t \geq \frac{(1-\alpha)(\eta+\zeta)}{1-\eta-\theta} \int_a^\infty \frac{z(t)h(t)[\overline{H}^\rho(t)]^{\eta+\zeta-1}}{[\overline{G}(t)]^{\eta+\theta-1}} \diamond_\alpha t \tag{8.93}$$

and

$$\int_a^\infty \frac{z(t)[\overline{H}^\rho(t)]^{\eta+\zeta}}{[\overline{G}^\rho(t)]^{\eta+\theta}} \diamond_\alpha t \geq \left[\frac{(1-\alpha)(\eta+\zeta)}{1-\eta-\theta}\right]^\zeta \int_a^\infty \frac{z(t)h^\zeta(t)[\overline{H}^\rho(t)]^\eta}{[\overline{G}^\rho(t)]^{\eta+\theta-\zeta}} \diamond_\alpha t. \tag{8.94}$$

(ii) *If* $\eta + \zeta \geq 1$, *then*

$$\int_a^\infty \frac{z(t)[\overline{H}(t)]^{\eta+\zeta}}{[\overline{G}^\rho(t)]^{\eta+\theta}} \diamond_\alpha t \geq \frac{(1-\alpha)(\eta+\zeta)}{1-\eta-\theta} \int_a^\infty \frac{z(t)h(t)[\overline{H}(t)]^{\eta+\zeta-1}}{[\overline{G}^\rho(t)]^{\eta+\theta-1}} \diamond_\alpha t \tag{8.95}$$

and

$$\int_a^\infty \frac{z(t)[\overline{H}(t)]^{\eta+\zeta}}{[\overline{G}^\rho(t)]^{\eta+\theta}} \diamond_\alpha t \geq \left[\frac{(1-\alpha)(\eta+\zeta)}{1-\eta-\theta}\right]^\zeta \int_a^\infty \frac{z(t)h^\zeta(t)[\overline{H}(t)]^\eta}{[\overline{G}^\rho(t)]^{\eta+\theta-\zeta}} \diamond_\alpha t. \tag{8.96}$$

Proof. It follows from (8.83) and formula (8.39) for $0 \leq \eta + \theta < 1$ that we have

$$\begin{aligned}
[\overline{G}^{1-\eta-\theta}(t)]^\nabla &= \int_0^1 \frac{(1-\eta-\theta)\overline{G}^\nabla(t)dw}{[w\overline{G}(t) + (1-w)\overline{G}^\rho(t)]^{\eta+\theta}} \\
&= \int_0^1 \frac{(1-\eta-\theta)[\alpha z^\rho(t) + (1-\alpha)z(t)]dw}{[w\overline{G}(t) + (1-w)\overline{G}^\rho(t)]^{\eta+\theta}} \\
&\leq \frac{(1-\eta-\theta)z(t)}{[\overline{G}^\rho(t)]^{\eta+\theta}},
\end{aligned} \tag{8.97}$$

where $\overline{G}(t) \geq \overline{G}(\rho(t))$ and the nondecreasingness property of z have been used.

(i) Using inequalities (8.81) and (8.97) in equation (8.85) yields

$$u^\nabla(t) \leq \frac{(1-\eta-\theta)z(t)}{[\overline{G}^\rho(t)]^{\eta+\theta}}[\overline{H}^\rho(t)]^{\eta+\zeta} - \frac{(1-\alpha)(\eta+\zeta)z(t)h(t)[\overline{H}^\rho(t)]^{\eta+\zeta-1}}{[\overline{G}(t)]^{\eta+\theta-1}},$$

or

$$\begin{aligned}
\int_a^\infty \frac{z(t)[\overline{H}^\rho(t)]^{\eta+\zeta}}{[\overline{G}^\rho(t)]^{\eta+\theta}} \diamond_\alpha t \geq{} & \frac{(1-\alpha)(\eta+\zeta)}{1-\eta-\theta} \int_a^\infty \frac{z(t)h(t)[\overline{H}^\rho(t)]^{\eta+\zeta-1}}{[\overline{G}(t)]^{\eta+\theta-1}} \diamond_\alpha t \\
& + \frac{1}{1-\eta-\theta} \int_a^\infty u^\nabla(t) \diamond_\alpha t.
\end{aligned}$$

If we employ inequality (8.87), we obtain from the above inequality that

$$\int_a^\infty \frac{z(t)[\overline{H}^\rho(t)]^{\eta+\zeta}}{[\overline{G}^\rho(t)]^{\eta+\theta}} \diamond_\alpha t \geq \frac{(1-\alpha)(\eta+\zeta)}{1-\eta-\theta} \int_a^\infty \frac{z(t)h(t)[\overline{H}^\rho(t)]^{\eta+\zeta-1}}{[\overline{G}(t)]^{\eta+\theta-1}} \diamond_\alpha t,$$

which is the desired inequality (8.93).

Since

$$\int_a^\infty \frac{z(t)[\overline{H}^\rho(t)]^{\eta+\zeta}}{[\overline{G}^\rho(t)]^{\eta+\theta}}\diamondsuit_a t \geq \frac{(1-\alpha)(\eta+\zeta)}{1-\eta-\theta}\int_a^\infty \frac{z(t)h(t)[\overline{H}^\rho(t)]^{\eta+\zeta-1}}{[\overline{G}(t)]^{\eta+\theta-1}}\diamondsuit_a t$$
$$\geq \frac{(1-\alpha)(\eta+\zeta)}{1-\eta-\theta}\int_a^\infty \frac{z(t)h(t)[\overline{H}^\rho(t)]^{\eta+\zeta-1}}{[\overline{G}^\rho(t)]^{\eta+\theta-1}}\diamondsuit_a t,$$

inequality (8.94) can be obtained by applying the reverse Hölder's inequality (8.40) with the constants $\zeta < 1$ and $\zeta/(\zeta-1) < 0$ to the right-hand side of inequality (8.93).

(ii) Using inequalities (8.82) and (8.97) in equation (8.88) yields

$$u^\nabla(t) \leq \frac{(1-\eta-\theta)z(t)[\overline{H}(t)]^{\eta+\zeta}}{[\overline{G}^\rho(t)]^{\eta+\theta}} - \frac{(1-\alpha)(\eta+\zeta)z(t)h(t)[\overline{H}(t)]^{\eta+\zeta-1}}{[\overline{G}^\rho(t)]^{\eta+\theta-1}},$$

or

$$\int_a^\infty \frac{z(t)[\overline{H}(t)]^{\eta+\zeta}}{[\overline{G}^\rho(t)]^{\eta+\theta}}\diamondsuit_a t \geq \frac{(1-\alpha)(\eta+\zeta)}{1-\eta-\theta}\int_a^\infty \frac{z(t)h(t)[\overline{H}(t)]^{\eta+\zeta-1}}{[\overline{G}^\rho(t)]^{\eta+\theta-1}}\diamondsuit_a t$$
$$+\frac{1}{1-\eta-\theta}\int_a^\infty u^\nabla(t)\diamondsuit_a t.$$

If we employ inequality (8.87), we obtain from the above inequality that

$$\int_a^\infty \frac{z(t)[\overline{H}(t)]^{\eta+\zeta}}{[\overline{G}^\rho(t)]^{\eta+\theta}}\diamondsuit_a t \geq \frac{(1-\alpha)(\eta+\zeta)}{1-\eta-\theta}\int_a^\infty \frac{z(t)h(t)[\overline{H}(t)]^{\eta+\zeta-1}}{[\overline{G}^\rho(t)]^{\eta+\theta-1}}\diamondsuit_a t,$$

which is the desired inequality (8.95).

Applying the reverse Hölder's inequality (8.40) with the constants $\zeta < 1$ and $\zeta/(\zeta-1) < 0$ to the right-hand side of inequality (8.95), one can obtain inequality (8.96). □

Remark 8.13. The diamond alpha Bennett–Leindler-type inequalities (8.93)–(8.96) obtained for $0 < \zeta < 1$, $\eta \geq 0$, and $0 \leq \eta + \theta < 1$ are derived for the first time due to the condition $0 \leq \eta + \theta < 1$. Moreover, these inequalities generalize the diamond alpha Bennett–Leindler-type inequality (8.26) given in [34, Theorem 16] established for $0 < \zeta < 1, \eta = 0$, and $\theta \leq 0$ and complement the diamond alpha Bennett–Leindler-type inequality (8.32) given in [38, Theorem 11] established for $\zeta > 1$, $\eta \geq 0$, and $0 \leq \eta + \theta < 1$.

Remark 8.14. The condition $0 \leq \eta + \theta < 1$ with $0 < \zeta < 1$ has not appeared in the literature before even in the special cases. This is one of the gaps in the literature that this theorem aims to fill. By the novel diamond alpha Bennett–Leindler inequalities (8.93)–(8.96) obtained for $0 < \zeta < 1$, $\eta \geq 0$, and $0 \leq \eta + \theta < 1$ for the first time, this aim is achieved.

Remark 8.15. Special cases of the diamond alpha Bennett–Leindler-type inequalities (8.93)–(8.96) can be seen below.

(i) Expressing inequalities (8.93)–(8.94) in terms of delta and nabla integrals for $0 < \zeta < 1$, $\eta \geq 0$, $0 \leq \eta + \theta < 1$, and $\eta + \zeta \leq 1$, and then choosing $\alpha = 0$ in those inequalities yields

$$\int_a^\infty \frac{z(t)[\overline{H}^\rho(t)]^{\eta+\zeta}}{[\overline{G}^\rho(t)]^{\eta+\theta}} \nabla t \geq \frac{\eta+\zeta}{1-\eta-\theta} \int_a^\infty \frac{z(t)h(t)[\overline{H}^\rho(t)]^{\eta+\zeta-1}}{[\overline{G}(t)]^{\eta+\theta-1}} \nabla t \tag{8.98}$$

and

$$\int_a^\infty \frac{z(t)[\overline{H}^\rho(t)]^{\eta+\zeta}}{[\overline{G}^\rho(t)]^{\eta+\theta}} \nabla t \geq \left[\frac{\eta+\zeta}{1-\eta-\theta}\right]^\zeta \int_a^\infty \frac{z(t)h^\zeta(t)[\overline{H}^\rho(t)]^\eta}{[\overline{G}^\rho(t)]^{\eta+\theta-\zeta}} \nabla t, \tag{8.99}$$

respectively, where $\overline{H}(t)$ and $\overline{G}(t)$ are defined in (8.15). Inequalities (8.98)–(8.99) are novel even in the nabla calculus. These novel inequalities generalize nabla Bennett–Leindler-type inequalities (8.16)–(8.17) in [29, Theorem 3.9] and [34, Remark 6] established for $0 < \zeta < 1$, $\eta = 0$ and $\theta \leq 0$ and complement nabla Bennett–Leindler-type inequalities (8.22)–(8.23) in [37, Theorem 8] and [38, Remark 7] established for $\zeta > 1$, $\eta \geq 0$, and $0 \leq \eta + \theta < 1$. We can conclude that (i) of Theorem 8.11 is a diamond alpha unification of Theorem 8.2 which is given in [29, Theorem 3.9] and [34, Remark 6] and is a generalization of the diamond alpha Bennett–Leindler-type inequality (8.26) established for $0 < \zeta < 1$, $\eta = 0$, and $\theta \leq 0$ given in [34, Theorem 16] and is a completion of the diamond alpha Bennett–Leindler-type inequality (8.32) established for $\zeta > 1$, $\eta \geq 0$, and $0 \leq \eta + \theta < 1$ given in [38, Theorem 11].

(ii) Expressing inequalities (8.95)–(8.96) in terms of delta and nabla integrals for $0 < \zeta < 1$, $\eta \geq 0$, $\eta + \theta \leq 0$, and $\eta + \zeta \geq 1$, and then choosing $\alpha = 0$ in those inequalities yields

$$\int_a^\infty \frac{z(t)[\overline{H}(t)]^{\eta+\zeta}}{[\overline{G}^\rho(t)]^{\eta+\theta}} \nabla t \geq \frac{\eta+\zeta}{1-\eta-\theta} \int_a^\infty \frac{z(t)h(t)[\overline{H}(t)]^{\eta+\zeta-1}}{[\overline{G}^\rho(t)]^{\eta+\theta-1}} \nabla t \tag{8.100}$$

and

$$\int_a^\infty \frac{z(t)[\overline{H}(t)]^{\eta+\zeta}}{[\overline{G}^\rho(t)]^{\eta+\theta}} \nabla t \geq \left[\frac{\eta+\zeta}{1-\eta-\theta}\right]^\zeta \int_a^\infty \frac{z(t)h^\zeta(t)[\overline{H}(t)]^\eta}{[\overline{G}^\rho(t)]^{\eta+\theta-\zeta}} \nabla t, \tag{8.101}$$

respectively, where $\overline{H}(t)$ and $\overline{G}(t)$ are defined in (8.15). Inequalities (8.100)–(8.101) are novel even in the nabla calculus. These novel inequalities generalize nabla Bennett–Leindler-type inequalities (8.16)–(8.17) in [29, Theorem 3.9] and [34, Remark 6] established for $0 < \zeta < 1$, $\eta = 0$ and $\theta \leq 0$ and complement nabla Bennett–Leindler-type inequalities (8.22)–(8.23) in [37, Theorem 8] and [38, Remark 7] established for $\zeta > 1$, $\eta \geq 0$, and $0 \leq \eta + \theta < 1$. We can conclude that (ii) of Theorem 8.11 is a diamond alpha unification of Theorem 8.2 which is given in [29, Theorem 3.9] and

[34, Remark 6] and is a generalization of the diamond alpha Bennett–Leindler-type inequality (8.26) established for $0 < \zeta < 1$, $\eta = 0$, and $\theta \le 0$ given in [34, Theorem 16] and is a completion of the diamond alpha Bennett–Leindler-type inequality (8.32) established for $\zeta > 1$, $\eta \ge 0$, and $0 \le \eta + \theta < 1$ given in [38, Theorem 11].

The following theorem asserts not only novel diamond alpha and nabla Bennett–Leindler-type inequalities when $0 < \zeta < 1$, $\eta \ge 0$, and $\eta + \theta > 1$, but also complements the diamond alpha Bennett–Leindler-type inequalities given in [38, Theorem 12] established for $\zeta > 1$, $\eta \ge 0$, and $\eta + \theta > 1$, and generalizations of the diamond alpha Bennett–Leindler-type inequalities given in [34, Theorem 18] established for $0 < \zeta < 1$, $\eta = 0$, and $\theta > 1$. This novelty is caused by the condition $\eta + \zeta \ge 1$, which has not been considered so far.

Theorem 8.12. *Suppose that z is a nonincreasing function on $[a,\infty)_{\mathbb{T}}$. For the functions $\overline{G}(t)$ and $\overline{H}(t)$ defined in (8.24), let $0 < \zeta < 1$, $\eta \ge 0$ and $\eta + \theta > 1$ be real constants.*

(i) *If $0 \le \eta + \zeta < 1$, then*

$$\int_a^\infty \frac{z(t)[\overline{H}^\rho(t)]^{\eta+\zeta}}{[\overline{G}(t)]^{\eta+\theta}} \diamond_\alpha t \ge \frac{(1-\alpha)(\eta+\zeta)}{\eta+\theta-1} \int_a^\infty \frac{z(t)h(t)[\overline{H}^\rho(t)]^{\eta+\zeta-1}}{[\overline{G}(t)]^{\eta+\theta-1}} \diamond_\alpha t \tag{8.102}$$

and

$$\int_a^\infty \frac{z(t)[\overline{H}^\rho(t)]^{\eta+\zeta}}{[\overline{G}(t)]^{\eta+\theta}} \diamond_\alpha t \ge \left[\frac{(1-\alpha)(\eta+\zeta)}{\eta+\theta-1}\right]^\zeta \int_a^\infty \frac{z(t)h^\zeta(t)[\overline{H}^\rho(t)]^{\eta}}{[\overline{G}(t)]^{\eta+\theta-\zeta}} \diamond_\alpha t. \tag{8.103}$$

(ii) *If $\eta + \zeta \ge 1$, then*

$$\int_a^\infty \frac{z(t)[\overline{H}(t)]^{\eta+\zeta}}{[\overline{G}(t)]^{\eta+\theta}} \diamond_\alpha t \ge \frac{(1-\alpha)(\eta+\zeta)}{\eta+\theta-1} \int_a^\infty \frac{z(t)h(t)[\overline{H}(t)]^{\eta+\zeta-1}}{[\overline{G}^\rho(t)]^{\eta+\theta-1}} \diamond_\alpha t \tag{8.104}$$

and

$$\int_a^\infty \frac{z(t)[\overline{H}(t)]^{\eta+\zeta}}{[\overline{G}(t)]^{\eta+\theta}} \diamond_\alpha t \ge \left[\frac{(1-\alpha)(\eta+\zeta)}{\eta+\theta-1}\right]^\zeta \int_a^\infty \frac{z(t)h^\zeta(t)[\overline{H}(t)]^{\eta}}{[\overline{G}(t)]^{\eta+\theta-\zeta}} \diamond_\alpha t. \tag{8.105}$$

Proof. It follows from (8.83) and formula (8.39) for $\eta + \theta > 1$ that we have

$$\begin{aligned}
[\overline{G}^{1-\eta-\theta}(t)]^\nabla &= \int_0^1 \frac{(1-\eta-\theta)\overline{G}^\nabla(t)dw}{[w\overline{G}(t)+(1-w)\overline{G}^\rho(t)]^{\eta+\theta}} \\
&= \int_0^1 \frac{(1-\eta-\theta)[\alpha z^\rho(t)+(1-\alpha)z(t)]dw}{[w\overline{G}(t)+(1-w)\overline{G}^\rho(t)]^{\eta+\theta}} \\
&\le \frac{(\eta+\theta-1)z(t)}{[\overline{G}(t)]^{\eta+\theta}},
\end{aligned} \tag{8.106}$$

where $\overline{G}(t) \geq \overline{G}(\rho(t))$ and the nonincreasingness property of z have been used.

(i) Using inequalities (8.81) and (8.106) in equation (8.85) yields

$$\begin{aligned} u^{\nabla}(t) &\leq -\frac{(\eta+\theta-1)z(t)}{[\overline{G}(t)]^{\eta+\theta}}[\overline{H}^{\rho}(t)]^{\eta+\zeta} - \frac{(1-\alpha)(\eta+\zeta)z(t)h(t)[\overline{H}^{\rho}(t)]^{\eta+\zeta-1}}{[\overline{G}(t)]^{\eta+\theta-1}} \\ &\leq \frac{(\eta+\theta-1)z(t)}{[\overline{G}(t)]^{\eta+\theta}}[\overline{H}^{\rho}(t)]^{\eta+\zeta} - \frac{(1-\alpha)(\eta+\zeta)z(t)h(t)[\overline{H}^{\rho}(t)]^{\eta+\zeta-1}}{[\overline{G}(t)]^{\eta+\theta-1}}, \end{aligned}$$

or

$$\begin{aligned} \int_a^{\infty} \frac{z(t)[\overline{H}^{\rho}(t)]^{\eta+\zeta}}{[\overline{G}(t)]^{\eta+\theta}} \diamond_\alpha t \geq{}& \frac{(1-\alpha)(\eta+\zeta)}{\eta+\theta-1} \int_a^{\infty} \frac{z(t)h(t)[\overline{H}^{\rho}(t)]^{\eta+\zeta-1}}{[\overline{G}(t)]^{\eta+\theta-1}} \diamond_\alpha t \\ &+ \frac{1}{\eta+\theta-1} \int_a^{\infty} u^{\nabla}(t) \diamond_\alpha t. \end{aligned}$$

If we employ inequality (8.87), we obtain from the above inequality that

$$\int_a^{\infty} \frac{z(t)[\overline{H}^{\rho}(t)]^{\eta+\zeta}}{[\overline{G}(t)]^{\eta+\theta}} \diamond_\alpha t \geq \frac{(1-\alpha)(\eta+\zeta)}{\eta+\theta-1} \int_a^{\infty} \frac{z(t)h(t)[\overline{H}^{\rho}(t)]^{\eta+\zeta-1}}{[\overline{G}(t)]^{\eta+\theta-1}} \diamond_\alpha t,$$

which is the desired inequality (8.102).

Inequality (8.103) can be obtained by applying the reverse Hölder's inequality (8.40) with the constants $\zeta < 1$ and $\zeta/(\zeta-1) < 0$ to the right-hand side of inequality (8.93).

(ii) Using inequalities (8.82) and (8.106) in equation (8.88) yields

$$u^{\nabla}(t) \leq -\frac{(\eta+\theta-1)z(t)[\overline{H}(t)]^{\eta+\zeta}}{[\overline{G}(t)]^{\eta+\theta}} - \frac{(1-\alpha)(\eta+\zeta)z(t)h(t)[\overline{H}(t)]^{\eta+\zeta-1}}{[\overline{G}^{\rho}(t)]^{\eta+\theta-1}},$$

or

$$\begin{aligned} \int_a^{\infty} \frac{z(t)[\overline{H}(t)]^{\eta+\zeta}}{[\overline{G}(t)]^{\eta+\theta}} \diamond_\alpha t \geq{}& \frac{(1-\alpha)(\eta+\zeta)}{\eta+\theta-1} \int_a^{\infty} \frac{z(t)h(t)[\overline{H}(t)]^{\eta+\zeta-1}}{[\overline{G}^{\rho}(t)]^{\eta+\theta-1}} \diamond_\alpha t \\ &+ \frac{1}{\eta+\theta-1} \int_a^{\infty} u^{\nabla}(t) \diamond_\alpha t. \end{aligned}$$

If we employ inequality (8.87), we obtain from the above inequality that

$$\int_a^{\infty} \frac{z(t)[\overline{H}(t)]^{\eta+\zeta}}{[\overline{G}(t)]^{\eta+\theta}} \diamond_\alpha t \geq \frac{(1-\alpha)(\eta+\zeta)}{\eta+\theta-1} \int_a^{\infty} \frac{z(t)h(t)[\overline{H}(t)]^{\eta+\zeta-1}}{[\overline{G}^{\rho}(t)]^{\eta+\theta-1}} \diamond_\alpha t,$$

which is the desired inequality (8.104).

Since

$$\int_a^\infty \frac{z(t)[\overline{H}(t)]^{\eta+\zeta}}{[\overline{G}(t)]^{\eta+\theta}} \diamond_\alpha t \geq \frac{(1-\alpha)(\eta+\zeta)}{\eta+\theta-1} \int_a^\infty \frac{z(t)h(t)[\overline{H}(t)]^{\eta+\zeta-1}}{[\overline{G}^\rho(t)]^{\eta+\theta-1}} \diamond_\alpha t$$
$$\geq \frac{(1-\alpha)(\eta+\zeta)}{\eta+\theta-1} \int_a^\infty \frac{z(t)h(t)[\overline{H}(t)]^{\eta+\zeta-1}}{[\overline{G}(t)]^{\eta+\theta-1}} \diamond_\alpha t,$$

applying the reverse Hölder's inequality (8.40) with the constants $\zeta < 1$ and $\zeta/(\zeta-1) < 0$ to the right-hand side of the above inequality, one can obtain inequality (8.105). □

Remark 8.16. The diamond alpha Bennett–Leindler-type inequalities (8.102)–(8.105) obtained for $0 < \zeta < 1$, $\eta \geq 0$, and $\eta + \theta > 1$ are generalizations of the diamond alpha Bennett–Leindler-type inequality (8.28) given in [34, Theorem 18] for $0 < \zeta < 1$, $\eta = 0$, and $\theta > 1$ and complements of the diamond alpha Bennett–Leindler-type inequality (8.34) given in [38, Theorem 12] for $\zeta > 1$, $\eta \geq 0$, and $\eta + \theta > 1$.

Remark 8.17. Although the special case of the condition $\eta + \zeta \leq 1$ is automatically satisfied in [34, Theorem 18], the other case, $\eta + \zeta \geq 1$ with $0 < \zeta < 1$, has not appeared in the literature before even for the special case. This is one of the gaps in the literature that this theorem aims to fill. By the novel diamond alpha Bennett–Leindler-type inequalities (8.104)–(8.105) obtained for $0 < \zeta < 1$, $\eta \geq 0$, $\eta + \theta > 1$, and $\eta + \zeta \geq 1$ for the first time, this aim is achieved.

Remark 8.18. Special cases of the diamond alpha Bennett–Leindler-type inequalities (8.102)–(8.105) can be seen below.

(i) Expressing inequalities (8.102)–(8.103) in terms of delta and nabla integrals for $0 < \zeta < 1$, $\eta \geq 0, \eta + \theta > 1$, and $\eta + \zeta \leq 1$, and then choosing $\alpha = 0$ in those inequalities yields

$$\int_a^\infty \frac{z(t)[\overline{H}^\rho(t)]^{\eta+\zeta}}{[\overline{G}(t)]^{\eta+\theta}} \nabla t \geq \frac{\eta+\zeta}{\eta+\theta-1} \int_a^\infty \frac{z(t)h(t)[\overline{H}^\rho(t)]^{\eta+\zeta-1}}{[\overline{G}(t)]^{\eta+\theta-1}} \nabla t \tag{8.107}$$

and

$$\int_a^\infty \frac{z(t)[\overline{H}^\rho(t)]^{\eta+\zeta}}{[\overline{G}(t)]^{\eta+\theta}} \nabla t \geq \left[\frac{\eta+\zeta}{\eta+\theta-1}\right]^\zeta \int_a^\infty \frac{z(t)h\zeta(t)[\overline{H}^\rho(t)]^{\eta}}{[\overline{G}(t)]^{\eta+\theta-\zeta}} \nabla t, \tag{8.108}$$

respectively, where $\overline{H}(t)$ and $\overline{G}(t)$ are defined in (8.15). Inequalities (8.107)–(8.108) generalize nabla Bennett–Leindler-type inequalities (8.18)–(8.19) established for $0 < \zeta < 1$, $\eta = 0$, and $\theta > 1$ given in [34, Remark 8] and complement nabla Bennett–Leindler-type inequalities given in [38, Remark 9] established for $\zeta > 1$, $\eta \geq 0$, and

$\eta + \theta > 1$. We can conclude that (i) of Theorem 8.12 is a diamond alpha unification of Theorem 8.2 given in [34, Remark 8] and a generalization of the diamond alpha Bennett–Leindler-type inequality (8.28) established for $0 < \zeta < 1$, $\eta = 0$, and $\theta > 1$ given in [34, Theorem 18] and a completion of the diamond alpha Bennett–Leindler-type inequality (8.34) established for $\zeta > 1$, $\eta \geq 0$, and $\eta + \theta > 1$ given in [38, Theorem 12].

(ii) Expressing inequalities (8.104)–(8.105) in terms of delta and nabla integrals for $0 < \zeta < 1$, $\eta \geq 0, \eta + \theta > 1$, and $\eta + \zeta \geq 1$, and then choosing $\alpha = 0$ in those inequalities yields

$$\int_a^\infty \frac{z(t)[\overline{H}(t)]^{\eta+\zeta}}{[\overline{G}(t)]^{\eta+\theta}} \nabla t \geq \frac{\eta+\zeta}{\eta+\theta-1} \int_a^\infty \frac{z(t)h(t)[\overline{H}(t)]^{\eta+\zeta-1}}{[\overline{G}^{\rho}(t)]^{\eta+\theta-1}} \nabla t \tag{8.109}$$

and

$$\int_a^\infty \frac{z(t)[\overline{H}(t)]^{\eta+\zeta}}{[\overline{G}(t)]^{\eta+\theta}} \nabla t \geq \left[\frac{\eta+\zeta}{\eta+\theta-1}\right]^{\zeta} \int_a^\infty \frac{z(t)h^{1/\zeta}(t)[\overline{H}(t)]^{\eta\zeta}}{[\overline{G}(t)]^{\eta+\theta-\zeta}} \nabla t, \tag{8.110}$$

respectively, where $\overline{H}(t)$ and $\overline{G}(t)$ are defined in (8.15). Inequalities (8.109)–(8.110) are novel even in the nabla calculus. These novel inequalities generalize nabla Bennett–Leindler-type inequalities (8.18)–(8.19) established for $0 < \zeta < 1$, $\eta = 0$, and $\theta > 1$ given in [34, Remark 8] and complement nabla Bennett–Leindler-type inequalities given in [38, Remark 9] established for $\zeta > 1$, $\eta \geq 0$, and $\eta + \theta > 1$. We can conclude that (ii) of Theorem 8.12 is a diamond alpha unification of Theorem 8.2 given in [34, Remark 8] and a generalization of the diamond alpha Bennett–Leindler-type inequality (8.28) established for $0 < \zeta < 1$, $\eta = 0$, and $\theta > 1$ given in [34, Theorem 18] and a completion of the diamond alpha Bennett–Leindler-type inequality (8.34) established for $\zeta > 1$, $\eta \geq 0$, and $\eta + \theta > 1$ given in [38, Theorem 12].

Bibliography

[1] R. Agarwal, M. Bohner and A. Peterson, Inequalities on time scales: A survey, *Math. Inequal. Appl.*, 4(4) (2001), 535–557.

[2] R. P. Agarwal, R. R. Mahmoud, S. Saker and C. Tunç, New generalizations of Németh–Mohapatra type inequalities on time scales, *Acta Math. Hung.*, 152(2) (2017), 383–403.

[3] R. Agarwal, D. O'Regan and S. Saker, Dynamic Inequalities on Time Scales, Springer, Cham, 2014.

[4] R. Agarwal, D. O'Regan and S. Saker, Hardy Type Inequalities on Time Scales, Springer, Cham, 2016.

[5] M. R. S. Ammi, R. A. C. Ferreira and D. F. M. Torres, Diamond-α Jensen's inequality on time scales, *J. Inequal. Appl.*, 2008 (2008), 576876.

[6] D. R. Anderson, Time-scale integral inequalities, *J. Inequal. Pure Appl. Math.*, 6(3) (2005), 66.

[7] N. Atasever, B. Kaymakçalan, G. Lešaja and K. Taş, Generalized diamond-α dynamic Opial inequalities, *Adv. Differ. Equ.*, 2012(109) (2012), 1–9.

[8] F. M. Atici and G. S. Guseinov, On Green's functions and positive solutions for boundary value problems on time scales, *J. Comput. Appl. Math.*, 141(1–2) (2002), 75–99.

[9] A. A. Balinsky, W. D. Evans and R. T. Lewis, The Analysis and Geometry of Hardy's Inequality, Springer, Switzerland, 2015.

[10] G. Bennett, Some elementary inequalities II, *Q. J. Math.*, 39 (1988), 385–400.

[11] M. Bohner and O. Duman, Opial-type inequalities for diamond alpha derivatives and integrals on time scales, *Differ. Equ. Dyn. Syst.*, 18(1–2) (2010), 229–237.

[12] M. Bohner, R. Mahmoud and S. H. Saker, Discrete, continuous, delta, nabla, and diamond alpha Opial inequalities, *Math. Inequal. Appl.*, 18(3) (2015), 923–940.

[13] M. Bohner, R. R. Mahmoud and S. H. Saker, Improvements of dynamic Opial-type inequalities and applications, *Dyn. Syst. Appl.*, 24 (2015), 229–242.

[14] M. Bohner and A. Peterson, Dynamic Equations on Time Scales. An Introduction With Applications, Birkhäuser, Boston, MA, 2001.

[15] M. Bohner and A. Peterson, Advances in Dynamic Equations on Time Scales, Birkhäuser, Boston, MA, 2003.

[16] E. T. Copson, Note on series of positive terms, *J. Lond. Math. Soc.*, 3(1) (1928), 49–51.

[17] E. T. Copson, Some integral inequalities, *Proc. R. Soc. Edinb., Sect. A*, 75(2) (1976), 157–164.

[18] A. A. El-Deeb, H. A. Elsennary and Z. A. Khan, Some reverse inequalities of Hardy type on time scales, *Adv. Differ. Equ.*, 2020(402) (2020), 1–18.

[19] A. A. El-Deeb, H. A. Elsennary and B. Dumitru, Some new Hardy-type inequalities on time scales, *Adv. Differ. Equ.*, 2020(441) (2020), 1–22.

[20] P. Gao and H. Y. Zhao, On Copson's inequalities for $0 < p < 1$, *J. Inequal. Appl.*, 2020(72) (2020), 1–13.

[21] G. S. Guseinov and B. Kaymakçalan, Basics of Riemann delta and nabla integration on time scales, *J. Differ. Equ. Appl.*, 8(11) (2002), 1001–1017.

[22] M. Gürses, G. S. Guseinov and B. Silindir, Integrable equations on time scales, *J. Math. Phys.*, 46(11) (2005), 113510.

[23] A. F. Güvenilir, B. Kaymakçalan and N. N. Pelen, Constantin's inequality for nabla and diamond alpha derivative, *J. Inequal. Appl.*, 2015(167) (2015), 1–17.

[24] G. H. Hardy, Note on a theorem of Hilbert, *Math. Z.*, 6(3–4) (1920), 314–317.

[25] G. H. Hardy, Notes on some points in the integral calculus, LX. An inequality between integrals, *Messenger Math.*, 54(3) (1925), 150–156.

[26] G. H. Hardy and J. E. Littlewood, Elementary theorems concerning power series with positive coefficients and moment constants of positive functions, *J. Reine Angew. Math.*, 157 (1927), 141–158.

[27] G. H. Hardy, J. E. Littlewood and G. Pólya, Inequalities, Cambridge University Press, London, 1934.

[28] S. Iqbal, M. J. S. Sahir and M. Samraiz, Symmetric Rogers–Hölder's inequalities on diamond α calculus, *Int. J. Nonlinear Anal. Appl.*, 9(2) (2018), 9–19.

[29] Z. Kayar, B. Kaymakçalan and N. N. Pelen, Bennett–Leindler type inequalities for time scale nabla calculus, *Mediterr. J. Math.*, 18(14) (2021), 1–18. https://doi.org/10.1007/s00009-020-01674-5.

[30] Z. Kayar and B. Kaymakçalan, Hardy–Copson type inequalities for nabla time scale calculus, *Turk. J. Math.*, 45(2) (2021), 1040–1064.

[31] Z. Kayar and B. Kaymakçalan, Some new extended nabla and delta Hardy–Copson type inequalities and their applications in oscillation theory, *Bull. Iranian Math. Soc.*, 48 (2022), 2407–2439. https://doi.org/10.1007/s41980-021-00651-2.

[32] Z. Kayar and B. Kaymakçalan, Complements of nabla and delta Hardy–Copson type inequalities and their applications, under review.

[33] Z. Kayar and B. Kaymakçalan, Diamond-alpha Hardy–Copson type dynamic inequalities, *Hacet. J. Math. Stat.*, 51 (2022), 48–73.

[34] Z. Kayar, B. Kaymakçalan and N. N. Pelen, Diamond-alpha Bennett–Leindler type dynamic inequalities, *Math. Methods Appl. Sci.*, 45(5) (2022), 2797–2819. https://doi.org/10.1002/mma.7955.

[35] Z. Kayar and B. Kaymakçalan, Extensions of diamond alpha Hardy–Copson type dynamic inequalities and their applications to oscillation theory, *Dyn. Syst. Appl.*, 30(7) (2021), 1180–1209.
[36] Z. Kayar and B. Kaymakçalan, Applications of the novel diamond alpha Hardy–Copson type dynamic inequalities to half linear difference equations, *J. Differ. Equ. Appl.*, 28(4) (2022), 457–484. https://doi.org/10.1080/10236198.2022.2042522.
[37] Z. Kayar and B. Kaymakçalan, The complementary nabla Bennett–Leindler type inequalities, *Commun. Fac. Sci. Univ. Ank. Sér. A1 Math. Stat.*, 71(2) (2022), 1–28. https://doi.org/10.31801/cfsuasmas.930138.
[38] Z. Kayar and B. Kaymakçalan, Novel diamond-alpha Bennett–Leindler type dynamic inequalities, *Bull. Malays. Math. Soc.*, 45(3) (2022), 1027–1054. https://doi.org/10.1007/s40840-021-01224-6.
[39] A. Kufner, L. Maligranda and L. E. Persson, The Hardy Inequality. About Its History and Some Related Results, Vydavatelský Servis, Pilsen, 2007.
[40] A. Kufner, L. E. Persson and N. Samko, Weighted Inequalities of Hardy Type, World Scientific Publishing Co. Pte. Ltd., Hackensack, NJ, 2017.
[41] L. Leindler, Some inequalities pertaining to Bennett's results, *Acta Sci. Math.*, 58(1–4) (1993), 261–279.
[42] L. Leindler, Further sharpening of inequalities of Hardy and Littlewood, *Acta Sci. Math.*, 54(3–4) (1990), 285–289.
[43] A. B. Malinowska and D. F. M. Torres, On the diamond alpha Riemann integral and mean value theorems on time scales, *Dyn. Syst. Appl.*, 18(3–4) (2009), 469–481.
[44] N. Masmoudi, About the Hardy Inequality, in *An Invitation to Mathematics. From Competitions to Research*, Springer, Heidelberg, 2011.
[45] T. Z. Mirković, Dynamic Opial diamond-α integral inequalities involving the power of a function, *J. Inequal. Appl.*, 2017(139) (2017), 1–10.
[46] D. Mozyrska and D. F. M. Torres, A study of diamond alpha dynamic equations on regular time scales, *Afr. Diaspora J. Math.*, 8(1) (2009), 35–47.
[47] U. M. Özkan, M. Z. Sarikaya and H. Yildirim, Extensions of certain integral inequalities on time scales, *Appl. Math. Lett.*, 21(10) (2008), 993–1000.
[48] N. N. Pelen, Hardy–Sobolev–Mazya inequality for nabla time scale calculus, *Eskişehir Tech. Univ. J. Sci. Technol. B - Theor. Sci.*, 7(2) (2019), 133–145.
[49] P. Řehák, Hardy inequality on time scales and its application to half-linear dynamic equations, *J. Inequal. Appl.*, 2005(5) (2005), 495–507.
[50] J. W. Rogers Jr. and Q. Sheng, Notes on the diamond-α dynamic derivative on time scales, *J. Math. Anal. Appl.*, 326(1) (2007), 228–241.
[51] S. H. Saker, Dynamic inequalities on time scales: A survey, *J. Fract. Calc. Appl.*, 3(S)(2) (2012), 1–36.
[52] S. H. Saker and R. R. Mahmoud, A connection between weighted Hardy's inequality and half-linear dynamic equations, *Adv. Differ. Equ.*, 2014(129) (2019), 1–15.
[53] S. H. Saker, R. R. Mahmoud and A. Peterson, Some Bennett–Copson type inequalities on time scales, *J. Math. Inequal.*, 10(2) (2016), 471–489.
[54] S. H. Saker, R. R. Mahmoud, M. M. Osman and R. P. Agarwal, Some new generalized forms of Hardy's type inequality on time scales, *Math. Inequal. Appl.*, 20(2) (2017), 459–481.
[55] S. H. Saker, D. O'Regan and R. P. Agarwal, Dynamic inequalities of Hardy and Copson type on time scales, *Analysis*, 34(4) (2014), 391–402.
[56] S. H. Saker, D. O'Regan and R. P. Agarwal, Generalized Hardy, Copson, Leindler and Bennett inequalities on time scales, *Math. Nachr.*, 287(5–6) (2014), 686–698.
[57] S. H. Saker, M. M. Osman, D. O'Regan and R. P. Agarwal, Inequalities of Hardy type and generalizations on time scales, *Analysis*, 38(1) (2018), 47–62.
[58] S. H. Saker, R. R. Mahmoud and A. Peterson, A unified approach to Copson and Beesack type inequalities on time scales, *Math. Inequal. Appl.*, 21(4) (2018), 985–1002.
[59] S. H. Saker, D. O'Regan and R. P. Agarwal, Converses of Copson's inequalities on time scales, *Math. Inequal. Appl.*, 18(1) (2015), 241–254.
[60] Q. Sheng, M. Fadag, J. Henderson and J. M. Davis, An exploration of combined dynamic derivatives on time scales and their applications, *Nonlinear Anal., Real World Appl.*, 7(3) (2006), 395–413.

Sibel Doğru Akgöl and Abdullah Özbekler

9 De La Vallée Poussin-type inequality for impulsive dynamic equations on time scales

Abstract: We derive a de La Vallée Poussin-type inequality for impulsive dynamic equations on time scales. This inequality is often used in conjunction with disconjugacy and/or (non)oscillation. Hence, it appears to be a very useful tool for the qualitative study of dynamic equations. In this work, generalizing the classical de La Vallée Poussin inequality for impulsive dynamic equations on arbitrary time scales, we obtain a disconjugacy criterion and some results on nonoscillation. We also present illustrative examples that support our findings.

9.1 Introduction

The celebrated de La Vallée Poussin inequality was given for the second-order linear homogeneous differential equation

$$x'' + p(t)x' + q(t)x = 0$$

with the Dirichlet boundary conditions

$$x(a) = x(b) = 0$$

as in the following theorem.

Theorem 9.1 ([11, 14]). *If $x(t) \neq 0$ in (a, b), then*

$$\frac{q_0}{2}(b-a)^2 + p_0(b-a) > 1$$

holds, where

$$p_0 = \max_{t\in[a,b]} |p(t)| \quad \textit{and} \quad q_0 = \max_{t\in[a,b]} |q(t)|.$$

Later, Hartman and Wintner [11] obtained the less restrictive inequality

$$\max\left\{\int_a^b (s-a)|p(s)|\,\mathrm{d}s, \int_a^b (b-s)|p(s)|\,\mathrm{d}s\right\} + \int_a^b (s-a)(b-s)|q(s)|\,\mathrm{d}s > b-a, \tag{9.1}$$

Sibel Doğru Akgöl, Abdullah Özbekler, Department of Mathematics, Atilim University, 06830 Incek, Ankara, Turkey, e-mails: sibel.dogruakgol@atilim.edu.tr, aozbekler@gmail.com

https://doi.org/10.1515/9783111182971-009

which implies that

$$\frac{q_0}{6}(b-a)^3 + \frac{p_0}{2}(b-a)^2 > b-a,$$

or equivalently that

$$\frac{q_0}{6}(b-a)^2 + \frac{p_0}{2}(b-a) > 1.$$

In 2018, some extensions of the de La Vallée Poussin-type inequalities were obtained for dynamic equations on time scales [8], fractional differential equations [9], and partial differential equations [1]. Recently, the inequality (9.1) has also been extended for impulsive differential equations, see [3].

In this work, we aim to obtain a de La Vallée Poussin-type inequality for impulsive dynamic equations on time scales. First, we recall some basic concepts regarding time scale calculus.

A time scale $\mathbb{T}$ is a nonempty arbitrary closed subset of $\mathbb{R}$. The set of real numbers $\mathbb{R}$, the set of integers $\mathbb{Z}$, the Cantor set $\mathscr{C}$, the set $q^{\mathbb{N}_0} := \{q^n : n \in \mathbb{N}_0\}$, where $q > 1$ and $h\mathbb{Z} := \{hz : z \in \mathbb{Z}\}$, where $h > 0$, are some of the well-known examples of time scales. The necessary definitions and elements of time scale calculus can be found in [2], and comprehensive and detailed information about time scales is given in the books [4–7, 10, 12, 13].

We let $\mathrm{PLC}_{rd}[a,b]$ be the set of functions φ such that $\varphi(t)$ is rd-continuous on $(t_k, t_{k+1}]_{\mathbb{T}} := (t_k, t_{k+1}] \cap \mathbb{T}$ and $\varphi(t_k^+)$ exists for each k for which $t_k \in [a,b)_{\mathbb{T}}$, $k = 1, 2, \ldots$.

Let $\mathbb{T}$ be an arbitrary time scale and suppose a and b ($a < b$) are two consecutive zeros of the linear impulsive dynamic equation

$$\begin{cases} x^{\Delta\Delta} + p(t)x^{\Delta} + q(t)x^{\sigma} = 0, & t \neq t_k; \\ \triangle x^{\Delta} + p_k x^{\Delta} + q_k x^{\sigma} = 0, & t = t_k \end{cases} \tag{9.2}$$

for $t \in [a,b]_{\mathbb{T}}$, $k = 1, 2, \ldots$, where $p, q \in \mathrm{C}_{\mathrm{rd}}[a,b]$, $\{p_k\}$ and $\{q_k\}$ are real sequences, $\{t_k\}$ is a strictly increasing sequence of real numbers such that $\lim_{k\to\infty} t_k = \infty$, and the impulse operator is defined by $\triangle\varphi(t_k) = \varphi(t_k^+) - \varphi(t_k^-)$ with $\varphi(t_k^{\pm}) = \lim_{t\to t_k^{\pm}} \varphi(t)$. Throughout this study, we assume that $\sigma(t_k) = t_k$, $k = 1, 2, \ldots$, i. e., each impulse point is right dense, and we use the notations $\varphi^+ = \max\{\varphi, 0\}$, $\underline{n}(t) := \inf\{k : t_k \geq t\}$ and $\overline{n}(t) := \sup\{k : t_k < t\}$; also, for brevity, we shall denote the intervals as $[a,b]$ instead of $[a,b]_{\mathbb{T}}$.

9.2 Main results

In this section, we derive a de La Vallée Poussin-type inequality for Eq. (9.2). Below is our main result.

Theorem 9.2. *Let $x \in C_{rd}[a, \rho(b)]$ be a nontrivial solution of Eq. (9.2) satisfying the boundary conditions*

$$x(a) = x(b) = 0. \tag{9.3}$$

If $x(t) \neq 0$ for $t \in (a, b)$, then the inequality

$$\begin{aligned} &\int_a^{\rho(b)} (\sigma(s) - a)(b - \sigma(s))q^+(s)\Delta s + \sum_{k=\underline{n}(a)}^{\overline{n}(b)} (t_k - a)(b - t_k)q_k^+ \\ &+ \max\left\{ \int_a^{\rho(b)} (\sigma(s) - a)p^+(s)\Delta s, \int_a^{\rho(b)} (b - \sigma(s))p^+(s)\Delta s \right\} \\ &+ \max\left\{ \sum_{k=\underline{n}(a)}^{\overline{n}(b)} (t_k - a)p_k^+, \sum_{k=\underline{n}(a)}^{\overline{n}(b)} (b - t_k)p_k^+ \right\} \geq b - a \end{aligned} \tag{9.4}$$

holds.

Proof. For clarity, we define $f(t) = p(t)x^\Delta + q(t)x^\sigma$ and $f_k = p_k x^\Delta(t_k) + q_k x(t_k)$. Thus, Eq. (9.2) turns into

$$\begin{cases} x^{\Delta\Delta}(t) = -f(t), & t \neq t_k; \\ \triangle x^\Delta(t) = -f_k, & t = t_k, \quad k = 1, 2, \ldots \end{cases} \tag{9.5}$$

Any solution of (9.5) satisfying the Dirichlet boundary conditions (9.3) is represented by

$$x(t) = \int_a^{\rho(b)} G(t, \sigma(s))f(s)\Delta s + \sum_{k=\underline{n}(a)}^{\overline{n}(b)} G(t, t_k)f_k, \tag{9.6}$$

where

$$G(t, \tau) = \frac{1}{b - a} \begin{cases} (\tau - a)(b - t), & a \leq \tau < t; \\ (t - a)(b - \tau), & t \leq \tau \leq b \end{cases}$$

is the Green function for Eq. (9.5) with the boundary conditions (9.3).

Indeed, observe that (9.6) can be expanded as

$$\begin{aligned} x(t) = \frac{1}{b - a} \Bigg\{ &\int_a^t (b - t)(\sigma(s) - a)f(s)\Delta s - \int_t^b (t - a)(b - \sigma(s))f(s)\Delta s \\ &+ \sum_{k=\underline{n}(a)}^{\overline{n}(t)} (t_k - a)f_k - \sum_{k=\underline{n}(t)}^{\overline{n}(b)} (b - t_k)f_k \Bigg\}. \end{aligned} \tag{9.7}$$

The delta differentiation of both sides of (9.7) with respect to t gives

$$x^{\Delta}(t) = -\frac{1}{b-a}\left\{\int_a^t (\sigma(s)-a)f(s)\Delta s - \int_t^{\rho(b)} (b-\sigma(s))f(s)\Delta s + \sum_{k=\underline{n}(a)}^{\overline{n}(t)} (t_k - a)f_k - \sum_{k=\underline{n}(t)}^{\overline{n}(b)} (b-t_k)f_k\right\}.$$

If we denote

$$I(t) := \int_a^t (\sigma(s)-a)|f(s)|\Delta s + \int_t^{\rho(b)} (b-\sigma(s))|f(s)|\Delta s, \quad t \in [a, \rho(b)],$$

and

$$S(t) := \sum_{k=\underline{n}(a)}^{\overline{n}(t)} (t_k - a)|f_k| + \sum_{k=\underline{n}(t)}^{\overline{n}(b)} (b-t_k)|f_k|, \quad t \in [a, \rho(b)],$$

then we can write

$$|x^{\Delta}(t)| \leq \frac{1}{b-a}\{I(t) + S(t)\}.$$

By taking the delta derivative of $I(t)$, one has

$$I^{\Delta}(t) = (2\sigma(t) - a - b)|f(t)|,$$

which shows that the maximum occurs either at $t = a$ or at $t = \rho(b)$. Thus, we can write

$$I(t) \leq \max\left\{\int_a^{\rho(b)} (\sigma(s)-a)|f(s)|\Delta s, \int_a^{\rho(b)} (b-\sigma(s))|f(s)|\Delta s\right\}. \tag{9.8}$$

On the other hand, we have

$$S(t) - \sum_{k=\underline{n}(a)}^{\overline{n}(b)} (t_k - a)|f_k| = \sum_{k=\underline{n}(t)}^{\overline{n}(b)} (a+b-2t_k)|f_k|$$

and

$$S(t) - \sum_{k=\underline{n}(a)}^{\overline{n}(b)} (b - t_k)|f_k| = \sum_{k=\underline{n}(a)}^{\overline{n}(t)} (2t_k - a - b)|f_k|$$

for $S(t)$. If $t_k < (a+b)/2$ for $t_k \in [t, b)$ and $t_k > (a+b)/2$ for $t_k \in [a, t)$, then both

$$\sum_{k=\underline{n}(a)}^{\overline{n}(b)} (a+b-2t_k)|f_k| \tag{9.9}$$

and

$$\sum_{k=\underline{n}(a)}^{\overline{n}(b)} (2t_k-a-b)|f_k| \tag{9.10}$$

would be positive, but this is impossible since $\{t_k\}$ is increasing. Thus, at least one of (9.9) or (9.10) must be nonpositive. This yields

$$S(t) \le \max\left\{ \sum_{k=\underline{n}(a)}^{\overline{n}(b)} (t_k-a)|f_k|, \sum_{k=\underline{n}(a)}^{\overline{n}(b)} (b-t_k)|f_k| \right\}. \tag{9.11}$$

Now, observe that

$$|f(t)| = |p(t)x^{\Delta}(t) + q(t)x^{\sigma}(t)| \le p^{+}(t)|x^{\Delta}(t)| + q^{+}(t)|x(t)|$$

and

$$|f_k| = |p_k x^{\Delta}(t_k) + q_k x(t_k)| \le p_k^{+}|x^{\Delta}(t_k)| + q_k^{+}|x(t_k)|.$$

At this point, we need the following lemma:

Lemma 9.1 (Mean value theorem [6]). *Let φ be continuous on $[a,b]$ and delta-differentiable on $[a,b)$. Then, there exist some constants $c, d \in [a,b)$ such that*

$$\varphi^{\Delta}(c) \le \frac{\varphi(b)-\varphi(a)}{b-a} \le \varphi^{\Delta}(d).$$

Since $x^{\Delta} \in \mathrm{PLC}_{\mathrm{rd}}[a,b]$, there exists $\alpha \in [a,b]$ such that

$$\max_{\tau\in[a,b]} |x^{\Delta}(\tau)| = |x^{\Delta}(\alpha)|.$$

On the other hand, since $x \in \mathrm{C}_{\mathrm{rd}}[a,b]$ and x^{Δ} exists for all $x \in [a,\rho(b)]$, Lemma 9.1 applies. Hence,

$$|x(\tau)| \le |x^{\Delta}(\alpha)|(\tau-a) \quad \text{and} \quad |x(\tau)| \le |x^{\Delta}(\alpha)|(b-\tau).$$

Let $m(\tau) := \min\{(\tau-a),(b-\tau)\}$, then clearly $|x(\tau)| \le m(\tau)|x^{\Delta}(\alpha)|$. If we define

$$M_1 := \frac{1}{b-a}\left(\max\left\{ \int_a^{\rho(b)} (\sigma(s)-a)p^{+}(s)\Delta s, \int_a^{\rho(b)} (b-\sigma(s))p^{+}(s)\Delta s \right\}\right.$$

$$+\max\left\{\sum_{k=\underline{n}(a)}^{\overline{n}(b)}(t_k-a)p_k^+,\sum_{k=\underline{n}(a)}^{\overline{n}(b)}(b-t_k)p_k^+\right\}\Bigg)$$

and

$$M_2:=\frac{1}{b-a}\Bigg(\max\left\{\int_a^{\rho(b)}(\sigma(s)-a)m(s)q^+(s)\Delta s,\int_a^{\rho(b)}(b-\sigma(s))m(s)q^+(s)\Delta s\right\}$$
$$+\max\left\{\sum_{k=\underline{n}(a)}^{\overline{n}(b)}(t_k-a)m(t_k)q_k^+,\sum_{k=\underline{n}(a)}^{\overline{n}(b)}(b-t_k)m(t_k)q_k^+\right\}\Bigg),$$

then it follows from (9.8) and (9.11) that

$$|x^\Delta(a)|\leq\frac{1}{b-a}\{|I(t)|+|S(t)|\}\leq M_1|x^\Delta(a)|+M_2|x^\Delta(a)|,$$

or equivalently

$$1\leq M_1+M_2. \tag{9.12}$$

Since $(b-\tau)m(\tau)\leq(\tau-a)(b-\tau)$ and $(\tau-a)m(\tau)\leq(\tau-a)(b-\tau)$ for $\tau\in[a,b]$, we can write

$$M_2\leq\frac{1}{b-a}\Bigg(\int_a^{\rho(b)}(\sigma(s)-a)(b-\sigma(s))q^+(s)\Delta s+\sum_{k=\underline{n}(a)}^{\overline{n}(b)}(t_k-a)(b-t_k)q_k^+\Bigg). \tag{9.13}$$

From (9.12) and (9.13), we obtain the dynamic de La Vallée Poussin-type inequality given by (9.4). □

To express some corollaries of the above theorem, we introduce the following definition.

Definition 9.1. Equation (9.2) is said to be disconjugate on (a,b) if every nontrivial solution of it has at most one zero on (a,b).

Corollary 9.1. *Suppose, on the contrary to* (9.4), *that the inequality*

$$\int_a^{\rho(b)}(\sigma(s)-a)(b-\sigma(s))q^+(s)\Delta s+\sum_{k=\underline{n}(a)}^{\overline{n}(b)}(t_k-a)(b-t_k)q_k^+$$
$$+\max\left\{\int_a^{\rho(b)}(\sigma(s)-a)p^+(s)\Delta s,\int_a^{\rho(b)}(b-\sigma(s))p^+(s)\Delta s\right\}$$
$$+\max\left\{\sum_{k=\underline{n}(a)}^{\overline{n}(b)}(t_k-a)p_k^+,\sum_{k=\underline{n}(a)}^{\overline{n}(b)}(b-t_k)p_k^+\right\}<b-a$$

holds. Then, Eq. (9.2) *is disconjugate on* (a, b).

Remark 9.1. If we take $\triangle x^{\Delta} = 0$, Eq. (9.2) turns into the nonimpulsive dynamic equation

$$x^{\Delta\Delta} + p(t)x^{\Delta} + q(t)x^{\sigma} = 0, \tag{9.14}$$

and hence the associated de La Vallée Poussin inequality reduces to

$$\int_a^{\rho(b)} (\sigma(s) - a)(b - \sigma(s))q^{+}(s)\Delta s + \max\left\{\int_a^{\rho(b)} (\sigma(s) - a)p^{+}(s)\Delta s, \int_a^{\rho(b)} (b - \sigma(s))p^{+}(s)\Delta s\right\} \geq b - a. \tag{9.15}$$

Since $p^{+}(s) \leq |p(s)|$ and $q^{+}(s) \leq |q(s)|$, inequality (9.15) is better than the inequality given in [8]. Thus, (9.15) is the best possible de La Vallée Poussin inequality for the dynamic equations of the form (9.14).

9.3 Examples

In this section, we provide some examples that support the above results.

Example. Let $\mathbb{T} = \mathbb{P}_{1,1} = \bigcup_{n=0}^{\infty}[2n, 2n+1]$, and consider the following impulsive dynamic equation on the time scale $\mathbb{P}_{1,1}$:

$$\begin{cases} x^{\Delta\Delta} + \cos(\frac{\pi t}{2})x^{\Delta} - tx^{\sigma} = 0, & t \neq 2k; \\ \triangle x^{\Delta} - x^{\Delta} + \frac{x}{(3t)^3} = 0, & t = 2k, \quad k = 1, 2, \ldots, 5, \end{cases} \tag{9.16}$$

with the boundary conditions

$$x(0) = x(9) = 0.$$

Since $p(t) = \cos(\pi t/2)$ is positive on the intervals $(0, 1)$, $(3, 5)$, $(7, 9)$ and negative on $(1, 3)$, $(5, 7)$, we have $p^{+}(t) = \cos(\pi t/2)$ for $t \in (0, 1) \cup (3, 5) \cup (7, 9)$ and $p^{+}(t) = 0$ for $t \in (1, 3) \cup (5, 7)$. For $q(s)$, p_k, and q_k, it is easy to see that $q^{+}(s) = 0 = p_k^{+}$ and $q_k^{+} = 1/(6k)^3$. Thus,

$$\begin{aligned} I_1 &:= \int_a^{\rho(b)} (\sigma(s) - a)p^{+}(s)\Delta s \\ &= \int_0^1 s\cos(\pi s/2)\Delta s + \int_3^5 s\cos(\pi s/2)\Delta s + \int_7^9 s\cos(\pi s/2)\Delta s \end{aligned}$$

$$= \int_0^1 s\cos(\pi s/2)\mathrm{d}s + \int_4^5 s\cos(\pi s/2)\mathrm{d}s + \int_8^9 s\cos(\pi s/2)\mathrm{d}s$$
$$= \frac{30}{\pi} - \frac{12}{\pi^2} < 8.4.$$

Similarly,

$$I_2 := \int_a^{\rho(b)} (b - \sigma(s))p^+(s)\Delta s$$
$$= \int_0^1 (9-s)\cos(\pi s/2)\mathrm{d}s + \int_4^5 (9-s)\cos(\pi s/2)\mathrm{d}s + \int_8^9 (9-s)\cos(\pi s/2)\mathrm{d}s$$
$$= \frac{24}{\pi} + \frac{12}{\pi^2} < 8.9$$

and

$$S_1 := \sum_{k=\underline{n}(a)}^{\overline{n}(b)} (t_k - a)(b - t_k)q_k^+ = \sum_{k=1}^{4} \frac{9-2k}{108k^2} = \frac{1245}{15552} < 0.1.$$

Since $\max\{I_1, I_2\} + S_1 < 9 = b - a$, and the remaining terms on the left-hand side of (9.4) are all zero, the inequality in Corollary 9.1 is satisfied, as opposed to Theorem 9.2. Thus, (9.16) is disconjugate on $(0, 9)$.

Example. Consider the impulsive dynamic equation

$$\begin{cases} x^{\Delta\Delta} + \frac{1}{s^3}x^{\Delta} - x^{\sigma} = 0, & t \neq 4k; \\ \triangle x^{\Delta} - t^3 x^{\Delta} + \sin(\frac{\pi t}{4}) = 0, & t = 4k, \quad k \in \mathbb{N} \end{cases} \tag{9.17}$$

on $\mathbb{T} = \mathbb{P}_{2,2} = \bigcup_{n=0}^{\infty}[4n, 4n+2]$ satisfying the Dirichlet boundary conditions

$$x(0) = x(b) = 0,$$

where $b > 0$. Observe that $p^+(s) = 1/s^3$ and $q^+(s) = p_k^+ = q_k^+ = 0$. Let j be the largest integer satisfying $\rho(b) \geq 4j + 4$. Then,

$$J_1 := \int_a^{\rho(b)} (\sigma(s) - a)p^+(s)\Delta s = \sum_{n=0}^{j} \int_{4n}^{4n+2} \frac{1}{s^2}\mathrm{d}s + \sum_{n=0}^{j} \frac{1}{(4n+2)^2} + \int_{4j+4}^{\rho(b)} \frac{1}{s^2}\mathrm{d}s$$
$$= \sum_{n=0}^{j}\left[\frac{1}{2n(4n+2)}\right] + \sum_{n=0}^{j} \frac{1}{(4n+2)^2} + \frac{1}{4j+4} - \frac{1}{\rho(b)}$$

and

$$\begin{aligned}
J_2 &:= \int_a^{\rho(b)} (b-\sigma(s))p^+(s)\Delta s \\
&= \sum_{n=0}^{j} \int_{4n}^{4n+2} \frac{b-s}{s^3} ds + \sum_{n=0}^{j} \frac{b-(4n+2)}{(4n+2)^3} + \int_{4j+4}^{\rho(b)} \frac{b-s}{s^3} ds \\
&= \sum_{n=0}^{j} \left[\frac{b(4n+1)}{4n^2(4n+2)^2} - \frac{1}{2n(4n+2)} \right] + \sum_{n=0}^{j} \frac{b-(4n+2)}{(4n+2)^3} \\
&\quad + \frac{b}{(4j+4)^2} - \frac{b}{(\rho(b))^2} - \frac{1}{4j+4} + \frac{1}{\rho(b)}.
\end{aligned}$$

Letting $b \to \infty$ and thus, $j \to \infty$, it is clearly seen that $\lim_{t\to\infty} J_1 < \infty$ and $\lim_{t\to\infty} J_2 < \infty$ which imply that $\max\{J_1, J_2\} < \infty$. Thus, by Corollary 9.1, the dynamic impulsive Eq. (9.17) is disconjugate on $(0,\infty)$. In other words, (9.17) is nonoscillatory.

Example. Consider the impulsive dynamic equation

$$\begin{cases} x^{\Delta\Delta} - 2x^{\Delta} + x^{\sigma} = 0, & t \neq k; \\ \triangle x^{\Delta} + 2(1+\cosh 1)x = 0, & t = k, \quad k \in \mathbb{N} \end{cases} \tag{9.18}$$

on $\mathbb{T} = \mathbb{R}$ with the boundary conditions

$$x\left(\frac{e}{e+1}\right) = x\left(1 + \frac{e}{e+1}\right) = 0. \tag{9.19}$$

By direct computation, it can be shown that

$$x_k(t) = (-1)^k e^{t-k+1}[(1+e)(k-t) - 1], \quad t \in (k-1, k]$$

is a solution of (9.18) with the boundary conditions (9.19). Moreover, it is not hard to see that $x_k(t) > 0$ if $x \in (e/(e+1), 1 + e/(e+1))$ which means that (9.18) is disconjugate on $(e/(e+1), 1 + e/(e+1))$. Thus, Theorem 9.2 applies, i. e., the inequality (9.4) must hold.

Since $p^+(t) = 0$, $q^+(t) = 1$, $p_k^+ = 0$ and $q_k^+ = 2(1+\cosh 1)$, we have

$$\begin{aligned}
&\int_a^{\rho(b)} (\sigma(s)-a)(b-\sigma(s))q^+(\sigma(s))\Delta s \\
&\quad = \int_{e/(e+1)}^{1+e/(e+1)} \left(\sigma(s) - \frac{e}{e+1}\right)\left(1 + \frac{e}{e+1} - \sigma(s)\right) ds \\
&\quad = \frac{1}{6}
\end{aligned}$$

and

$$\sum_{k=\underline{n}(a)}^{\overline{n}(b)} (t_k - a)(b - t_k)q_k^+ = 2(1 + \cosh 1) \sum_{k=1}^{1} \left(t_k - \frac{e}{e+1}\right)\left(1 + \frac{e}{e+1} - t_k\right) = 1.$$

Thus,

$$\int_a^{\rho(b)} (\sigma(s) - a)(b - \sigma(s))q^+(s)\Delta s + \sum_{k=\underline{n}(a)}^{\overline{n}(b)} (t_k - a)(b - t_k)q_k^+ = \frac{7}{6} > b - a = 1.$$

Hence, the de La Vallée Poussin-type inequality (9.4) is indeed valid.

Bibliography

[1] R. P. Agarwal, M. Jleli and B. Samet, On de La Vallée Poussin-type inequalities in higher dimension and applications, *Appl. Math. Lett.*, 86 (2018), 264–269.

[2] S. D. Akgöl, Asymptotic equivalence of impulsive dynamic equations on time scales, *Hacet. J. Math. Stat.*, 52(2) (2023), 277–291.

[3] S. D. Akgöl and A. Özbekler, De La Vallée Poussin inequality for impulsive differential equations, *Math. Slovaca*, 71(4) (2021), 881–888.

[4] M. Bohner and A. Peterson, Advances in Dynamic Equations on Time Scales, Birkhäuser, Boston, 2003.

[5] M. Bohner and A. Peterson, Dynamic Equations on Time Scales: An Introduction with Applications, Birkhäuser, Basel, 2001.

[6] M. Bohner and S. G. Georgiev, Multivariable Dynamic Calculus on Time Scales, Springer, Switzerland, 2016.

[7] İ. M. Erhan and S. G. Georgiev, Nonlinear Integral Equations on Time Scales, Nova Science Publishers, 2019.

[8] R. A. C. Ferreira, A de La Vallée Poussin Type Inequality on Time Scales, *Results Math.*, 73(3) (2018), 88.

[9] R. A. C. Ferreira, *Fract. Vallée Poussin Inequal.*, 22(3) (2019), 917–930.

[10] S. G. Georgiev, Integral Equations on Time Scales, Atlantis Studies in Dynamical Systems, Springer, New York, 2016.

[11] P. Hartman and A. Wintner, On an oscillation criterion of de La Vallée Poussin, *Q. Appl. Math.*, 13 (1955), 330–332.

[12] S. Hilger, Analysis on Measure Chains – A Unified Approach to Continuous and Discrete Calculus, *Results Math.*, 18 (1990), 18–56.

[13] B. Kaymakçalan, V. Lakshmikantham and S. Sivasundaram, Dynamic Systems on Measure Chains, Kluwer, Dordrecht, 1996.

[14] C. de La Vallée Poussin, Sur l'équation différentielle linéaire du second order. Détermination d'une intégrale par deux valuers assignés. Extension aux équasions d'ordre *n*, *J. Math. Pures Appl.*, 8 (1929), 125–144.

Najlaa Jaddoa, Rezan Sevinik-Adıgüzel, and İnci M. Erhan

10 Divided and σ-divided differences on time scales

Abstract: In this chapter, the divided differences and σ-divided differences on time scales are introduced. The Newton and σ-Newton interpolation polynomial are constructed. In addition, the Hermite interpolation polynomial on time scales is constructed by using the divided differences table. Examples are presented to illustrate the theoretical results.

10.1 Introduction

The theory of time scales was commenced by Hilger in his PhD thesis [9]. He unified the continuous and discrete analysis [1, 10]. Afterwards it has been studied by many researchers (see [2, 4, 5] and the references therein). The development of the theory of time scales is still in progress. In particular, numerical analysis and numerical methods on time scales is one of the very recent subjects of interest [3, 6–8].

Polynomial interpolation is a very practical method to represent a discrete set of data by a polynomial. Lagrange interpolation is one of the most widely used techniques to construct a polynomial which interpolates a given function, or a given discrete data set of nodes and function values. Another approach to construct the interpolation polynomial is the Newton interpolation which is based on using divided differences. On the other hand, the polynomial interpolation problem extends further to constructing a polynomial which takes not only the given function values but also the derivative values at the interpolation points. This problem uses the Hermite interpolation polynomial. In a recent book [6], the authors investigated the interpolation problem on arbitrary time scales and defined Lagrange and σ-Lagrange interpolation polynomials, as well as Hermite and σ-Hermite interpolation polynomials, on time scales.

In this study we intend to give the construction of Newton and σ-Newton interpolation polynomial by introducing the divided and σ-divided differences on arbitrary time scales. This approach provides an alternative way to compute the interpolation polynomial for a given set of data. Moreover, we propose an alternative and simpler way to

Najlaa Jaddoa, Department of Mathematics, Gazi University, 06500 Beşevler, Ankara, Turkey, e-mail: nabd.jaddoa@gazi.edu.tr
Rezan Sevinik-Adıgüzel, Department of Mathematics, Atilim University, 06830 Incek, Ankara, Turkey, e-mail: rezan.adiguzel@atilim.edu.tr
İnci M. Erhan, Department of Computer Engineering, Aydın Adnan Menderes University, 09010 Aydın, Turkey, e-mail: incierhan@gmail.com

https://doi.org/10.1515/9783111182971-010

write the Hermite interpolation polynomial using modified divided differences for an arbitrary time scale.

The chapter is organized as follows. In Section 10.2, we propose the construction of divided differences on an arbitrary time scale and define the Newton interpolation polynomial via the divided difference table. In Section 10.3, we construct the σ-divided differences on an arbitrary time scale and the σ-Newton interpolation polynomial by using them. Section 10.4 contains the construction of the Hermite interpolation polynomial using modified divided differences on an arbitrary time scale.

10.2 Divided differences

Let $\mathbb{T}$ be a time scale with the forward jump operator σ and graininess function μ. Suppose that $x_i \in \mathbb{T}$, $i \in \{0, 1, \ldots, n\}$ are distinct points in the time scale and $y_i = f(x_i)$, $i \in \{0, 1, \ldots, n\}$ are given real numbers. The Newton interpolation polynomial is constructed in the following way.

Using the notation $f[x_0] = f(x_0) = y_0, f[x_1] = f(x_1) = y_1, \ldots, f[x_n] = f(x_n) = y_n$, one can generate the polynomials $P_n(x)$ recursively by taking

$$P_0(x) = f(x_0) = f[x_0]$$

and defining

$$P_1(x) = P_0(x) + f[x_0, x_1](x - x_0) = f[x_0] + f[x_0, x_1](x - x_0). \tag{10.1}$$

The next polynomial has the form

$$\begin{aligned} P_2(x) &= P_1(x) + f[x_0, x_1, x_2](x - x_0)(x - x_1) \\ &= f[x_0] + f[x_0, x_1](x - x_0) + f[x_0, x_1, x_2](x - x_0)(x - x_1). \end{aligned} \tag{10.2}$$

The nth degree polynomial is generalized as

$$P_n(x) = \sum_{i=0}^{n} f[x_0, \ldots, x_i] \prod_{j=0}^{i-1} (x - x_j), \tag{10.3}$$

and is called the Newton interpolation polynomial, where $f[x_0, \ldots, x_i]$ represents the divided difference of order i. Here we take $\prod_{j=0}^{-1}(x - x_j) = 1$.

In order to evaluate the divided differences $f[x_0, \ldots, x_i]$, for $i = 0, 1, \ldots, n$, we first employ (10.1) and the interpolation conditions

$$P_n(x_i) = f(x_i), \quad \text{for any } n \in \mathbb{N}_0 \text{ and } i = 0, 1, \ldots, n,$$

leading to

$$f[x_0, x_1] = \frac{f(x_1) - f(x_0)}{x_1 - x_0}. \tag{10.4}$$

Next by using (10.2), we derive

$$f[x_0, x_1, x_2] = \frac{f(x_0)}{(x_0 - x_1)(x_0 - x_2)} + \frac{f(x_1)}{(x_1 - x_0)(x_1 - x_2)} + \frac{f(x_2)}{(x_2 - x_0)(x_2 - x_1)}.$$

Substituting the identity

$$\frac{f(x_1)}{(x_1 - x_0)(x_1 - x_2)} = \frac{f(x_1)}{(x_1 - x_0)(x_0 - x_2)} - \frac{f(x_1)}{(x_1 - x_2)(x_0 - x_2)},$$

we deduce

$$f[x_0, x_1, x_2] = \frac{\left\{\frac{f(x_2)-f(x_1)}{x_2-x_1} - \frac{f(x_1)-f(x_0)}{x_1-x_0}\right\}}{x_2 - x_0} = \frac{f[x_1, x_2] - f[x_0, x_1]}{x_2 - x_0}. \tag{10.5}$$

Notice here that the form of $f[x_0, x_1, x_2]$ obtained in terms of $f[x_1, x_2]$ and $f[x_0, x_1]$ provides an easy way of generating divided differences recursively. Then, the general nth divided difference can be derived recursively as follows:

$$f[x_0, \ldots, x_n] = \frac{f[x_1, \ldots, x_n] - f[x_0, \ldots, x_{n-1}]}{x_n - x_0}. \tag{10.6}$$

The formula (10.6) allows us to generate all the divided differences needed for the Newton interpolation polynomial in a simple manner by using a divided difference table. We illustrate such a table for the case $n = 4$ (see Table 10.1).

Table 10.1: Divided difference table.

x_k	$f(x_k)$	$f[x_k, x_{k+1}]$	$f[x_k, x_{k+1}, x_{k+2}]$	$f[x_k, x_{k+1}, x_{k+2}, x_{k+3}]$	$f[x_k, x_{k+1}, x_{k+2}, x_{k+3}, x_{k+4}]$
x_0	$f(x_0)$				
		$f[x_0, x_1]$			
x_1	$f(x_1)$		$f[x_0, x_1, x_2]$		
		$f[x_1, x_2]$		$f[x_0, x_1, x_2, x_3]$	
x_2	$f(x_2)$		$f[x_1, x_2, x_3]$		$f[x_0, x_1, x_2, x_3, x_4]$
		$f[x_2, x_3]$		$f[x_1, x_2, x_3, x_4]$	
x_3	$f(x_3)$		$f[x_2, x_3, x_4]$		
		$f[x_3, x_4]$			
x_4	$f(x_4)$				

The divided differences in the table are calculated columnwise by using the formula

$$f[x_i, \ldots, x_{i+k}] = \frac{f[x_{i+1}, \ldots, x_{i+k}] - f[x_i, \ldots, x_{i+k-1}]}{x_{i+k} - x_i}, \tag{10.7}$$

for i, depending on n and k. The coefficients used in the Newton interpolation polynomial are the first entries in each column.

In the next theorem we prove that the Newton interpolation polynomial defined in (10.3) interpolates the given set of points (x_i, y_i) for $i = 0, 1, \ldots, n$.

In the following, we use the notation $y[x_0, x_1, \ldots, x_k]$ instead of $f[x_0, x_1, \ldots, x_k]$, $k = 0, 1, \ldots, n$, for the divided differences.

Theorem 10.1. *Consider the polynomial*

$$P_n(x) = y[x_0] + y[x_0, x_1](x - x_0) + \cdots + y[x_0, x_1, \ldots, x_n](x - x_0) \cdots (x - x_{n-1}),$$

where

$$y[x_0, \ldots, x_k] = \frac{y[x_1, \ldots, x_k] - y[x_0, \ldots, x_{k-1}]}{x_k - x_0}, \quad k = 1, \ldots, n.$$

Then $P_n(x)$ interpolates the points (x_i, y_i) for $i = 0, 1, \ldots, n$.

Proof. We will prove the theorem by using induction. For $n = 1$, the Newton polynomial is

$$P_1(x) = y[x_0] + y[x_0, x_1](x - x_0).$$

Clearly,

$$P_1(x_0) = y[x_0] = y_0$$

and

$$P_1(x_1) = y[x_0] + y[x_0, x_1](x_1 - x_0) = y[x_1] = y_1$$

gives

$$y[x_0, x_1] = \frac{y[x_1] - y[x_0]}{x_1 - x_0},$$

which indeed is the form of $y[x_0, x_1]$.

Assume that the hypothesis holds for $n = k$, that is,

$$P_k(x) = y[x_0] + \cdots + y[x_0, \ldots, x_k](x - x_0) \cdots (x - x_{k-1})$$

interpolates the points (x_i, y_i) for $i = 0, 1, \ldots, k$. Consider the polynomial

$$q_k(x) = y[x_1] + y[x_1, x_2](x - x_1) + \cdots + y[x_1, \ldots, x_{k+1}](x - x_1) \cdots (x - x_k).$$

It is obvious that the polynomial q_k interpolates the points (x_i, y_i) for $i = 1, \ldots, k + 1$. Define the $(k + 1)$th degree polynomial

$$r(x) = \frac{(x - x_0)q_k(x) + (x_{k+1} - x)P_k(x)}{x_{k+1} - x_0}.$$

Then for x_i, $i = 1, 2, \dots, k$, we have

$$\begin{aligned} r(x_i) &= \frac{(x_i - x_0)q_k(x_i) + (x_{k+1} - x_i)P_k(x_i)}{x_{k+1} - x_0} \\ &= \frac{(x_i - x_0)y_i + (x_{k+1} - x_i)y_i}{x_{k+1} - x_0} \\ &= \frac{(x_{k+1} - x_0)y_i}{x_{k+1} - x_0} \\ &= y_i. \end{aligned}$$

For $i = 0$, we have

$$r(x_0) = \frac{0.q_k(x_0) + (x_{k+1} - x_0)P_k(x_0)}{x_{k+1} - x_0} = P_k(x_0) = y_0,$$

and, for $i = k + 1$,

$$r(x_{k+1}) = (x_{k+1} - x_0)q_k(x_{k+1}) = q_k(x_{k+1}) = y_{k+1}.$$

Thus, r_k is a polynomial of degree $k + 1$ which interpolates the points (x_i, y_i) for $i = 0, 1, \dots, k + 1$ and hence, $r_k = P_{k+1}$. The leading coefficient, that is, the coefficient of the highest degree term x^{k+1} of P_{k+1}, is then

$$\frac{y[x_1, \dots, x_{k+1}] - y[x_0, \dots, x_k]}{x_{k+1} - x_0},$$

which completes the proof. □

10.3 σ-Divided differences

In order to define the σ-divided differences, we need to give the definition of σ-distinct points on a time scale.

Definition 10.1 ([6]). Let $n \in \mathbb{N}_0$. The points $x_j \in [a, b)$, $j \in \{0, 1, \dots, n\}$, will be called σ-distinct if $\sigma(x_n) \le b$ and

$$\sigma(x_0) < \sigma(x_1) < \cdots < \sigma(x_n).$$

Suppose that $x_i \in \mathbb{T}$, $i \in \{0, 1, \dots, n\}$ are σ-distinct points on a time scale and $y_i = f(x_i)$, $i \in \{0, 1, \dots, n\}$ are given real numbers. We will define the σ-Newton interpolation polynomial via σ-divided differences.

Using the notation $f_\sigma[x_0] = f(x_0) = y_0, f_\sigma[x_1] = f(x_1) = y_1, \ldots, f_\sigma[x_n] = f(x_n) = y_n$, the polynomials $P_{\sigma_n}(x)$ can be generated recursively as follows.

Let

$$P_{\sigma_0}(x) = f(x_0) = f_\sigma[x_0],$$

and define

$$P_{\sigma_1}(x) = P_{\sigma_0}(x) + f_\sigma[x_0, x_1](\sigma(x) - \sigma(x_0)) = f_\sigma[x_0] + f_\sigma[x_0, x_1](\sigma(x) - \sigma(x_0)). \quad (10.8)$$

Then,

$$\begin{aligned} P_{\sigma_2}(x) &= P_{\sigma_1}(x) + f_\sigma[x_0, x_1, x_2](\sigma(x) - \sigma(x_0))(\sigma(x) - \sigma(x_1)) \\ &= f_\sigma[x_0] + f_\sigma[x_0, x_1](\sigma(x) - \sigma(x_0)) \\ &\quad + f_\sigma[x_0, x_1, x_2](\sigma(x) - \sigma(x_0))(\sigma(x) - \sigma(x_1)). \end{aligned} \quad (10.9)$$

The general formula for $P_{\sigma_n}(x)$ is generalized as

$$P_{\sigma_n}(x) = \sum_{i=0}^{n} f_\sigma[x_0, \ldots, x_i] \prod_{j=0}^{i-1} (\sigma(x) - \sigma(x_j)), \quad (10.10)$$

where we take $\prod_{j=0}^{-1}(\sigma(x) - \sigma(x_j)) = 1$. The notation $f_\sigma[x_0, \ldots, x_i]$ will be referred to as σ-divided difference of order i.

Definition 10.2. The polynomial given by the formula (10.10) will be called σ-Newton interpolation polynomial.

Remark 10.1. Note that if, on a given time scale the forward jump operator σ is a nonlinear function, the σ-Newton interpolation polynomial may not be a polynomial of degree n. Therefore, in [6] such functions are called σ-polynomials, since they are polynomials in $\sigma(x)$ but not in x.

The σ-divided differences $f_\sigma[x_0, \ldots, x_i]$ used in the formula (10.10) are evaluated recursively below. From (10.8) we have

$$f_\sigma[x_0, x_1] = \frac{f(x_0)}{\sigma(x_0) - \sigma(x_1)} + \frac{f(x_1)}{\sigma(x_1) - \sigma(x_0)} = \frac{f(x_1) - f(x_0)}{\sigma(x_1) - \sigma(x_0)},$$

and from (10.9) we can calculate

$$f_\sigma[x_0, x_1, x_2] = \frac{\frac{f(x_2)-f(x_1)}{\sigma(x_2)-\sigma(x_1)} - \frac{f(x_1)-f(x_0)}{\sigma(x_1)-\sigma(x_0)}}{\sigma(x_2) - \sigma(x_0)} = \frac{f_\sigma[x_1, x_2] - f_\sigma[x_0, x_1]}{\sigma(x_2) - \sigma(x_0)}.$$

The form of $f_\sigma[x_0, x_1, x_2]$ indicates an easy way of generating σ-divided differences recursively. The general form can be derived by the following formula:

$$f_\sigma[x_0,\dots,x_k] = \frac{f_\sigma[x_1,\dots,x_k] - f_\sigma[x_0,\dots,x_{k-1}]}{\sigma(x_k) - \sigma(x_0)},$$

for $k = 1, 2, \dots, n$. This formula allows us to generate all the σ-divided differences needed for the σ-Newton interpolation polynomial in a simple manner by using a divided difference table. In Table 10.2, we show the σ-divided differences for $n = 4$. The calculation is done columnwise using the recursion formula

$$f_\sigma[x_i,\dots,x_{i+k}] = \frac{f_\sigma[x_{i+1},\dots,x_{i+k}] - f_\sigma[x_i,\dots,x_{i+k-1}]}{\sigma(x_{i+k}) - \sigma(x_i)},$$

for i depending on n, and k.

The coefficients of the σ-Newton interpolation polynomial appear at the top of each column.

Table 10.2: σ-Divided difference table.

$\sigma(x_k)$	$f(x_k)$	$f_\sigma[x_k,x_{k+1}]$	$f_\sigma[x_k,x_{k+1},x_{k+2}]$	$f_\sigma[x_k,x_{k+1},x_{k+2},x_{k+3}]$	$f_\sigma[x_k,x_{k+1},x_{k+2},x_{k+3},x_{k+4}]$
$\sigma(x_0)$	$f(x_0)$				
		$f_\sigma[x_0,x_1]$			
$\sigma(x_1)$	$f(x_1)$		$f_\sigma[x_0,x_1,x_2]$		
		$f_\sigma[x_1,x_2]$		$f_\sigma[x_0,x_1,x_2,x_3]$	
$\sigma(x_2)$	$f(x_2)$		$f_\sigma[x_1,x_2,x_3]$		$f_\sigma[x_0,x_1,x_2,x_3,x_4]$
		$f_\sigma[x_2,x_3]$		$f_\sigma[x_1,x_2,x_3,x_4]$	
$\sigma(x_3)$	$f(x_3)$		$f_\sigma[x_2,x_3,x_4]$		
		$f_\sigma[x_3,x_4]$			
$\sigma(x_4)$	$f(x_4)$				

Theorem 10.2. *Consider the σ- polynomial*

$$\begin{aligned} P_{\sigma_n}(x) &= y_\sigma[x_0] + y_\sigma[x_0,x_1](\sigma(x) - \sigma(x_0)) \\ &\quad + \cdots + y_\sigma[x_0,x_1,\dots,x_n](\sigma(x) - \sigma(x_0))\cdots(\sigma(x) - \sigma(x_{n-1})), \end{aligned} \tag{10.11}$$

where

$$y_\sigma[x_0,\dots,x_k] = \frac{y_\sigma[x_1,\dots,x_k] - y_\sigma[x_0,\dots,x_{k-1}]}{\sigma(x_k) - \sigma(x_0)}. \tag{10.12}$$

Then $P_{\sigma_n}(x)$ interpolates the σ-distinct points $\{x_0,\dots,x_n\}, \{y_0,\dots,y_n\}$.

Proof. We consider the proof by induction. For $n = 1$, we have

$$P_{\sigma_1}(x_0) = y_\sigma[x_0] + y_\sigma[x_0,x_1](\sigma(x) - \sigma(x_0)) = y_\sigma[x_0]$$

and

$$P_{\sigma_1}(x_1) = y_\sigma[x_0] + y_\sigma[x_0, x_1](\sigma(x_1) - \sigma(x_0)) = y_\sigma[x_1],$$

provided that

$$y_\sigma[x_0, x_1] = \frac{y_\sigma[x_1] - y_\sigma[x_0]}{\sigma(x_1) - \sigma(x_0)}.$$

Let $n = k$ and assume that

$$P_{\sigma_k}(x) = y_\sigma[x_0] + \cdots + y_\sigma[x_0, \ldots, x_k](\sigma(x) - \sigma(x_0)) \cdots (\sigma(x) - \sigma(x_{k-1}))$$

interpolates the points $\{x_0, \ldots, x_k\}$, $\{y_0, \ldots, y_k\}$. Let

$$\begin{aligned} q_{\sigma_k}(x) &= y_\sigma[x_1] + y_\sigma[x_1, x_2](\sigma(x) - \sigma(x_1)) \\ &\quad + \cdots + y_\sigma[x_1, \ldots, x_{k+1}](\sigma(x) - \sigma(x_1)) \cdots (\sigma(x) - \sigma(x_k)). \end{aligned} \tag{10.13}$$

Clearly, q_{σ_k} interpolates the points $\{x_1, \ldots, x_{k+1}\}$, $\{y_1, \ldots, y_{k+1}\}$. Define

$$r_\sigma(x) = \frac{(\sigma(x) - \sigma(x_0))q_{\sigma_k}(x) + (\sigma(x_{k+1}) - \sigma(x))p_{\sigma_k}(x)}{\sigma(x_{k+1}) - \sigma(x_0)}.$$

It can be easily verified that r_σ interpolates the σ-distinct points $\{x_1, \ldots, x_{k+1}\}$, $\{y_1, \ldots, y_{k+1}\}$. Indeed, we have

$$\begin{aligned} r_\sigma(x_i) &= \frac{(\sigma(x_i) - \sigma(x_0))q_{\sigma_k}(x_i) + (\sigma(x_{k+1}) - \sigma(x_i))p_{\sigma_k}(x_i)}{\sigma(x_{k+1}) - \sigma(x_0)} \\ &= \frac{(\sigma(x_i) - \sigma(x_0))y_i + (\sigma(x_{k+1}) - \sigma(x_i))y_i}{\sigma(x_{k+1}) - \sigma(x_0)} \\ &= y_i. \end{aligned}$$

For $i = 0$, we get

$$r_\sigma(x_0) = \frac{0.q_{\sigma_k}(x_0) + (\sigma(x_{k+1}) - \sigma(x_0))P_{\sigma_k}(x_0)}{\sigma(x_{n+1}) - \sigma(x_0)} = P_{\sigma_k}(x_0) = y_0.$$

For $i = k + 1$, we have

$$r_\sigma(x_{k+1}) = (\sigma(x_{k+1}) - \sigma(x_0))q_{\sigma_k}(x_{k+1}) = q_{\sigma_k}(x_{k+1}) = y_{k+1}.$$

Then r_σ is a σ-polynomial of degree $(k + 1)$ which interpolates the σ-distinct points $\{x_0, \ldots, x_{k+1}\}$, $\{y_0, \ldots, y_{k+1}\}$. Hence, $r_\sigma = P_{\sigma,k+1}$ and its leading coefficient is

$$\frac{y_\sigma[x_1, \ldots, x_{k+1}] - y_\sigma[x_0, \ldots, x_k]}{\sigma(x_{k+1}) - \sigma(x_0)},$$

which completes the proof. □

Next, we present some examples of σ-divided differences on different time scales and the resulting σ-Newton interpolation polynomial. We also compute the σ-Lagrange interpolation polynomial as described in [8].

Example. Let $\mathbb{T} = \{\sqrt{2n+1} : n = 0, 1, 2, \ldots\}$ and

$$x_0 = 1, \quad x_1 = \sqrt{5}, \quad x_2 = \sqrt{11},$$
$$y_0 = -2, \quad y_1 = 0, \quad y_2 = 4.$$

We will compute both σ-Lagrange and σ-Newton interpolation polynomials and verify that they are equal.

On the given time scale $\mathbb{T} = \{\sqrt{1}, \sqrt{3}, \sqrt{5}, \ldots\}$ we have $\sigma(x) = \sqrt{x^2+2}$, so that

$$\begin{aligned} \sigma(x_0) &= \sigma(1) = \sqrt{3}, \\ \sigma(x_1) &= \sigma(\sqrt{5}) = \sqrt{7}, \\ \sigma(x_2) &= \sigma(\sqrt{11}) = \sqrt{13}. \end{aligned}$$

The σ-divided differences are computed in the following Table 10.3.

Table 10.3: The divided difference table for the first example.

x	$\sigma(x)$	$f(x)$	$f[x_k, x_{k+1}]$	$f[x_k, x_{k+1}, x_{k+2}]$
1	$\sqrt{3}$	-2		
			$\frac{0-(-2)}{\sqrt{7}-\sqrt{3}} = \frac{2}{\sqrt{7}-\sqrt{3}}$	
$\sqrt{5}$	$\sqrt{7}$	0		$\frac{\frac{4}{\sqrt{13}-\sqrt{7}} - \frac{2}{\sqrt{7}-\sqrt{3}}}{\sqrt{13}-\sqrt{3}}$
			$\frac{4-0}{\sqrt{13}-\sqrt{7}} = \frac{4}{\sqrt{13}-\sqrt{7}}$	
$\sqrt{11}$	$\sqrt{13}$	4		

Then the σ-Newton interpolation polynomial can be derived as

$$\begin{aligned} N_{\sigma_2}(x) &= f(x) + f[x_0, x_1](\sigma(x) - \sigma(x_0)) \\ &\quad + f[x_0, x_1, x_2](\sigma(x) - \sigma(x_0))(\sigma(x) - \sigma(x_1)) \\ &= -2 + \frac{2}{\sqrt{7}-\sqrt{3}}(\sigma(x) - \sqrt{3}) + \frac{\frac{4}{\sqrt{13}-\sqrt{7}} - \frac{2}{\sqrt{7}-\sqrt{3}}}{\sqrt{13}-\sqrt{3}}(\sigma(x) - \sqrt{3})(\sigma(x) - \sqrt{7}), \end{aligned}$$

which is simplified as

$$N_{\sigma_2}(x) = 1.056\,\sigma^2(x) - 2.435\,\sigma(x) - 0.951. \tag{10.14}$$

To verify the uniqueness of an interpolation polynomial, we compute the σ-Lagrange interpolation polynomial which is defined in [8]. Following the construction procedure presented in [8], we obtain

$$\begin{aligned}
L_{\sigma_0}(x) &= \frac{(\sigma(x)-\sigma(x_1))(\sigma(x)-\sigma(x_2))}{(\sigma(x_0)-\sigma(x_1))(\sigma(x_0)-\sigma(x_2))} \\
&= \frac{(\sigma(x)-\sqrt{7})(\sigma(x)-\sqrt{13})}{(\sqrt{3}-\sqrt{7})(\sqrt{3}-\sqrt{13})} \\
&= \frac{\sigma^2(x)+(-\sqrt{13}-\sqrt{7})\sigma(x)+\sqrt{91}}{3-\sqrt{39}-\sqrt{21}+\sqrt{91}}, \\
L_{\sigma_1}(x) &= \frac{(\sigma(x)-\sigma(x_0))(\sigma(x)-\sigma(x_2))}{(\sigma(x_1)-\sigma(x_0))(\sigma(x_1)-\sigma(x_2))} \\
&= \frac{(\sigma(x)-\sqrt{3})(\sigma(x)-\sqrt{13})}{(\sqrt{7}-\sqrt{3})(\sqrt{7}-\sqrt{13})} \\
&= \frac{\sigma^2(x)+(-\sqrt{13}-\sqrt{3})\sigma(x)+\sqrt{39}}{7-\sqrt{91}-\sqrt{21}+\sqrt{39}}, \\
L_{\sigma_2}(x) &= \frac{(\sigma(x)-\sigma(x_0))(\sigma(x)-\sigma(x_1))}{(\sigma(x_2)-\sigma(x_0))(\sigma(x_2)-\sigma(x_1))} \\
&= \frac{(\sigma(x)-\sqrt{3})(\sigma(x)-\sqrt{7})}{(\sqrt{13}-\sqrt{3})(\sqrt{13}-\sqrt{7})} \\
&= \frac{\sigma^2(x)+(-\sqrt{7}-\sqrt{3})\sigma(x)+\sqrt{21}}{13-\sqrt{91}-\sqrt{39}+\sqrt{21}}.
\end{aligned}$$

Then, the σ-Lagrange interpolation polynomial has the form

$$\begin{aligned}
P_{\sigma_2}(x) &= y_0L_{\sigma_0}(x)+y_1L_{\sigma_1}(x)+y_2L_{\sigma_2}(x) \\
&= -2L_{\sigma_0}(x)+4L_{\sigma_2}(x) \\
&= -2\frac{\sigma^2(x)+(-\sqrt{13}-\sqrt{7})\sigma(x)+\sqrt{91}}{3-\sqrt{39}-\sqrt{21}+\sqrt{91}} \\
&\quad +4\frac{\sigma^2(x)+(-\sqrt{7}-\sqrt{3})\sigma(x)+\sqrt{21}}{13-\sqrt{91}-\sqrt{39}+\sqrt{21}}
\end{aligned}$$

and simplifies as

$$P_{\sigma_2}(x) = 1.056\sigma^2(x) - 2.435\,\sigma(x) - 0.951. \tag{10.15}$$

Observe that the σ-Lagrange polynomial (10.15) and σ-Newton polynomial (10.14) are equal.

The next example presents a computation of a third-degree σ-interpolation polynomial.

Example. Let $\mathbb{T} = \{\frac{1}{n} : n \in \mathbb{N}\} = \{\dots, \frac{1}{n}, \frac{1}{n-1}, \dots, \frac{1}{3}, \frac{1}{2}, 1\}$ and

$$\begin{aligned}
x_0 &= \frac{1}{16}, \quad x_1 = \frac{1}{12}, \quad x_2 = \frac{1}{10}, \quad x_3 = \frac{1}{7}, \\
y_0 &= \frac{1}{272}, \quad y_1 = \frac{1}{156}, \quad y_2 = \frac{1}{110}, \quad y_3 = \frac{1}{56}.
\end{aligned}$$

We will find the σ-Lagrange and the σ-Newton interpolating polynomials for the given points (x_i, y_i), $i = 0, 1, 2, 3$.

On the given time scale, the forward jump operator is

$$\sigma(x) = \frac{x}{1-x},$$

so that we obtain

$$\begin{aligned} \sigma(x_0) &= \sigma\left(\frac{1}{16}\right) = \frac{1}{15}, \\ \sigma(x_1) &= \sigma\left(\frac{1}{12}\right) = \frac{1}{11}, \\ \sigma(x_2) &= \sigma\left(\frac{1}{10}\right) = \frac{1}{9}, \\ \sigma(x_3) &= \sigma\left(\frac{1}{7}\right) = \frac{1}{6}. \end{aligned}$$

The σ-divided differences are computed in Table 10.4.

Table 10.4: The divided difference table for the second example.

x	$\sigma(x)$	$f(x)$	$f[x_k, x_{k+1}]$	$f[x_k, x_{k+1}, x_{k+2}]$	$f[x_k, x_{k+1}, x_{k+2}, x_{k+3}]$
$\frac{1}{16}$	$\frac{1}{15}$	$\frac{1}{272}$			
			$\frac{\frac{1}{156}-\frac{1}{272}}{\frac{1}{11}-\frac{1}{15}} = \frac{257}{2279}$		
$\frac{1}{12}$	$\frac{1}{11}$	$\frac{1}{156}$		$\frac{\frac{69}{520}-\frac{257}{2279}}{\frac{1}{9}-\frac{1}{15}} = \frac{2531}{5646}$	
			$\frac{\frac{1}{110}-\frac{1}{156}}{\frac{1}{9}-\frac{1}{11}} = \frac{69}{520}$		$\frac{\frac{603}{1820}-\frac{2531}{5646}}{\frac{1}{6}-\frac{1}{15}} = \frac{-986}{843}$
$\frac{1}{10}$	$\frac{1}{9}$	$\frac{1}{110}$		$\frac{\frac{243}{1540}-\frac{69}{520}}{\frac{1}{6}-\frac{1}{11}} = \frac{603}{1820}$	
			$\frac{\frac{1}{56}-\frac{1}{110}}{\frac{1}{6}-\frac{1}{9}} = \frac{243}{1540}$		
$\frac{1}{7}$	$\frac{1}{6}$	$\frac{1}{56}$			

From Table 10.4, the σ-Newton polynomial follows easily as

$$\begin{aligned} N_{\sigma_3}(x) &= f(x) + f[x_0, x_1](\sigma(x) - \sigma(x_0)) \\ &\quad + f[x_0, x_1, x_2](\sigma(x) - \sigma(x_0))(\sigma(x) - \sigma(x_1)) \\ &\quad + f[x_0, x_1, x_2, x_3](\sigma(x) - \sigma(x_0))(\sigma(x) - \sigma(x_1))(\sigma(x) - \sigma(x_2)) \\ &= \frac{1}{272} + \frac{257}{2279}\left(\sigma(x) - \frac{1}{15}\right) + \frac{2531}{5646}\left(\sigma(x) - \frac{1}{15}\right)\left(\sigma(x) - \frac{1}{11}\right) \\ &\quad + \frac{-986}{843}\left(\sigma(x) - \frac{1}{15}\right)\left(\sigma(x) - \frac{1}{11}\right)\left(\sigma(x) - \frac{1}{9}\right), \end{aligned}$$

which becomes

$$N_{\sigma_3}(x) = \frac{-986}{843}\sigma^3(x) + \frac{1284}{1745}\sigma^2(x) + \frac{314}{12951}\sigma(x) - \frac{7}{16371}.$$

For the σ-Lagrange interpolation polynomial, we first compute the σ-Lagrange basis polynomials as in [6, 8]:

$$\begin{aligned}
L_{\sigma_0}(x) &= \frac{(\sigma(x)-\sigma(x_1))(\sigma(x)-\sigma(x_2))(\sigma(x)-\sigma(x_3))}{(\sigma(x_0)-\sigma(x_1))(\sigma(x_0)-\sigma(x_2))(\sigma(x_0)-\sigma(x_3))}\\
&= \frac{(\sigma(x)-\frac{1}{11})(\sigma(x)-\frac{1}{9})(\sigma(x)-\frac{1}{6})}{(\frac{1}{15}-\frac{1}{11})(\frac{1}{15}-\frac{1}{9})(\frac{1}{15}-\frac{1}{6})},\\
L_{\sigma_1}(x) &= \frac{(\sigma(x)-\sigma(x_0))(\sigma(x)-\sigma(x_2))(\sigma(x)-\sigma(x_3))}{(\sigma(x_1)-\sigma(x_0))(\sigma(x_1)-\sigma(x_2))(\sigma(x_1)-\sigma(x_3))}\\
&= \frac{(\sigma(x)-\frac{1}{15})(\sigma(x)-\frac{1}{9})(\sigma(x)-\frac{1}{6})}{(\frac{1}{11}-\frac{1}{15})(\frac{1}{11}-\frac{1}{9})(\frac{1}{11}-\frac{1}{6})},\\
L_{\sigma_2}(x) &= \frac{(\sigma(x)-\sigma(x_0))(\sigma(x)-\sigma(x_1))(\sigma(x)-\sigma(x_3))}{(\sigma(x_2)-\sigma(x_0))(\sigma(x_2)-\sigma(x_1))(\sigma(x_2)-\sigma(x_3))}\\
&= \frac{(\sigma(x)-\frac{1}{15})(\sigma(x)-\frac{1}{11})(\sigma(x)-\frac{1}{6})}{(\frac{1}{9}-\frac{1}{15})(\frac{1}{9}-\frac{1}{11})(\frac{1}{9}-\frac{1}{6})},\\
L_{\sigma_3}(x) &= \frac{(\sigma(x)-\sigma(x_0))(\sigma(x)-\sigma(x_1))(\sigma(x)-\sigma(x_2))}{(\sigma(x_3)-\sigma(x_0))(\sigma(x_3)-\sigma(x_1))(\sigma(x_3)-\sigma(x_2))}\\
&= \frac{(\sigma(x)-\frac{1}{15})(\sigma(x)-\frac{1}{11})(\sigma(x)-\frac{1}{9})}{(\frac{1}{6}-\frac{1}{15})(\frac{1}{6}-\frac{1}{11})(\frac{1}{6}-\frac{1}{6})}.
\end{aligned}$$

This results in the σ-Lagrange polynomial

$$\begin{aligned}
P_{\sigma_3}(x) &= f(x_0)L_{\sigma_0}(x) + f(x_1)L_{\sigma_1}(x) + f(x_2)L_{\sigma_2}(x) + f(x_3)L_{\sigma_3}(x)\\
&= \frac{1}{272}\left[\frac{(\sigma(x)-\frac{1}{11})(\sigma(x)-\frac{1}{9})(\sigma(x)-\frac{1}{6})}{(\frac{1}{15}-\frac{1}{11})(\frac{1}{15}-\frac{1}{9})(\frac{1}{15}-\frac{1}{6})}\right]\\
&\quad + \frac{1}{156}\left[\frac{(\sigma(x)-\frac{1}{15})(\sigma(x)-\frac{1}{9})(\sigma(x)-\frac{1}{6})}{(\frac{1}{11}-\frac{1}{15})(\frac{1}{11}-\frac{1}{9})(\frac{1}{11}-\frac{1}{6})}\right]\\
&\quad + \frac{1}{110}\left[\frac{(\sigma(x)-\frac{1}{15})(\sigma(x)-\frac{1}{11})(\sigma(x)-\frac{1}{6})}{(\frac{1}{9}-\frac{1}{15})(\frac{1}{9}-\frac{1}{11})(\frac{1}{9}-\frac{1}{6})}\right]\\
&\quad + \frac{1}{56}\left[\frac{(\sigma(x)-\frac{1}{15})(\sigma(x)-\frac{1}{11})(\sigma(x)-\frac{1}{9})}{(\frac{1}{6}-\frac{1}{15})(\frac{1}{6}-\frac{1}{11})(\frac{1}{6}-\frac{1}{9})}\right],
\end{aligned}$$

simplifying into

$$P_{\sigma_3}(x) = \frac{-986}{843}\sigma^3(x) + \frac{1284}{1745}\sigma^2(x) + \frac{314}{12951}\sigma(x) - \frac{7}{16371}.$$

The equality of the two σ-polynomials is obvious, which demonstrates the consistency with the theoretical results.

10.4 Hermite interpolation polynomial via divided differences on time scales

In this section, the construction of a modified divided difference table is discussed, which simplifies the computation of the Hermite interpolation polynomial considerably.

Let $\mathbb{T}$ be a time scale with the forward jump operator σ, delta derivative Δ, and graininess function μ. Let $x_0, x_1, \ldots, x_n$ be given σ-distinct points such that $x_j \neq \sigma(x_i)$ for $i, j \in \{0, \ldots, n\}$. Let $y_0, y_1, \ldots, y_n$ and $z_0, z_1, \ldots, z_n$ be the points such that $y_i = f(x_i)$, $z_i = f^{\Delta}(x_i)$ for some f.

The idea is based on doubling the number of points in the following way. Define

$$t_{2k} = x_k, \quad t_{2k+1} = \sigma(x_k), \quad k = 0, \ldots, n.$$

Then

$$y(t_{2k}) = y_k, \quad y(t_{2k+1}) = y(\sigma(x_k)), \quad k = 0, \ldots, n,$$

where $y(\sigma(x_i))$ are in general unknown. The first divided differences for $t_0, \ldots, t_{2n}$ will be defined as

$$y[t_i] = \begin{cases} y_k, & i = 2k, \\ w_k, & i = 2k+1, \end{cases}$$

where w_k is not given if the function itself is unknown.

The second divided differences are defined as

$$y[t_j, t_{j+1}] = \frac{y(t_{j+1}) - y(t_j)}{t_{j+1} - t_j}, \quad j = 0, \ldots, 2n.$$

Now for $j = 2k$ we have

$$\begin{aligned} y[t_{2k}, t_{2k+1}] &= \frac{y(t_{2k+1}) - y(t_{2k})}{t_{2k+1} - t_{2k}} \\ &= \frac{y(\sigma(x_k)) - y(x_k)}{\sigma(x_k) - x_k} = y^{\Delta}(x_k) = z_k \end{aligned}$$

and

$$y[t_{2k-1}, t_{2k}] = \frac{y(x_k) - y(\sigma(x_{k-1}))}{x_k - \sigma(x_{k-1})}$$

$$= \frac{y(x_k) - y(x_{k-1}) - (y(\sigma(x_{k-1})) - y(x_{k-1}))}{x_k - \sigma(x_{k-1})}$$
$$= \frac{y(x_k) - y(x_{k-1})}{x_k - \sigma(x_{k-1})} - \frac{y(\sigma(x_{k-1})) - y(x_{k-1})}{\sigma(x_{k-1}) - x_{k-1}} \frac{\sigma(x_{k-1}) - x_{k-1}}{x_k - \sigma(x_{k-1})}$$
$$= \frac{y(x_k) - y(x_{k-1})}{x_k - \sigma(x_{k-1})} - z_{k-1} \frac{\sigma(x_{k-1}) - x_{k-1}}{x_k - \sigma(x_{k-1})}.$$

The other divided differences are computed in the usual way.

Then, Table 10.5 enables us to construct the Hermite interpolation polynomial

$$\begin{aligned} P_{2k+1}(x) &= y[t_0] + y[t_0, t_1](x - x_0) + y[t_0, t_1, t_2](x - x_0)(x - \sigma(x_0)) \\ &\quad + \cdots + y[t_0, \ldots, t_{2n+1}](x - x_0)(x - \sigma(x_0)) \cdots (x - x_n). \end{aligned} \tag{10.16}$$

Since $P_{2k+1}(x)$ interpolates the points $(t_j, f(t_j))$ for $j = 0, 1, \ldots, 2k$, we have

$$P_{2k+1}(x_i) = P_{2k+1}(t_{2i}) = y_i,$$
$$P_{2k+1}(\sigma(x_i)) = P_{2k+1}(t_{2i+1}) = w_i,$$

so that

$$P^{\Delta}_{2k+1}(x_i) = \frac{P_{2k+1}(\sigma(x_i)) - P_{2k+1}(x_i)}{\sigma(x_i) - x_i} = \frac{w_i - y_i}{\sigma(x_i) - x_i} = \frac{f(\sigma(x_i)) - f(x_i)}{\sigma(x_i) - x_i} = f^{\Delta}(x_i).$$

Therefore, the values of the polynomial in (10.16) and its Δ-derivative coincide with the values of the function and its delta derivative at the given points. In the examples, the divided difference table is employed in the construction of the Hermite interpolation polynomial.

Table 10.5: Divided difference table for Hermite interpolation polynomial.

t_k	$y[t_k]$	$y[t_k, t_{k+1}]$	$y[t_k, t_{k+1}, t_{k+2}]$	$y[t_k, t_{k+1}, t_{k+2}, t_{k+3}]$	$y[t_k, t_{k+1}, t_{k+2}, t_{k+3}, t_{k+4}]$
$t_0 = x_0$	y_0				
		z_0			
$t_1 = \sigma(x_0)$	w_0		$y[t_0, t_1, t_2]$		
		$y[t_1, t_2]$		$y[t_0, t_1, t_2, t_3]$	
$t_2 = x_1$	y_1		$y[t_1, t_2, t_3]$		$y[t_0, t_1, t_2, t_3, t_4]$
		z_1		$y[t_1, t_2, t_3, t_4]$	$\vdots$
				$\vdots$	
$t_3 = \sigma(x_1)$	w_1		$y[t_2, t_3, t_4]$		
		$y[t_3, t_4]$	$\vdots$		
		$\vdots$			
$t_4 = x_2$	y_2				
$\vdots$	$\vdots$				
$t_{2k+1} = \sigma(x_k)$	w_k				

Example. Let $\mathbb{T} = \mathbb{Z}$, $f(x) = \frac{1}{x+5}$, and $x_0 = 1$, $x_1 = 4$ be given points on $\mathbb{T}$. We will compute the Hermite polynomial in two different ways and compare the results.

Note that on this time scale we have $\sigma(x) = x + 1$ and hence,

$$f^{\Delta}(x) = \frac{f(x+1) - f(x)}{1} = \frac{-1}{(x+5)(x+6)}.$$

Then, it follows that

$$\begin{aligned} x_0 &= 1, & x_1 &= 4, \\ \sigma(x_0) &= 2, & \sigma(x_1) &= 5, \\ y_0 &= \frac{1}{6}, & y_1 &= \frac{1}{9}, \\ z_0 &= \frac{-1}{42}, & z_1 &= \frac{-1}{90}. \end{aligned}$$

We first compute the divided difference table for the given points (see Table 10.6).

Table 10.6: Divided difference table for the first example.

t_k	$y(t_k)$	$y[t_k, t_{k+1}]$	$y[t_k, t_{k+1}, t_{k+2}]$	$y[t_k, t_{k+1}, t_{k+2}, t_{k+3}]$
$t_0 = 1$	$\frac{1}{6}$			
		$\frac{\frac{1}{7} - \frac{1}{6}}{2-1} = \frac{-1}{42}$		
$t_1 = \sigma(x_0) = 2$	$\frac{1}{7}$		$\frac{\frac{-1}{63} + \frac{1}{42}}{4-1} = \frac{1}{378}$	
		$\frac{\frac{1}{9} - \frac{1}{7}}{4-2} = \frac{-1}{63}$		$\frac{\frac{1}{630} - \frac{1}{378}}{5-1} = \frac{\frac{-1}{945}}{4} = \frac{-1}{3780}$
$t_2 = 4$	$\frac{1}{9}$		$\frac{\frac{-1}{90} + \frac{1}{63}}{5-2} = \frac{1}{630}$	
		$\frac{\frac{1}{10} - \frac{1}{9}}{5-4} = \frac{-1}{90}$		
$t_3 = \sigma(x_1) = 5$	$\frac{1}{10}$			

This enables us to write easily the Hermite interpolation polynomial as

$$\begin{aligned} P_3 &= \frac{1}{6} - \frac{1}{42}(x-1) + \frac{1}{378}(x-1)(x-2) - \frac{1}{3780}(x-1)(x-2)(x-4) \\ &= \frac{1}{6} - \frac{1}{42}x + \frac{1}{42} + \frac{1}{378}(x^2 - 3x + 2) - \frac{1}{3780}(x^3 - 7x^2 - 10x - 8) \\ &= \frac{1}{6} - \frac{1}{42}x + \frac{1}{42} + \frac{1}{378}x^2 - \frac{3}{378}x + \frac{2}{378} - \frac{1}{3780}x^3 + \frac{7}{3780}x^2 + \frac{10}{3780}x + \frac{8}{3780} \\ &= \frac{-1}{3780}x^3 + \frac{17}{3780}x^2 - \frac{67}{1890}x + \frac{187}{945}. \end{aligned}$$

The Hermite interpolation polynomial can also be computed using the formulation given in [6]. First, we compute

$$
\begin{aligned}
L_0(x) &= \frac{x - x_1}{x_0 - x_1} = \frac{x-4}{1-4} = \frac{-1}{3}(x-4),\\
L_1(x) &= \frac{x - x_0}{x_1 - x_0} = \frac{x-1}{4-1} = \frac{1}{3}(x-1),\\
M_0(x) &= \frac{x - \sigma(x_1)}{x_0 - \sigma(x_1)} = \frac{x-5}{1-5} = \frac{-1}{4}(x-5),\\
M_1(x) &= \frac{x - \sigma(x_0)}{x_1 - \sigma(x_0)} = \frac{x-2}{4-2} = \frac{1}{2}(x-2),\\
L_0^{\Delta}(x_0) &= \frac{-1}{3},\\
L_1^{\Delta}(x_1) &= \frac{1}{3},\\
M_0^{\Delta}(x_0) &= \frac{-1}{4},\\
M_1^{\Delta}(x_1) &= \frac{1}{2},\\
L_0(\sigma(x_0)) = L_0(2) &= \frac{-1}{3}(2-4) = \frac{2}{3},\\
L_1(\sigma(x_1)) = L_1(5) &= \frac{1}{3}(5-1) = \frac{4}{3},\\
M_0(\sigma(x_0)) = M_0(2) &= \frac{-1}{4}(2-5) = \frac{3}{4},\\
M_1(\sigma(x_1)) = M_1(5) &= \frac{1}{2}(5-2) = \frac{3}{2}.
\end{aligned}
$$

Then, the Hermite interpolation polynomial has the form

$$
\begin{aligned}
P_3 &= \left[\left(1 - \frac{\frac{-1}{4} + \frac{3}{4}\cdot\frac{-1}{3}}{\frac{3}{4}\cdot\frac{2}{3}}(x-1)\right)\frac{1}{6} + \frac{\frac{-1}{42}}{\frac{3}{4}\cdot\frac{2}{3}}(x-1)\right]\frac{-1}{3}(x-4)\frac{-1}{4}(x-5)\\
&\quad + \left[\left(1 - \frac{\frac{1}{2} + \frac{3}{2}\cdot\frac{1}{3}}{\frac{3}{2}\cdot\frac{4}{3}}(x-4)\right)\frac{1}{9} + \frac{\frac{-1}{90}}{\frac{3}{2}\cdot\frac{4}{3}}(x-4)\right]\frac{1}{3}(x-1)\frac{1}{2}(x-2)\\
&= \left[(1 + (x-1))\frac{1}{6} + \frac{-1}{21}(x-1)\right]\frac{-1}{3}(x-1)\frac{1}{2}(x-2)\\
&\quad + \left[\left(1 - \frac{1}{2}(x-4)\right)\frac{1}{9} + \frac{-1}{180}(x-4)\right]\frac{1}{3}(x-1)\frac{1}{2}(x-2)\\
&= \frac{-1}{3780}x^3 + \frac{17}{3780}x^2 - \frac{67}{1890}x + \frac{187}{945}.
\end{aligned}
$$

Notice that the polynomials obtained in the two different ways are identical.

Example. Let $\mathbb{T} = 2^{\mathbb{N}}, f(x) = \frac{1}{x}$, and $x_0 = 1, x_1 = 4$. The Hermite interpolation polynomial is constructed by the two methods and the results are compared.

First, note that on the given time scale the forward jump operator is $\sigma(x) = 2x$ and we evaluate

$$f^{\Delta}(x) = \frac{f(2x) - f(x)}{x} = \frac{-1}{2x^2}.$$

Then

$$\begin{aligned} x_0 &= 1, & x_1 &= 4, \\ y_0 &= 1, & y_1 &= \frac{1}{4}, \\ z_0 &= \frac{-1}{2}, & z_1 &= \frac{-1}{32}, \\ \sigma(1) &= 2, & \sigma(4) &= 8. \end{aligned}$$

The modified divided difference table is derived as follows (see Table 10.7).

Table 10.7: Divided difference table for the second example.

t_k	$y(t_k)$	$y[t_k, t_{k+1}]$	$y[t_k, t_{k+1}, t_{k+2}]$	$y[t_k, t_{k+1}, t_{k+2}, t_{k+3}]$
$t_0 = 1$	1			
		$\frac{\frac{1}{2}-1}{2-1} = \frac{-1}{2}$		
$t_1 = \sigma(x_0) = 2$	$\frac{1}{2}$		$\frac{\frac{-1}{8}+\frac{1}{2}}{4-1} = \frac{1}{8}$	
		$\frac{\frac{1}{4}-\frac{1}{2}}{4-2} = \frac{-1}{8}$		$\frac{\frac{1}{64}-\frac{1}{8}}{8-1} = \frac{\frac{-7}{64}}{7} = \frac{-1}{64}$
$t_2 = 4$	$\frac{1}{4}$		$\frac{\frac{-1}{32}+\frac{1}{8}}{8-2} = \frac{1}{64}$	
		$\frac{\frac{1}{8}-\frac{1}{4}}{8-4} = \frac{-1}{32}$		
$t_3 = \sigma(x_1) = 8$	$\frac{1}{8}$			

Using this table, we can get the Hermite polynomial

$$\begin{aligned} P_3 &= 1 - \frac{1}{2}(x-1) + \frac{1}{8}(x-1)(x-2) - \frac{1}{64}(x-1)(x-2)(x-4) \\ &= 1 - \frac{1}{2}x + \frac{1}{2} + \frac{1}{8}(x^2 - 3x + 2) - \frac{1}{64}(x^3 - 7x^2 - 10x - 8) \\ &= 1 - \frac{1}{2}x + \frac{1}{2} + \frac{1}{8}x^2 - \frac{3}{8}x + \frac{1}{4} - \frac{1}{64}x^3 + \frac{7}{64}x^2 + \frac{10}{64}x + \frac{1}{8} \\ &= \frac{-1}{64}x^3 + \frac{15}{64}x^2 - \frac{23}{32}x + \frac{15}{8}. \end{aligned}$$

Considering the general form of the Hermite interpolation polynomial defined in [6], we also compute

$$\begin{aligned} L_0(x) &= \frac{x - x_1}{x_0 - x_1} = \frac{x-4}{1-4} = \frac{-1}{3}(x-4), \\ L_1(x) &= \frac{x - x_0}{x_1 - x_0} = \frac{x-1}{4-1} = \frac{1}{3}(x-1), \end{aligned}$$

$$
\begin{aligned}
M_0(x) &= \frac{x - \sigma(x_1)}{x_0 - \sigma(x_1)} = \frac{x-8}{1-8} = \frac{-1}{7}(x-8),\\
M_1(x) &= \frac{x - \sigma(x_0)}{x_1 - \sigma(x_0)} = \frac{x-2}{4-2} = \frac{1}{2}(x-2),\\
L_0^{\Delta}(x_0) &= \frac{-1}{3},\\
L_1^{\Delta}(x_1) &= \frac{1}{3},\\
M_0^{\Delta}(x_0) &= \frac{-1}{7},\\
M_1^{\Delta}(x_1) &= \frac{1}{2},\\
L_0(\sigma(x_0)) = L_0(2) &= \frac{-1}{3}(2-4) = \frac{-1}{3}(-2) = \frac{2}{3},\\
L_1(\sigma(x_1)) = L_1(8) &= \frac{1}{3}(8-1) = \frac{7}{3},\\
M_0(\sigma(x_0)) = M_0(2) &= \frac{-1}{7}(2-8) = \frac{6}{7},\\
M_1(\sigma(x_1)) = M_1(8) &= \frac{1}{2}(8-2) = 3,
\end{aligned}
$$

which yields the Hermite polynomial as

$$
\begin{aligned}
P_3 &= \left[\left(1 - \frac{\frac{-1}{7} + \frac{6}{7}\cdot\frac{-1}{3}}{\frac{6}{7}\cdot\frac{2}{3}}(x-1)\right).1 + \frac{\frac{-1}{2}}{\frac{6}{7}\cdot\frac{2}{3}}(x-1)\right]\frac{-1}{3}(x-4)\frac{-1}{7}(x-8)\\
&\quad + \left[\left(1 - \frac{\frac{1}{2} + 3.\frac{1}{3}}{3.\frac{7}{3}}(x-4)\right)\frac{1}{4} + \frac{\frac{-1}{32}}{3.\frac{7}{3}}(x-4)\right]\frac{1}{3}(x-1)\frac{1}{2}(x-2)\\
&= \left[\left(1 - \frac{\frac{-3}{7}}{\frac{4}{7}}(x-1)\right) + \frac{-7}{8}(x-1)\right]\frac{-1}{3}(x-4)\frac{-1}{7}(x-8)\\
&\quad + \left[\left(1 - \frac{\frac{3}{2}}{7}(x-4)\right) + \frac{-1}{224}(x-4)\right]\frac{1}{3}(x-1)\frac{1}{2}(x-2)\\
&= \frac{-1}{64}x^3 + \frac{15}{64}x^2 - \frac{23}{32}x + \frac{15}{8}.
\end{aligned}
$$

Clearly, both polynomials are the same.

10.5 Conclusion

In this chapter, we introduced the divided and σ-divided differences on time scales. They provide an alternative form for the interpolation polynomial for a given data set. Moreover, the modified divided differences make the construction of the Hermite interpolation polynomial very simple. As a further study, one can consider the derivation of a modified σ-divided differences table to be used to simplify the construction of the σ-Hermite interpolation polynomials, which is also complicated in its present form given in [6].

Bibliography

[1] B. Aulbach and S. Hilger, A unified approach to continuous and discrete dynamics, in *Qualitative Theory of Differential Equations, Szeged, 1988, Colloq.*, Math. Soc. Janos Bolyai, vol. 53, pp. 37–56, North-Holland, Amsterdam, 1990.

[2] M. Bohner, Calculus of variations on time scales, in *Dynamic systems and applications*, vol. 13, pp. 339–349, 2004.

[3] M. Bohner, İ. M. Erhan and S. Georgiev, The Euler method for dynamic equations on general time scales, *Nonlinear Stud.*, 27(2) (2020), 415–431.

[4] M. Bohner and S. G. Georgiev, Multivariable Dynamic Calculus on Time Scales, Springer, Switzerland, 2016.

[5] M. Bohner and A. Peterson, A survey of exponential functions on time scales, *Rev. Cubo Mat. Educ.*, 3(2) (2001), 285–301.

[6] S. Georgiev and İ. M. Erhan, Numerical Analysis on Time Scales, De Gruyter, 2022.

[7] S. Georgiev and İ. M. Erhan, The Taylor series method and trapezoidal rule on time scales, *Appl. Math. Comput.*, 378 (2020), 125200. https://doi.org/10.1016/j.amc.2020.125200.

[8] S. Georgiev and İ. M. Erhan, Lagrange interpolation on time scales, *J. Appl. Anal. Comput.*, 12(4) (2022), 1294–1307. https://doi.org/10.11948/20200461.

[9] S. Hilger, Ein Maßkettenkalkül mit Anwendung auf Zentrumsmannigfaltigkeiten, PhD thesis, Universität Würzburg, 1988.

[10] S. Hilger, Differential and difference calculus – unified, *Nonlinear Anal.*, 30(5) (1997), 2683–2694.

Index

https://doi.org/10.1515/9783111182971-011

Bei Fragen zur Produktsicherheit wenden Sie sich bitte an:
If you have any questions regarding product safety,
please contact:

Walter de Gruyter GmbH
Genthiner Straße 13
10785 Berlin
productsafety@degruyterbrill.com